LIFE HISTORIES
OF NORTH AMERICAN
WILD FOWL

by Arthur Cleveland Bent

IN TWO PARTS

PART I

Dover Publications, Inc., New York

Published in the United Kingdom by Constable
and Company Limited, 10 Orange Street, London
W. C. 2.

This new Dover edition, first published in 1962,
is an unabridged and unaltered republication of the
work first published by the United States Government Printing Office. Part I was originally published
in 1923 as Smithsonian Institution United States
National Museum *Bulletin 126;* Part II was originally published in 1925 as Smithsonian Institution
United States National Museum *Bulletin 130.*

International Standard Book Number: 0-486-20285-2
Library of Congress Catalog Card Number: 62-51885

Manufactured in the United States of America

Dover Publications, Inc.
180 Varick Street
New York 14, N.Y.

INTRODUCTION

Nearly all those who contributed material for former volumes have rendered similiar service in this case. In addition to those whose contributions have been previously acknowledged, our thanks are due to the following new contributors:

Photographs have been contributed by A. A. Allen, J. H. Bowles, A. D. Dubois, J. Labarthe, C. W. Townsend, and W. Raine.

Notes and data have been contributed by A. A. Allen, G. M. Allen, H. W. Brandt, L. Griscom, J. F. Honecker, W. J. Hoxie, J. C. Phillips, and G. M. Sutton. Mr. Wharton Huber has written the life history and the distribution of his new species, the New Mexican duck.

Dr. John C. Phillips has furnished the references for the life history of the falcated teal and very kindly placed at the author's disposal his entire unpublished manuscript on this species.

The information previously furnished by Dr. T. S. Palmer, on bird reservations belonging to the United States, and a lot of data recently furnished by Mr. Hoyes Lloyd, on Canadian bird reservations and sanctuaries, form such a voluminous mass that it seems best to leave it for future publication by itself, rather than attempt to do but scant justice to it in this volume.

The distributional part of this volume is the work of the author; but it has been examined by Mr. James H. Fleming, Mr. P. A. Taverner, and Mr. F. Seymour Hersey. Mr. Hersey also figured the egg measurements, after collecting a lot of additional measurements from some of the large egg collections, notably those of the California Academy of Sciences (Mailliard collection), the Colorado Museum of Natural History (Bradbury collection), Mr. Richard C. Harlow, Col. John E. Thayer, and the University of California (Grinnell collection).

In outlining the breeding ranges of the ducks, no attempt has been made to mention all of the many cases recorded of northern breeding species which have lingered for the summer and bred far south of their normal breeding ranges, as such birds are often cripples; only a few cases have been mentioned which seemed to be casual breeding records of normal birds.

The author regrets that, in a general work of this kind, he did not feel justified in giving detailed descriptions of the interesting plumage changes of the ducks; so the sequences of molts and plumages have been only briefly indicated with sufficient descriptions to make them recognizable. The reader, who wishes to make a more thorough study of this subject, is referred to the splendid monographs of the ducks by Mr. John G. Millais and to the full and accurate descriptions of the various plumages by Miss Annie C. Jackson, in A Practical Handbook of British Birds, edited by Mr. H. F. Witherby.

The old check list is now so far out of date that it seems unwise to follow it in the use of scientific names. Therefore, Dr. Charles W. Richmond and Dr. Harry C. Oberholser have kindly furnished the scientific names for this volume, which, according to their best judgment, most satisfactorily represent our present knowledge.

THE AUTHOR.

TABLE OF CONTENTS.

LIFE HISTORIES OF NORTH AMERICAN WILDFOWL, ORDER ANSERES (PART).

By Arthur Cleveland Bent,
Of Taunton, Massachusetts.

Family ANATIDAE, Ducks, Geese, and Swans.

MERGUS MERGANSER AMERICANUS Cassin.

AMERICAN MERGANSER.

HABITS.

Spring.—This large and handsome duck has always been associated in my mind with the first signs of the breaking up of winter. Being a hardy species, it lingers on the southern border of ice and snow and is the very first of our waterfowl to start on its spring migration. We may confidently look for it in New England during the first warm days in February or as soon as the ice has begun to break up in our rivers and lakes. We are glad to greet these welcome harbingers of spring, for the sight of the handsome drakes flying along our water courses or circling high in the air over our frozen lakes, with their brilliant colors flashing in the winter sunshine, reminds us of the migratory hosts that are soon to follow. They are looking for open water in the rivers, for rifts in the ice or open borders around the shores of the lakes, where the first warm sunshine has tempted the earliest fish to seek the genial shallows, but they are often doomed to disappointment, for winter lingers in the lap of spring and again locks the lakes with solid ice driving the hardy pioneers back to winter quarters. The drakes are always the first to arrive and the females follow a few weeks later. Mr. Fred A. Shaw writes to me that in Maine—"The males generally make their appearance in March and in a short time select their mates, leaving early in April for their breeding grounds."

Courtship.—Perhaps the best account of the courtship of this species is given by Dr. Charles W. Townsend (1916), as follows:

A group of five or six male mergansers may be seen swimming energetically back and forth by three or four passive females. Sometimes the drakes swim in a compact mass or in a file for six or seven yards or even farther, and then each turns abruptly

1

and swims back. Again they swim in and out among each other, and every now and then one with swelling breast and slightly raised wings spurts ahead at great speed by himself or in the pursuit of a rival. The birds suggest swift motor boats by the waves which curl up on either side, and by the rapidity with which they turn and swash around. Again they suggest polo ponies, as one in rapid course pushes sidewise against a rival, in order to keep him away from the object of the quest. They frequently strike at each other with their bills, and I have seen two splendid drakes rise up in the water breast to breast, and, amid a great splashing, during which it was impossible to see details, fight like gamecocks. The pursuit is varied by sudden, momentary dives and much splashing of water.

The smooth iridescent green heads, the brilliant carmine bills tipped with black nails, the snowy white of flanks and wing patches and the red feet, which flash out in the dive, make a wonderful color effect, contrasting well with the dark water and white ice. The smaller females, with their shaggy brown heads, their neat white throat bibs, their quaker blue-gray backs and modest wing patches, which are generally hidden, are fitting foils to their mates. I have reserved for the last the mention of the delicate salmon yellow tint of the lower breast and the belly of the male, a coloration of which he is deservedly proud, for, during courtship, he frequently raises himself up almost onto his tail with or without a flapping of the wings and reveals this color, in the same way that the eider displays his jet-black shield. Most of the time he keeps his tail cocked up and spread, so that it shows from behind a white center and blue border. Every now and then he points his head and closed bill up at an angle of 45° or to the zenith. Again he bows or bobs his head nervously and often at the same time tilts up the front of his breast from which flashes out the salmon tint. From time to time he emits a quickly repeated purring note, "dorr-dorr" or "krr-krr."

The most surprising part of the performance is the spurt of water fully 3 or 4 feet long which every now and then is sent backward into the air by the powerful kick of the drake's foot. It is similar to the performance of the whistler but much greater, and while the foot of the whistler is easily seen and is plainly a part of the display, it is difficult to see the red foot of the merganser in the rush of water, although it is evident, doubtless, to the females. The display of the brilliantly colored foot in both species is probably the primary sexual display, and the splash, at first incidental and secondary, has now become of primary importance.

During all this time the female swims about unconcernedly, merely keeping out of the way of the ardent and belligerent males, although she sometimes joins in the dance and bobs in a mild way. At last she succumbs to the captivating display and submerges herself so that only a small part of her body with a bit of the crest appear above the water, and she swims slowly beside or after her mate, sometimes even touching him with her bill. Later she remains motionless, flattens herself still more, the crest disappears, and she sinks so that only a line like that made by a board floating on the water is seen. One would never imagine it to be a live duck. The drake slowly swims around her several times, twitches his head and neck, picks at the water, at his own feathers, and at her before he mounts and completely submerges her holding tightly with his bill to her neck meanwhile. Then she bathes herself, washes the water vigorously through her feathers and flaps her wings; the drake stretches himself and flaps his wings likewise. From the beginning of submergence by the female the process is the same in all the duck family that I have observed.

Nesting.—On the nesting habits of the American merganser there has been much discussion and many conflicting opinions, some asserting that it always nests in hollow trees or that it never nests on the ground. As a matter of fact it does both, for its nesting habits vary greatly in different localities or even with different individuals in the same locality. Major Bendire carried on an extensive correspondence with Mr. Manly Hardy on this subject and the latter was finally convinced that the goosander, as he called it, does occasionally nest on the ground. On June 12, 1891, Mr. Hardy wrote that, in passing through Caribou Lake, June 8, he found three nests, containing 26 eggs, on ledges under low fir bushes, which settled the controversy. I believe, however, that in the Eastern States and Provinces this merganser prefers to nest in hollow trees where it can find suitable cavities, which are usually scarce. Mr. Fred A. Shaw, who has had 30 years' experience with this species in the vicinity of Sebago Lake, Maine, contributes the following notes on its nesting habits:

A few breed around Sebago, especially near the mouth of Songo River, the principal tributary, where there is a large area of bog, flooded in spring, through which are scattered large hollow trees providing safe nesting places for them. The nest of this species is commonly placed in a hollow tree standing near the water and is composed of feathers and down from the breast of the parent bird. A nest of this bird near Whites Bridge at the outlet of Sebago Lake was shown me in May, 1897. It was in a white birch stub which was broken off about 15 feet from the ground and was hollow for about 10 feet from the top and contained 10 eggs, which were laid at the bottom of the hole on a warm bed of soft down from the breast of the mother bird.

Mr. Hardy wrote that it nested in hollow trees, usually hardwood trees, such as maples and ashes, and often in green trees. Mr. John H. Sage (1881) found a nest on an island in Moosehead Lake, Maine, on June 19, 1881—

in a hollow under the roots of a standing tree, roots, earth, and moss forming a perfect roof, so that the nest, after the heavy shower of that day, seemed well protected and was quite dry. The eggs were covered with leaves, moss, and feathers— mostly feathers. The old bird was seen to leave the nest.

On Lake Winnipegosis and Waterhen Lake, Manitoba, we found the American merganser very common and nesting on the numerous islands, wherever suitable nesting sites could be found among the piles of loose bowlders along the shores. Mr. Walter Raine found as many as 30 nests of this species on Gun Island, a large island in Lake Winnipegosis. On one island that we visited the Indians had collected about 60 eggs of the " saw bills," as they call them, a short time previously. The nests were very well hidden in remote crevices under the piles of large bowlders; many of them were quite inaccessible, where the bowlders were too large for us to move them. We often saw mergansers flying away from islands where we felt sure that they were nesting, but where we were unable to find or reach the nests. A few nests were found in the dense tangles of gooseberry bushes and nettles on the tops of the islands where hunting for them was difficult and painful unless a telltale path, strewn with feathers and droppings, told us just where to look.

Near the north end of Lake Winnipegosis, on Whiskey Jack Island, we visited, on June 18, 1913, a deserted ice house where we were told that we might find the "little saw bills," hooded mergansers, nesting. It was an old tumbled-down affair, with the roof nearly gone and partly filled with loosely piled bales of hay; there did not seem to be any suitable nesting sites for "saw bills" anywhere in the vicinity, so we sat down to eat our luncheon. While so occupied we were surprised to see a female American merganser fly up and alight on the landing and gaze longingly into the ice house. We then began an exhaustive search by moving the bales of hay and crawling into the crevices between them. While peering into a dark cavity I thought I saw something moving regularly like a breathing duck; we pulled away some more bales and there sat a female merganser on her nest within 2 feet of my face; I reached in to catch her but she slipped away and escaped through another passage way. There were 15 eggs under her in a nest profusely lined with white down, mixed with hay. Further search revealed another nest near by, similarly located; the bird had left the nest and had carefully covered the 12 eggs which it contained with a soft blanket down. Lieut. I. T. Van Kammen (1915) found two nests in an old, abandoned lighthouse tower; the nests were about 3 feet apart and "each nest was placed in a depression, perhaps 5 inches deep, scraped out of the soft dirt of the lighthouse floor."

Audubon (1840) gives an attractive account of finding the nest of a goosander on a marshy island. He describes the nesting site and nest as follows:

The islands on which the goosander is wont to breed are mostly small, as if selected for the purpose of allowing the sitting bird to get soon to the water in case of danger. The nest is very large, at times raised 7 or 8 inches on the top of a bed of all the dead weeds which the bird can gather in the neighborhood. Properly speaking, the real nest, however, is not larger than that of the dusky duck, and is rather neatly formed externally of fibrous roots and lined round the edges with the down of the bird. The interior is about 7½ inches in diameter and 4 inches in depth.

Mr. W. L. Dawson (1909) says of its nesting habits in Washington:

Now and then a crevice in the face of a cliff does duty, and old nests of hawk or crow have been pressed into service. Moderate elevations are favored, but Mr. Bowles once found a nest near Puget Sound in a decayed fir stub at a height of over a hundred feet. The cavity, wherever found, is warmly lined with weeds, grasses, and rootlets, and plentifully supplied with down from the bird's breast.

Mr. Fred H. Andrus (1896) thus describes a nest which he found in Oregon:

May 26, 1895, I collected a set of 10 American merganser's eggs from a hole in the rocks about 100 feet above the Umpqua River. The nest was about 15 feet from the top of a nearly perpendicular cliff about 50 feet in height, and was found by watching the bird. In going to the nest the bird would fly up and down the river in an oval course several times, and finally, coming close to the water as if to light, would rise to the nest. The entrance to the hole was 6 inches by 12, and the inside dimensions 4 feet long, 2 feet deep, and 18 inches high. The nest was about 1 foot in diameter, of down mixed with moss, one-half inch thick in the center and thicker around the edges.

Eggs.—The American merganser raises but one brood in a season and lays from 6 to 17 eggs; some writers say that it lays from 6 to 10 eggs, but I think these small sets must have been incomplete or second attempts; I should say that the commonest numbers would run between 9 and 12, but I have personally taken sets of 15 and 16. The eggs are usually distinctive, and typical eggs are not easily mistaken for anything else. The female is difficult to distinguish from the female red-breasted merganser when she flies from the nest, though she is a decidedly heavier looking bird and has more white in the wings. The down, however, in the American merganser's nest is much whiter than that in the nest of the red-breasted and the color of the eggs is different. The down is grayish white in color, about No. 10 gray of Ridgway, or "pale gull gray," and it is usually mixed with numerous pure white breast feathers and considerable rubbish or bits of straw. The eggs of the American merganser are very pale buff or "ivory yellow." The shell is thick and strong, with little, if any, luster. The shape varies from elliptical ovate to elliptical oval.

The measurements of 93 eggs in various collections average 64.3 by 44.9 millimeters; the eggs showing the four extremes measure **72** by 46, 64 by **50, 55.4** by 38.5 and 56.7 by **37** millimeters.

The drakes desert the ducks and usually disappear from the breeding grounds entirely as soon as the eggs are laid, leaving the females to perform the duties of incubation and care for the young alone. In Newfoundland we saw only females on the lakes, where they were busy with family cares, but we saw plenty of males on the swift-water rivers, playing in the rapids and fishing in the pools. Several observers in Maine have said that the males are not seen during the summer, but this may be due to the fact that the males are in eclipse plumage at this time and are very shy and retiring. Mergansers which nest in hollow trees are usually very close sitters, and it is often impossible to drive a sitting bird from her nest by pounding the tree; on the other hand those which nest on the ground on islands usually slip away long before the intruder reaches the vicinity of the nest, often before a boat lands on the island; a deviation from either of these habits would, in either case, tend to reveal the nest. Fresh eggs, taken from incomplete sets in Manitoba and hatched in our incubators, showed an incubation period of 28 days.

Young.—Several writers state that the young mergansers are carried from a nest in a hollow tree to the water, in the bill of the parent bird; Millais (1913) says that Mr. Oswin Lee has seen a female goosander carry down nine young ones out of the nest, and that she "carried them partly in her beak, partly between the beak and the breast."

Mr. Shaw, however, offers the following evidence to the contrary:

An interesting occurrence in connection with the breeding of this bird was related to me by Mr. G. H. Moses, who while in camp at Songo River had exceptional opportunity to observe them. In the spring of 1896 a nest was located in a tall hollow tree where it could be readily seen from his camp door. After the young were hatched Mr. Moses saw the mother bird alight in the water at the foot of the stub in which the nest was located and commence to call to the young birds in the nest. Immediately the little ones came tumbling down one after another from the hole in the tree top to the water and at once swam away with their mother.

Mr. William S. Post (1914) has twice witnessed a similar performance; the following is his account of it:

It was my good fortune to witness twice the emerging of a young brood of mergansers from an extreme situation of this kind, an old pileated woodpecker's hole about 40 feet high in the limb of a live elm, standing about 15 feet from the edge of the Tobique River in New Brunswick.

On June 18, 1910, I fished the famous salmon pool at the fork of the river, and having incidentally run the canoe close to the shore near where this old elm stands, I landed and rapped several times sharply on the tree with a stick, for I had been told that a wood duck—which on the Tobique means a golden eye—nested there the previous spring. The female merganser immediately flew out and having circled about over the river, alighted on the water. After assuring myself of the identification, which caused me some astonishment on account of the size of the bird in proportion to the entrance of the hole, I returned to my fishing.

In a few moments I noticed a small bird drop down apparently from the hole, and in a few more seconds another and then a third. My first thought was that a bank swallow, of which there are many on the river, had flown up near to the hole and down again three times in succession. This caused me to stop fishing and to watch, when to my astonishment a small bird with white breast appeared in the hole, jumped out, and was followed by another, and again another. I then lost no time in reaching a point in the river opposite the tree, where I saw in the water against the bank, swimming around, a brood of 11 young ducks. I was much surprised, as I had been under the impression from what I had read that the old duck would certainly carry down the young from such an inaccessible position, and though I believe the young birds must have landed in the water, I was yet astonished that they could withstand the shock of such a drop, and I presumed that by rapping on the tree I had caused the old bird to leave in such fright that her fear had been communicated to the young and they had followed her example, and that the whole procedure was therefore an unnatural one.

The clubhouse is situated directly across the river, and on June 12, 1913, two years later, I was sitting on the piazza when my attention was attracted by seeing something large drop from the top of this same elm into the water. I immediately saw that it was the old sheldrake and that she was swimming around close to the shore.

In a few seconds another dropped from the hole to the ground and I could see it run down the bank and join its mother who was calling loudly and turning round and round in the water. This one was quickly followed by others in succession until there were seven. By this time I had called my guide and in company of one of the members of the club was crossing the river, provided with trout-landing nets.

The old bird seeing us immediately swam upstream and around the point with her brood and this was the last we saw of her. We landed and stood under the tree where we could hear distinctly more young ducks peeping in the hole. Looking up we saw one tottering on the edge, and before we could take stations where we could properly observe the actual drop he had struck the ground close to my friend and made such rapid progress toward the water that he escaped in spite of landing nets. In a few seconds another, which proved to be the last, followed, falling on the other side of the tree, and I promptly made him captive. The first bird was in the water and had immediately dived. It is strange that he should have known enough to seek the water, and also to dive immediately.

After a day or two of rest in the nest, probably longer in tree nests than in those on the ground, the young have dried off their down and gained sufficient strength to take to the water, where they are very precocious. The downy young are very handsome and attractive. It is a beautiful sight to see a female merganser swimming in the clear calm water of some mountain lake or wilderness stream, where the mirrored reflections of picturesque scenery and forest trees make a splendid setting for the picture of a swiftly gliding, graceful duck followed by a procession of pretty little balls of down, with perhaps one or two of them riding on her back. If danger threatens she quickens her pace, but the little fellows are good swimmers and keep right at her heels; even if she dives they can follow her under water, working their little paddles vigorously and darting along like so many fish. If too hard pressed she rises and flaps along the surface, half flying; they can almost keep up with her at this pace, for they can run along the surface as fast as we can paddle our canoe. They soon become exhausted with some exertion, so she leads them into some sheltered cove, where they can run up on the shore and hide in the grass, or even up into the thick woods, where it is almost hopeless to hunt for them; it is surprising to see how quickly the young and even the mother bird can disappear. Millais (1913) relates the following interesting incident:

When rushing down the swift rivers of Newfoundland in my canoe, I have often wondered at the resource or natural instinct of the broods of goosander and their mothers which remain perfectly still when suddenly confronted with danger. As the little boat flies down a rapid, swiftly passing silent pools in the rock eddies at the sides, I have often turned my head and noticed a female goosander and her nearly full-grown young. On a lake or open stretch of the river, knowing that concealment was impossible, the mother would have dashed out in the open, and either hurried by flapping along the surface to the middle of the lake, or, in the case of the river, downstream, and so endeavor to escape. When suddenly confronted within a few yards in the eddies of the rapids, she felt that such a method of escape was useless, and with swift intuition remained perfectly still, each member of the brood keeping the neck held stiffly, so that the whole party looked like the stiff twigs of an upturned tree. This sudden assimilation to surroundings, so wonderfully exhibited in the common or little bittern amid the rushes, seems to be a natural instinct in all birds, and they often adopt it as a last resort.

Plumages.—Downy young mergansers are beautiful creatures; the upper parts, including the crown, down to the lores and eyes, hind neck, and back, are rich deep "bister" or "warm sepia," relieved by the white edging of the wing and a large white spot on each side of the rump; the sides of the head and neck are "mikado brown" or "pecan brown", shading off on the neck to "light vinaceous cinnamon" or "buff pink;" a pure white stripe extends from the lores to a point below the eyes and it is bordered above and below by dark-brown stripes; the rest of the lower parts are pure white. The nostril is in the central third of the bill, instead of in the basal third, as in the red-breasted merganser.

In the juvenal plumage, which is acquired at about the time that the young bird attains its growth, the sexes are practically indistinguishable except that the male is slightly larger. This plumage is similar to that of the adult female except that, in the young bird, the white throat extends down to the chest, whereas, in the adult female, the lower throat is brown. The wing pattern is also different in young males, which have the outer secondaries white and the inner secondaries gray. During the fall and winter an almost continual molt is in progress, black feathers appearing in the head and neck, producing a mottled effect, and vermiculated feathers appearing in the flanks. The tail is renewed in the spring but not the wings. The first postnuptial molt, which can hardly be said to involve an eclipse plumage, takes place in August and September; this molt is complete and is prolonged through October at least; by November, when the bird is nearly a year and a half old, the adult plumage is complete.

Millais (1913) says:

In May the adult male goosander begins to assume its eclipse plumage. The adult male in August has the crown reddish brown, with a gray tinge; chin white, and the rest of the head and upper neck rich red brown. There is a black mark in front of the eye, and a whitish line from this to the lower angle of the upper mandible. The lower neck is blue gray, interspersed with creamy white; mantle, flanks, scapulars, back, and tail blue gray; the flanks have a few white feathers on the outer sides, vermiculated with brownish gray; the last inner secondaries only change from black to black; wings as in winter, and now changing as usual only once; under parts not so rosy as in winter. In early September the wings and tail are renewed, and the black feathers on the mantle come in. After this the whole plumage proceeds to molt slowly, the full winter dress not being assumed until early December.

Subsequent molts of adult males consist of a postnuptial molt of the contour feathers early in the summer into the eclipse plumage, a molt of the flight feathers in August or September, and a complete molt of the contour feathers out of the eclipse plumage in the fall. Females probably do not make the double molt of the contour feathers but have a complete molt in the late summer.

Food.—The merganser is primarily a fishing duck, at which it is very skillful and a voracious feeder. It pursues under water and catches successfully the swiftest fish. Often a party of sheldrakes may be seen fishing together, driving the panic-stricken fish into the shallows or into some small pool where they may be more easily caught. Mr. Hardy writes:

They fish in companies; as fast as they come up the hind ones run ahead of those in front of them and dive again, being in turn succeeded by others. I have seen them fishing on quick water in very cold weather until January 7. They feed exclusively on fish, several often uniting to capture one of large size. Last year we took a pickerel from a party of them which measured 14 inches in length; also took from one's throat a chub which, with head decomposed, measured 10 inches.

One of these gluttonous birds will often attempt to swallow a larger fish than it can dispose of, leaving the tail of the fish protruding from its mouth while the head is digesting. Mr. Shaw found in the "stomach" of one bird "13 perch, a few of which were nearly 3 inches in length."

Audubon (1840) says:

I have found fishes in its stomach 7 inches in length, and of smaller kinds so many as to weigh more than half a pound. Digestion takes place with great rapidity, insomuch that some which I have fed in captivity devoured more than two dozen of fishes about 4 inches in length, four times daily, and yet always seemed to be desirous of more.

Mr. Harry S. Swarth (1911) describes the following manner of feeding, which is unusual for this species:

I was concealed in the shrubbery at the water's edge examining a large flock of ducks for possible rarities, when a dozen or more mergansers (both *M. americanus* and *M. serrator*) began swimming back and forth but a very short distance from my blind. They swam slowly, with neck outstretched, and with the bill held just at the surface of the water, and at a slight angle, so that the head was submerged about to the level of the eyes. The water was evidently filtered through the bill, as a slight "gabbling" noise was quite audible, and obviously something was being retained as food, though just what it was I could not tell. This is rather remarkable, as it is exactly the manner of feeding usually employed by the shoveller (*Spatula clypeata*), a species which, as regards bill structure, is further removed from the mergansers than any other member of the Anatidae.

Mr. Ora W. Knight (1908) says:

Along the coast in winter they eat many mussels and allied species of mollusks, swallowing them shell and all. The shells are soon ground to pieces in their intestines and stomachs, and in dead birds dissected out I have traced the entire process from entire mussel shells down to impalpable mud at the lower end of the intestinal tract.

In the early spring, when live fish are difficult to obtain, they seem to enjoy frozen, rotten fish with the same gusto as fresh, picking them out of the floating ice. They also feed to some extent on frogs, small eels, aquatic salamanders, crawfish, and other small crustaceans, various bivalve mollusks and snails, leeches, worms, water insects and larvae, and the stems and roots of aquatic plants.

Behavior.—The American merganser is a heavy-bodied bird and sometimes experiences considerable difficulty in rising from the water; if the circumstances are not favorable, it has to patter along the surface for a considerable distance; when flying off an island it often does the same thing unless it gets a good start from some high place, so that it can swoop downward.   In swift water it has to rise downstream, as it can make no headway against the current; but it generally prefers to fly upstream if it can.   Mr. Aretas A. Saunders writes to me, in regard to the flight of mated pairs, noted in Montana, "that they flew off, with the male in the lead in each case," also "that they left the water flying in a long, low slant upstream, not rising high enough to see them above the willows that lined the stream until they had flown a considerable distance."   When well under way the flight of this species is strong, swift, and direct; on its breeding grounds it usually flies low, along the courses of rivers or about the shores of lakes, seldom rising above the tree tops; but on its migrations it flies in small flocks, high in the air with great velocity. The drake may be easily recognized in flight by its large size, loonlike shape, its black and white appearance above, dark green head and white underparts; its flight is said to resemble that of the mallard. The female closely resembles the female red-breasted merganser, but it is a more heavily built bird, has a more continuous white patch in the wings, the white tips of the greater coverts overlapping the black bases of the secondaries, giving the appearance of a large white speculum, whereas in the red-breasted merganser the black bases of the secondaries show below the greater coverts, forming a black stripe through the middle of the white speculum.   When flying to its nest cavity in a tree or cliff it rises in a long upward curve and enters the hole with speed and precision.   Mr. Harry S. Swarth (1911) refers to—

a peculiar habit which made this species quite conspicuous throughout the summer, was that of individuals rising high in the air and circling about for hours at a time, uttering at frequent and regular intervals a most unmelodious squawk. Both sexes were observed doing this, and the habit was kept up until about the end of August.

This sheldrake is probably the most expert diver of its tribe, being built somewhat like a loon and approaching it in aquatic ability.   It can sink quietly down into the water like a grebe or dive quickly with a forward curving plunge, clearing the water for a foot or more, as it does so.   It swims swiftly on the surface, but can attain even higher speed below it, where few fish can escape it.   Dr. Charles W. Townsend (1909) infers from its method of diving, that the wings are not used when swimming under water and he quotes a statement from Selous to the same effect; he says:

The American and the red-breasted merganser both dive like the cormorant. They often leap clear of the water, in graceful curves, with their wings cleaving closely to the sides.   At other times the leap is much curtailed, or they sink beneath the surface without apparent effort.   I should infer, therefore, that the wings were not used under water.

On the other hand I quote from Mr. Walter H. Rich (1907) as follows:

Seen under the water in pursuit of a breakfast or dodging about to escape capture when wounded the resemblance to some finny dweller of the sea is very marked—head and neck outstretched, every feather hugged closely to the body, the half-opened wings like large fins aiding the feet in their work, he goes shooting through the water like a flash.

Probably both observers are correct, for birds are not bound by hard and fast rules. The rapidity with which this species can dive from the air is remarkable. While in full flight it plunges into the water, swims below the surface for a distance and then suddenly emerges and continues its flight. Millais (1913) says of its behavior on land:

The walk is very heavy and rolling, and the feet are placed on the ground deliberately, whilst the bill is pointed downward, and each step taken as if the bird was afraid of tripping or falling. They seldom go more than a yard or two from the water's edge, but can run quite swiftly for a few yards if suddenly surprised. In winter it is a very rare event to see goosanders ashore, but in spring they often leave the water, and will spend hours sleeping and preening on some small island or point of land. No birds are more industrious in their toilet than the mergansers in spring, and most of their time, when not feeding, flying, or sleeping, is spent in polishing up their plumage and bathing.

Of the vocal performances of this species I know very little; I have never heard a sheldrake utter a sound, so far as I can remember, and very little seems to have been written on the subject. Audubon (1840) describes the notes of the goosander as "harsh, consisting of hoarse croaks, seldom uttered unless the bird be suddenly startled or when courting."

Game.—As they live almost exclusively on fish, sheldrakes are not considered good table birds and so are not much sought after by gunners. But young sheldrakes are not unpalatable, and many gunners shoot them regularly for food. They do not come to decoys, but, as they are swift fliers and hard to kill, shooting them is good sport. Many sportsmen feel justified in killing them on account of the large numbers of trout which they consume; but this is hardly justifiable, for they also destroy many predatory fish, such as pickerel and thus help to preserve the balance of nature.

Fall.—The name "pond sheldrake" has been applied to this species because it shows more preference for fresh water than its relative, the red-breasted merganser, its fall migration is more inland, where it flies along our larger water courses and frequents our lakes and ponds until it is forced coastwise by the freezing of its favorite resorts.

Winter.—But even in winter it still lingers wherever it can find open water near its summer home and its migration is one of the shortest. Mr. Shaw says that, at Sebago Lake, Maine—

a few remain through the winter in the coldest weather spending the day in the open water at the foot of the lake and in the upper part of Presumpscot River, its outlet, and at night leaving for the salt water.

It is more common on the coast of New England than in the interior in the winter, but it winters in large numbers in some of the Great Lakes and on the large rivers of the interior, especially in the rapids and about the cascades of clear-water streams. The icy waters of our northern streams have no terrors for this hardy fisherman provided it can find open water and plenty of food. Rev. Manley B. Townsend writes to me:

Every winter during my residence in Nashua, N. H., 1912–1918, I noted considerable numbers of these "fish ducks" as they are locally called, between Manchester and Nashua. Even in the coldest, most inclement weather, when the Merrimack was frozen several feet deep, these birds could be seen sitting on the ice about the rapids, where the swiftly flowing water kept the river free from ice, or swimming and diving in careless abandon. I once counted 40 on the ice or in the water at a single rapid. Fish seemed to be abundant. The birds apparently wintered well.

DISTRIBUTION.

Breeding range.—Northern States and Canada entirely across the continent. South to west central Nova Scotia (Gaspereaux Lakes), southern Maine (Washington to Cumberland Counties), central New Hampshire (White Mountain region), central Vermont (Windsor County), central New York (Adirondacks and Cayuga County), southern Ontario (Parry Sound and Red Bay), central Michigan (Josco County), northeastern Wisconsin (Dorr County), southwestern Minnesota (Heron Lake), southwestern South Dakota (Black Hills), northern New Mexico (near Santa Fe), north central Arizona (Fort Verde), and central California (Tulare County). South formerly, and perhaps casually now, to the mountain regions of western Massachusetts, central Pennsylvania and in Ohio. North to the base of the Alaska Peninsula (Iak Lake), southern Yukon (Lake Tagish), southern Mackenzie (Great Slave Lake), Hudson Bay (York Factory and Norway House), southern Ungava, central Labrador (Hamilton River), and Newfoundland (Humber River).

Winter range.—Mainly within the United States, including practically all of them. South to central western Florida (Tampa Bay); the Gulf coasts of Alabama, Louisiana, and Texas; northern Mexico (Sonora and Chihuahua); and northern Lower California (Colorado delta). North to the Aleutian Islands, rarely to the Pribilof Islands, regularly to southern British Columbia (Chilliwack and Comox), the Great Lakes, the St. Lawrence valley, and Prince Edward Island.

Spring migration.—Early dates of arrival: Ontario, Ottawa, February 25; Labrador, Hamilton River, May 28; Minnesota, Heron Lake, March 17; Wisconsin, Milwaukee, March 1. Average dates of arrival: Vermont, St. Johnsbury, March 17; Quebec, Montreal, April 5; Ontario, Ottawa, April 16; Prince Edward Island, April 21; Iowa, Hillsboro, March 28; Minnesota, Heron Lake, March 26; Manitoba, Aweme, April 11. Dates of departure: Maryland, March 29; Pennsylvania, Erie, April 7; Connecticut, April 17; Rhode Island, Newport, April 25.

Fall migration.—Average dates of arrival: Massachusetts, October 5; Chesapeake Bay, October 15 (earliest September 29). Average dates of departure: Prince Edward Island, November 1; Quebec, Montreal, November 6; Ontario, Ottawa, November 21 (latest November 26).

Casual records.—Accidental in the Aleutian Islands (Unalaska, May 26, 1906).

Egg dates.—Maine and Nova Scotia: Five records, May 22 to June 8. Michigan: Six records, May 13 to 30. Manitoba: Four records, June 16 to 19. California and Oregon: Three records, April 2, May 21 and June 20.

MERGUS SERRATOR Linnaeus.

RED-BREASTED MERGANSER.

HABITS.

Contributed by Charles Wendell Townsend.

The red-breasted merganser, or sheldrake, as it is commonly called in New England, the "bec-scie" or "saw bill" of the Acadians, although often hunted, is generally classed as a fish duck and considered almost worthless. But there are other things in life besides bread and meat and dollars and cents, and the esthetic appreciation of this, as well as of many other "worthless" birds, is surely increasing.

The drake in his newly acquired nuptial plumage is resplendent with a metallic green headdress and waving crest, the whole set off by a long coral-red bill. The white ring about his neck, the reddish brown and speckled breast, the snowy flanks and wing patches, and the dark back all go to create a picture of great beauty as he swims or dives or restlessly flies to and fro among the breakers. The females and young in their more modest suits of drab and brown are not to be despised from an esthetic point of view. They, too, like the drakes, are furnished with crests.

The great multitudes of these birds off the New England coast in winter is a wonderful sight and most satisfying to the bird lover, especially as there seems to be no doubt that the numbers have increased of late years. This increase is doubtless due partly to the better enforcement of game laws and to the stopping of spring shooting, but also to the fact that the great island of Anticosti in the Gulf of St. Lawrence has become a veritable haven for breeding birds since M. Meunier, the French chocolate king, has debarred all guns from this, his domain. During the last of October and the first part of November for several miles off Ipswich Beach the water is covered with these birds, and I have no doubt that the multitude at times numbers 25,000.

Spring.—The spring migration of this bird is at its height on the New England coast in March and April, but it also continues through May. Although it does not, as a rule, breed south of southern Maine, it is not uncommon to find two or three nonbreeding birds from place to place along the coast in summer as far south as Cape Cod.

Courtship.—The courtship of the red-breasted merganser is a spectacular performance. I (1911) have described it as observed at Ipswich as follows:

The nuptial performance is always at its best when several drakes are displaying their charms of movement, voice, and plumage, before a single duck, and each vies with the other in the ardor of the courtship. The drake begins by stretching up his long neck so that the white ring is much broadened, and the metallic green head, with its long crest and its narrow red bill, makes a conspicuous object. At once the bill is opened wide and the whole bird stiffly bobs or teeters, as if on a pivot, in such a way the breast and the lower part of the neck are immersed, while the tail and posterior part of the body swing upward. This motion brings the neck and head from a vertical position to an angle of 45°. All the motions are stiffly executed, and suggest a formal but ungraceful courtesy.

This song, emitted when the bill is opened, is a difficult one to describe, but easily recognized when once heard, and remains long in the memory after one has heard it repeated over and over again by a number of merganser suitors. It is a loud, rough, and purring, slightly double note which I wrote down "da-ah," but the note is probably insusceptible of expression by syllables.

The bobbing and the love note may be given twice in rapid succession, although at times the performance is a single one, or may consist of an extensive bob, preceded by a slighter but similar one. The performance is, however, repeated at frequent or infrequent intervals, depending on the ardor and number of the suitors, and, no doubt, on the attitude of the modestly dressed lady.

Although the female merganser may remain passive and coyly indifferent, as is the habit of her sex, she sometimes responds by a bobbing which is similar to that of the male, but of considerably less range. That is to say, the neck is not stretched so straight up, and the breast is not so much depressed during the bob. She emits a single note at this time, which is somewhat louder than that of the male and is of a different quality as it is decidedly rasping. As nearly as I can remember this note is similar to the rough croaks I have heard given by these birds in Labrador when they were flying to and from their nests.

When the female responds in this manner she appears to be very excited, and the ardor of the drakes is correspondingly increased, if one may judge by the frequent repetition of the love antics and notes, and by the fact that they crowd about the duck. Every now and then she darts out her neck and dashes at the ring of suitors, just as the female English sparrow does under similar circumstances.

The bobbing up of the stern of the male is the more conspicuous as the wings are then apparently slightly arched upwards, so that the white secondary feathers are very prominent. These show at all times as the male swims in the water, but in the female they are generally, but not always, invisible.

The drakes, in their eagerness, often rush through the water with slightly opened wings making the water foam about them. Again they rise in the water with wings close to the side until they almost seem to stand on tip-toe.

Nesting.—The nest of the red-breasted merganser is built on the ground, and, although the bird is marine in its haunts, the nest is generally situated in the borders of fresh-water ponds, pools, or rivers, often, however, in close proximity to the seacoast. Occasionally it is found on the shore of the ocean itself or on coastal islands. The bird also breeds throughout the interior at long distances from the sea. According to Macoun (1909) "it does not breed in the prairie region, but prefers the clear lakes and streams of the north." The nest is generally built within 25 yards of the water.

The nest, although sometimes built in the open, is generally placed under some shelter, as the overhanging and prostrate branches of dwarfed spruces, firs, or willows, or among the roots of trees or in a pile of driftwood, and is so well concealed and the female lies so close that the intruder often nearly steps into the nest before he is aware of its presence. Macfarlane (1908) mentions a nest near the Anderson River "on the border of the 'Barrens' to the east, under a fallen tree, close to a small lake. It was a scooped-out hole lined with feathers and down, and it contained six eggs."

[Author's note: The red-breasted merganser breeds abundantly in the Magdalen Islands, Quebec, where numerous nests have been found by others, as well as by me. An island near Grosse Isle, known as Seal Island, is a famous breeding resort for this species. It is a high island of red sandstone, nearly covered with a dense forest of spruces and firs, under which the nests are concealed. A typical nest, shown in the accompanying photograph, was located in the thick woods, about 1 rod from the edge and about 40 yards from the shore; it was perfectly concealed under a dense thicket of balsam firs and would never have been discovered except that we saw the bird fly out and a few pieces of down indicated where to look. The nest was a hollow in the ground, profusely lined with gray down and a few white breast feathers; it measured 14 by 12 inches in outside and 8 by 7 inches in inside diameter. Several pieces of dry egg membrane in the nest suggested the idea that it might have been used the previous season also. It contained eight fresh eggs on June 21, 1904.

On the previous day we had found a nest of this species in a very different situation, on what is locally known as the Gully Flats, a long stretch of beaches and sand dunes with numerous marshy or grassy hollows scattered among the sand hills. The nest was in one of these small marshy hollows, which was overgrown with coarse sedges or marsh grasses; it was well concealed in the thickest grass and was made of the dry stalks of this grass, scantly arranged under, around, and partially over the eggs; very little down had been added, as the six eggs were perfectly fresh and the set was probably incomplete.

The down in the nest of the red-breasted is much darker than that found in the nest of the American merganser; it is "mouse gray" with paler centers and usually pure white breast feathers and more or less rubbish are mixed with it. When the set is complete a thick blanket of down and rubbish is provided in sufficient quantity to entirely conceal the eggs when the bird has time to cover them before leaving the nest.

Eggs.—This merganser usually lays from 8 to 10 eggs, sometimes as many as 16. The eggs are quite different from those of the American merganser. The shape varies from elliptical ovate or elliptical oval to elongate ovate. The shell is smooth but without much luster. The color varies from a rich "olive buff" or "pale olive buff" to "cartridge buff"; the olive shades are commoner than the lighter shades. The measurements of 85 eggs in the United States National Museum average 64.5 by 45 millimeters; the eggs showing the four extremes measure **67.5** by 46, 67 by **46.5**, **56.5** by 43 and 60.5 by **41** millimeters.]

Young.—Incubation lasts from 26 to 28 days and is performed entirely by the female; the drakes are rarely seen in the neighborhood during this period. P. L. Hatch (1892), who has found this bird breeding within a few miles of both Minneapolis and St. Paul, says:

Only a very few individuals have seen these ducks during the summer, for the obvious reason that, like all other locally breeding ducks, they are rarely found on the wing.

The young are active within a few hours of hatching, as has been well described by R. M. Strong (1912), and wriggle in a prostrate manner over the ground like a snake. They are soon able to run about on their feet and climb easily to the mother's back.

The food of the young consists of small fish, water insects and larvae, worms, crustaceans, and sometimes frogs. Both parents are assiduous in caring for the young. The young mergansers are carefully fed and guarded by their parents, and the family group keeps together until the young are fully grown. At the slightest sign of danger the young conceal themselves under the bushes and among the reeds of the banks of the river or pond, while the adults do their best to entice the intruder away. When suddenly disturbed in the open the young are able to make their way over the surface of the water with surprising rapidity by the combined action of the wings and legs. The noise of such a flight often confuses the enemy. On open shores I have known the young to flee from the approaching canoe, creep ashore, and, trusting to their protecting coloring, crouch motionless among the rocks and small plants.

Rev. Manley B. Townsend contributes the following pretty picture of a family party:

One summer day, toward evening, as I sat upon the shore of a wilderness lake, drinking in the beauty of the forest and the mountain, a flock of red-breasted mergansers came sailing around a rocky point, close inshore. There were 10, led by a wary old male in full adult plumage. The other nine were much duller of color. I took them for the mother and her eight children. How alert! How wary! How incomparably wild! Suspiciously they scanned me, but I sat immovable. Plainly they were nonplussed. Yet they were taking no chances. Silently they submerged until only their heads and upper necks were above the surface, and turning swam quietly off out into the lake. A calculated movement on my part, and off went the whole family, led by the father, leaving a foamy wake to mark their tumultuous passage.

Plumages.—[Author's note: The downy young red-breasted merganser is exactly like the young American merganser except for two very slight differences in the head; the nostrils in the red-breasted are in the basal third of the bill, whereas in the American they are in the central third; and the white loral stripe is tinged with brownish or fluffy but with a more or less distinct white spot under the eye.

The down is worn for a long time. The first of the plumage appears on the under parts, then comes the tail, the flanks, and the scapulars in the order named; the remainder of the body plumage follows, then that of the head and neck; the wings appear last, and the bird is fully grown before it can fly. The last of the down is on the hind neck or central back.

Millais (1913) says that in its first plumage the young male—

resembles the adult female, but the crest is less, the bill much shorter, and the plumage of the upper parts more slaty and not nearly so brown, and the cheeks more red with less white. The ends of the tail are also worn. By the end of October young males are easily recognized by their superior size and bill. It is not until December that much change takes place. The red-brown crest is then abundant, and black feathers begin to appear on the sides of the crown and cheeks, chin, mantle, and scapulars. The tail and rump also begin to molt to blue gray, and many vermiculated feathers mixed with slaty-brown ones come in on the thighs and flanks. By the end of March some white feathers appear on the scapulars and the first white, broadly black-edged feathers come in on the sides of the breast overlapping the wings. These prominent feathers are, however, never complete as in the case of the adult males, but are always divided in color, the lower halves being red and vermiculated with black from the broad black edge to the white above. The nape is now very dark-brown edged with worn blue gray, and not a clear rich red brown, as in the female. The long inner secondaries, similar to adult males, now also appear.

I have seen young males in this plumage, with immature backs and wings, and with more or less black mottling in the heads and necks, in March, April, May, and June, during which time the old males are, of course, in full nuptial plumage.

Millais (1913) says:

The young male during May and June molts all signs of the brilliant spring plumage, and passes into an eclipse similar to the adult male. It can, however, always be identified by the immature wing, which is brown and slate on all its upper parts, instead of being black with a large white area in the center, as in the adult male. During August, September, and October the general molt toward complete winter plumage is in progress, and the young male does not come into full dress until the end of November. It may then be considered adult at 17 months.

The fully adult plumage is worn during the winter and spring until the molt into the eclipse begins; this sometimes begins in March and proceeds very slowly, but more often it does not begin until late spring or early summer; it is complete in August. Millais (1913) thus describes a specimen in full eclipse plumage taken on August 20:

Head, neck, and upper breast almost exactly similar to adult female, but with only a very short area of white on the chin; mantle and scapulars blackish brown, edged with gray; wings, which have just been renewed, as in winter; rump and lower back a mixture, brownish ash-gray feathers like the female, and white vermiculated with black (as in spring); flanks and sides of the chest brownish gray like the female. There are a few slate and brown vermiculated feathers at the sides of the vent. Under parts white, and soft parts as in spring, only not so bright.

The molt of the eclipse plumage begins early in September and continues through October and November, or with some individuals much longer; I have a specimen in my collection, taken November 7, in which this molt is only fairly started; by January at the latest most of the males are in full plumage again.

Of the immature females Millais (1913) says—

in first plumage the young female is similar to the adult female, except for the less abundant crest and small area of black round the eye. Tail feathers are worn and wing markings less distinct. The scapular and mantle feathers, too, which remain unchanged until March are like nearly all immature female ducks pale and worn on their outer edges and generally gray or sandy and unlike the clean rich feathers of adults. By April it is difficult to distinguish between immature and adult females, except that the young never possess the large area of black round the eye nor the black feathers at the sides of the chin, and only the throat. The wings are as usual the main character in distinguishing age. I do not think these young birds breed nor are they adult until the following November.]

Food.—The red-breasted merganser is chiefly a fish eater, but it does not disdain to gather up crustaceans and mollusks. In fresh water it is fond of crawfish. Its long serrated bill with the teeth pointing backward is well adapted to holding its slippery prey. Nelson (1887) remarks that it feeds on sticklebacks, which abound in the brackish ponds of Alaska. It delights in the rapids of rivers, in tidal estuaries, and in the shallow places off sand beaches and at the mouth of rivers, where small fish most do congregate. Strenuous must be the life of the small fry in these regions when a large flock of mergansers are diving together.

Behavior.—The flight of the shelldrake is lacking in the initial power shown by the black duck. Unlike the latter bird, it can not spring straight up into the air. Rising from the water or ground is indeed always a laborious process, but especially so in calm weather, when there is no wind to oppose its airplanes. There is a noisy flapping of the wings and a strenuous pushing away of the water or sand with the feet for some distance before the surface can be cleared. A pair that I disturbed from the beach on a calm day showed the marks of their feet for 29 yards before they succeeded in getting away from the sand. Once on the wing their flight is noiseless and is generally close to the water, differing in this way from that of the golden eye which frequents the same shores, but which usually rises to a considerable height. When flying in pairs in the spring, the female generally precedes. It is a rapid swimmer and perfectly at home in the roughest water. As a diver the bird is truly an expert, and it disappears under water with wings close to its sides, making use of its powerful feet alone except on rare occasions when its wings are also brought into play. At times it leaps clear of the surface, describes a graceful arc and enters the water like a curved arrow, while at other times it disappears with scarcely a sign of effort. It often swims with its head and neck stretched out in front, as if it were skimming the water and straining it with its serrated bill for food. Again it advances with the head, all but the crest, below the surface apparently on the lookout for fish, and, at such times, it is constantly diving. At the moment of diving the crest is flattened down; when the birds swim before a strong wind the crests often blow up and over the head.

On the land the red-breasted merganser is an awkward walker. It often rests flat on its belly or stands up with its body at an angle of 45°. Again, it stands with its body parallel with the ground like an ordinary duck.

The courtship note, or love song, has already been described. This may be heard not only in the spring but occasionally also in the autumn, as in the case of so many, if not all, birds, a phenomenon known as the "autumnal recrudescence of the amatory instinct." The rough croak of the female at this time has also been mentioned. A similar rough croak is emitted by these birds during the breeding season; I have heard it as they flew back and forth from their nests, and once on a small stream in Labrador, in early August, a female flew close to the water ahead of the canoe croaking hoarsely. She probably wished to entice me away from her young, which may have been concealed under the bushes.

Game.—Although very "fishy eating" the red-breasted merganser is assiduously hunted along the New England coast. It is a shy bird but comes in well to wooden decoys anchored off points, along the shore or in tidal estuaries. The gunner takes his station near at hand in a blind made of brush or sea weed or sometimes of ice cakes, and is most successful in the early morning when the birds are coming in from their night's rest on the ocean. Gunning punts covered with marsh grass are also used, but one must be a skillful sculler to be able to approach within gunshot of the wary birds. Birds that are merely winged are almost impossible to recover, as they are wonderful divers and generally elude pursuit. They often swim away with only the bill above water.

Winter.—In the latter part of September in New England, the species begins to arrive from the north and becomes exceedingly numerous during October and November. In December the numbers diminish, but it is one of our most abundant waterfowl on the Massachusetts coast throughout the winter. In the spring migration of March and April the numbers increase, but it is not until the last of May, or even the first of June that they have all left for the north. But the story of the migration of this bird is not so simple as the above statement would imply, for there is a sexual as well as an age difference to be considered. The large flocks in the early fall appear to be all in brown dress, and this is the dress not only of the females and young but also of the adult males, who are then in the eclipse plumage. In November this plumage is molted, and the males appear resplendent in their courtship dress, while the females and young of both sexes leave for the south, so that during the winter months the vast majority are in full male plumage. Thus, one January day, out of 500 sheldrakes off Ipswich Beach I could count only 6 in the dull plumage. Whether these were adult females or young or both I cannot say. In March the females put in an appearance and courting begins, and by the last of April and in May the birds are largely paired, although flocks of either or both sexes are common as well as those of immature males who have not molted into full plumage are common. Some at least of the immature males are slow in changing to adult plumage, and males in nearly complete immature dress with only a few greenish feathers about the head are often to be found in April and May. On the other hand, I have seen a bird that was half molted into adult male plumage as early as the 16th of February; this was probably an adult changing from the eclipse plumage, the others immature birds.

The southern side of this picture, which rounds out and corroborates my northern observations has been given me by Mr. William Brewster, who says that in Florida, in winter, he has seen large flocks of female and immature red-breasted mergansers, and by Mr. Arthur T. Wayne (1910) who, says of this species:

From the time when these fish-eating ducks arrive until the first week in February, the adult drakes are seldom, if ever, seen, but toward the second week in February they make their appearance in large numbers.

It would seem, therefore, that some of the drakes go south to escort their partners back to the breeding home.

On the New England coast in winter this bird is to be found in largest numbers off the beaches and in the coves and harbors. It frequents also the tidal estuaries among the salt marshes into which it enters at dawn and from which it flies at sunset in order to sleep with more safety on the ocean; in this respect its habits are the opposite of those of the black duck. It is evident that the merganser is not as common of late years in tidal estuaries, as it is more apt to be disturbed by gunners and motor boats. It occasionally visits the fresh-water ponds and rivers during the migrations along the sea coast, but does not prefer them to salt water, as does its cousin the goosander. Its habits during the winter have been described above. Courting takes place all along the New England coast even that part far from the breeding range, and begins in good weather as early as February.

The long neck, head, and bill of the sheldrake, its flat body, and conspicuous white-marked wing makes its recognition in the air usually an easy one. The adult drake is easily distinguished by its reddish breast and by its crest from the goosander; the females and young can often be distinguished, even at a distance, from the very similar females and young of the goosander by the more clearly defined white throat of the latter bird. In the red-breasted species the white is less in extent and shades gradually into the brown of the neck. This is an important field mark and is often overlooked, for most authorities write that the female and young of these two species can only be distinguished in the hand by the position of the nostril, which is in the middle of the bill in the case of the goosander and nearer the base in the red-breasted species. The back of the female goosander is of pearl-blue color while that of the red-breasted species is dark ashy with a brownish tinge. These differences are also noticeable in the field.

DISTRIBUTION.

Breeding range.—Northern portions of the Northern Hemisphere. In North America, south to Newfoundland, Nova Scotia (Kings County), New Brunswick (Grand Manan), coast of Maine (Jericho Bay), northern New York (Adirondacks), southern Ontario (Parry Sound), central Michigan and Wisconsin (Green Bay), central Minnesota (near St. Paul), southern Manitoba (Lake Manitoba), central Alberta (Buffalo Lake), southeastern British Columbia (Columbia River), southern Alaska (Chichagof Island), and the Aleutian Islands. North to the Arctic coast of Alaska (Icy Cape), Mackenzie (Fort Anderson), southern Baffin Land (Cumberland Sound) and central Greenland (Upérnivik and Scoresby Sound). In the Eastern Hemisphere, the breeding range includes Iceland, Ireland, Scotland, Scandinavia, northern Russia, northern Siberia, and the Kurile Islands.

Winter range.—Mainly on the coasts of the United States. On the Atlantic coast from Maine to Florida, on the Gulf coast from Florida to Texas. On the Pacific coast from British Columbia to Lower California (La Paz). In the interior from the Great Lakes southward. In the Eastern Hemisphere it is generally distributed over Europe, the Mediterranean, north Africa, the Black and Caspian Seas, Persia, northwest India, China and Japan.

Spring migration.—Average dates of arrival: Quebec, Montreal, April 16 (earliest April 6); Prince Eward Island, North River, April 21 (earliest April 15); Ungava, Lake Mistassini, May 11; Minnesota, Heron Lake, April 3; Manitoba, Aweme, April 22; Alaska, Chilcat, May 8, St. Michael, May 24, and Kowak River, middle of June. Late dates of departure: Missouri, Kansas City, May 4; Ohio, Oberlin, May 22; Pennsylvania, Erie, May 30; Rhode Island, Newport, May 16; Massachusetts, Essex County, May 20; California, Monterey, May 25.

Fall migration.—Early dates of arrival: Massachusetts, Essex County, September 23; Pennsylvania, Erie, September 6; California, Monterey, October 9. Late dates of departure: Mackenzie River, latitude 63°, October 16; Quebec, Montreal, November 1.

Casual records.—Accidental in Cuba (Habana, December, 1891), Bermuda, and Hawaiian Islands.

Egg dates.—Labrador: Fourteen records, June 4 to July 16; seven records, June 26 to July 7. Magdalen Islands: Eleven records, June 17 to 26. Alaska: Five records, June 26 to July 9. Iceland: Five records, May 20 to June 23.

LOPHODYTES CUCULLATUS (Linnaeus).

HOODED MERGANSER.

HABITS.

In the overflowed, heavily wooded bottoms of our great interior rivers, where rising waters have half submerged and killed the forest trees, this pretty little timberland duck finds a congenial home among the half-sunken snags, stumps, and dead trees, which offer suitable nesting hollows and where its striking color pattern matches its surroundings so well that it is easily overlooked. It is a widely distributed species, found in suitable localities almost anywhere in the wooded portions of North America; it breeds more or less regularly throughout this range from Florida and Arkansas northward to northern Canada. Its center of abundance extends from the northern half of the Mississippi Valley into central Canada. The male, with his showy crest and neat color pattern, is one of the handsomest of our ducks, a fit companion for the gaudy wood duck with which it is often associated in the watery woodlands where it breeds.

Spring.—As some individuals are present both in winter and in summer over so much of its range, its migratory movements are not easily traced. The birds which have wintered just below the frost line begin to move northward before the ice has disappeared from our large lakes and streams, frequenting the smaller and swifter open streams; these birds move on as soon as conditions are favorable in their northern breeding grounds. Others come later and spread out over the country wherever they can find suitable breeding grounds.

Courtship.—The courtship of this species must be a beautiful performance. I have never seen it and can not find any account of it by American writers. Mr John G. Millais (1913) gives the following brief description of it:

The courtship, according to my friend Mr. Francklyn, consists of a sudden rise of the body with depressed crest. On coming to the water again the crest is fully expanded. The males also stretch their necks forward with fully expanded crest.

Nesting.—The birds are probably mated when they arrive on their breeding grounds and soon begin the search for a suitable cavity for a nest, but they are not particular as to the size and shape of the cavity, the kind of a tree in which they find it, or the height from the ground; almost any hole or a hollow tree trunk will do, provided it is large enough to admit the bird and of the proper shape to hold and protect the eggs; even the open hollowed top of a stump or a fallen hollow log will do; and sometimes a hole in the ground is occupied.

Mr. Herbert Massey has sent me some notes regarding two sets of eggs in his collection. A set of 12 eggs, collected by Rev. P. B. Peabody, near Hallock, Minnesota, on May 9, 1899, was taken from a cavity in an elm tree about 100 feet from a wooded creek; the cavity was in a knot hole 15 or 20 feet from the ground and was 2 feet deep. The birds had used this tree for three years and had previously nested in an exactly similar hollow in an old elm stub half a mile below. The hole was so small that the bird could hardly squeeze into it. There was a scanty supply of trash at the bottom of the cavity, apparently brought in by squirrels, and there were a few of the breast feathers of the merganser mixed with the down; the eggs were nearly fresh. Mr. Edwin S. Bryant collected the other set of nine eggs, on May 28, 1899, near White Fish Lake, Montana; the nest of moss and down was in a hollow close to the top of a leaning tamarack stub 50 feet from the ground; the tree stood on a high ridge in a dense forest half a mile back from a small lake; the moss in the nest, apparently *Usnea*, Mr. Bryant thought had been brought in by flying squirrels. The female remained on the nest while he made the dangerous climb by nailing on cleats and he had to cut through from the upper side to pull her off the eggs.

Mr. J. Hooper Bowles has sent me the following interesting notes on his experience in inducing hooded mergansers to nest in boxes, near Tacoma, Washington:

I have never found a naturally located occupied nest of the hooded merganser, my rather limited experience being confined to nesting boxes that I put up for them. This was done through the kindness of Dr. G. D. Shaver, of Tacoma, Washington, who very kindly gave me entire use of his country estate near that city. The locality selected is a lake about half a mile in diameter, entirely surrounded by dense fir and deciduous woods, with a stream running in at one end and out again at the opposite end. At the head of the lake the stream runs through a large and heavily wooded swamp, in which I put up two of my boxes. A third was put up on a dead tree standing in the middle of the lake, a fourth on a tree at the outlet, a fifth on the side of the lake, and a sixth on a lone, giant fir tree that stands on a bare hillside some 300 yards from the water at the end of the lake. All are about 18 feet above the ground, or water, and seemed to cover as well as possible the nesting sites that might be suitable for these birds. Not to take up too much space, I will say that a set of 10 eggs was taken from one of the boxes in the swamp at the head of the lake, a brood being reared in the other box there. A set of 11 eggs was taken from the box on the tree in the lake, the bird using the box on the lone fir on the hillside for her second, and this time successful, attempt at rearing a brood for the season. The box on the side of the lake showed no signs of being visited, but down feathers on the entrance of the box at the outlet gave evidence that it had been thoroughly examined, although considered unsuitable for some reason. The birds are so exceedingly shy that I have never been able to see them enter their nests, but when leaving they come out at full flight, which would seem almost an impossibility under the circumstances. The eggs are just about the size, shape, and color of white billiard balls, and every bit as hard in their composition.

Mr. Glen Rinker (1899) describes two nests, found near Unionville, Missouri, as follows:

On the day mentioned I was sitting by the side of the lake watching a pair of females, when one of them raised and flew within 20 feet of my head. I was "all eyes" when she alighted on a snag about 50 yards back from me, and I noticed it looked down the hollow several times. I attempted to get closer, but she saw me and flew away. I then proceeded to examine the snag. It was about 2 feet in diameter and 10 feet high; the top was hollowed out to a depth of about 2 feet, and looked charred as though it had been burned. The nest was composed of leaves and some grass and a little moss, and had a complete lining of down. The eggs, six in number, were white, and were more round than most duck eggs.

June 13, 1899, found me near the lake again, but farther off in a thicket, watching a cardinal whose nest I knew was near. To my left was a tall bank where a lot of trees had fallen and which was overgrown with hazel bushes. I heard a whistle of wings, and looked up just in time to see a merganser settle down near on old stump. I waited about 5 or 10 minutes, and then walked quietly up to where I saw her light. When about 5 feet from the place, she jumped up with a quack, and started for the lake.

Now, I have several keys and other books, and they all say the hooded merganser nests in hollow trees and stumps, but this nest was on the ground under the roots of the stump, in a sort of a cave that was about 14 inches back under the stump. The nest was composed of about the same material as the other, but did not have as much down. There were only four eggs in the nest, so I left it until the 17th, when I collected them as only one more had been laid. On blowing them, incubation was just perceptible. The nest was in such a dark place that to photograph it was impossible without overexposing the outside. The down made the eggs hard to distinguish.

One of Major Bendire's correspondents, Mr. T. H. P. Lamb, writes that the Cree Indians of Saskatchewan call this bird the "beaver duck" and claim that it lays its eggs in deserted beaver houses, using the entrance under water, also occasionally in old muskrat houses. This seems hardly likely, however.

Where suitable nesting sites are scarce, the hooded merganser sometimes contends with other species of tree-nesting ducks for the possession of a coveted home and occasionally they share the home between them. Mr. George A. Boardman has been several times quoted as having witnessed such a contest between a wood duck and this species, which resulted in the two females laying in the same nest and occupying it by turns. Mr. George D. Peck (1896) writes:

I believe it is well known that the wood duck often drives the merganser from her nest, and in one nest I found 30 eggs of wood duck and 5 eggs of merganser. The hollow in the tree in which the nest was placed was not very large and the eggs were several layers deep.

In Maine, Mr. William Brewster (1900) says that several of the rounded, pure white, thick-shelled eggs of the hooded merganser are sometimes included in a set of the green, thin-shelled eggs of the whistler.

Eggs.—The hooded merganser is credited with laying anywhere from 6 to 18 eggs; probably from 10 to 12 would cover the usual numbers. The eggs are oval or subspherical in shape. The shell is thick and hard, smooth and usually quite glossy. The color is pure white, but they are often nest stained. The measurements of 116 eggs in various collections average 53.5 by 44.9 millimeters; the eggs showing the four extremes measure **57.5** by 45.2, 55.5 by **45.5**, **50** by 43, and 50.5 by **41.5** millimeters.

Mr. William Evans (1891) says that the period of incubation is 31 days: it is wholly performed by the female. The male is said by most observers, to desert the female as soon as the eggs are laid, but the following note by Mr. J. W. Preston (1892) is of interest in this connection:

While camping on Little Twin Lakes, northern Iowa, some years since, I noticed a male hooded merganser circling around a grove so often that it seemed certain that he was feeding his mate, which they do at incubating time. I concealed myself and watched for a long time, and finally was rewarded by seeing the fellow fly plump into a hollow in a gigantic oak. It would seem to be a piece of recklessness; certanly, if he had not aimed well he would have suffered for the error.

Young.—Several writers have written that the female conveys the young in her bill from the nest down into the water, soon after they are hatched, and Dr. P. L. Hatch (1892) says he has seen it done; he writes:

In one instance, a lady sharing my interest in birds and game, while rowing with me, noticed what we supposed to be a wood duck carrying her chick by the neck from a tree into the water. We waited in vain some time to see if the bird would not bring another young one. Reaching the middle of the small lake, we saw the duck, by the aid of the field glass, resume the loving task and discovered the bird to be a female of the species under consideration.

I suspect, however, that this instance was exceptional and that the usual method of precedure is for the mother to coax the young to climb up the edge of the cavity and then drop down into the water, or onto the soft ground, if circumstances are favorable, as is customary with most tree-nesting ducks. Their little bodies are so light and so elastic that the fall does not hurt them. Audubon (1840) says:

The affectionate mother leads her young among tall rank grasses which fill the shallow pools or the borders of creeks and teaches them to procure snails, tadpoles, and insects.

Mr. E. A. Samuels (1883) writes:

When the female is suddenly surprised, while with her young in a stream or pond, she gives a guttural, chattering cry, when the whole brood dives and swims off under the water to the shore, where they conceal themselves in the aquatic herbage. While they are thus retreating, the mother simulating lameness, almost exactly like some of the shore birds on the beach, flutters before the intruder, using every artifice to decoy him from the neighborhood of her young, when she takes wing and flies off. If, however, she has sufficient notice of the approach of a person before he reaches gunshot she swims rapidly off, with her whole brood paddling behind her, until she turns a point or neck in the pond or stream where she happens to be, when, silently creeping into shore, she, with her brood, hides herself in the herbage on the land until the danger is past. When about two-thirds grown, these young mergansers, like the young of most of the other fowls, are excellent eating. They are called "flappers" because of their habit of flapping their wings on the water to aid their escape from pursuers.

Plumages.—The downy young is thickly and warmly clothed with soft down in deep, rich shades of "bister" or "sepia" above, including the upper half of the head, the hind neck, and the flanks; the sides of the head, neck, and cheeks, up to the eyes, are "buff pink" or "light vinaceous cinnamon," the chin, throat, and under parts are pure white; and there is an obscure dusky band across the chest and an indistinct white spot on each side of scapular region and rump.

In the first plumage the sexes are alike and much resemble the adult female, but they are browner on the back and have undeveloped crests. Young males wear this immature plumage all through the first year, with only a slight change toward maturity during ths first spring and the following summer. The summer molt leaves them still in immature plumage and with but little change in the new wings, which still lack the pearl-gray lesser coverts and in which the greater coverts are only slightly white-tipped. In November and

December of this, their second, winter they begin to assume a plumage resembling that of the adult; the molt begins with the appearance of black feathers and white feathers in the head, spreading downward to the breast, flanks, and scapulars until by March or April a nearly adult plumage is assumed. In this plumage the colors are all duller than in old males; the crown, back, and rump are browner; the gray lesser wing-coverts are acquired, but the wings are otherwise immature. A partial eclipse plumage is assumed during the next summer, when the bird is 2 years old and late in the fall, November or December, the fully adult plumage is acquired. Young females can be distinguished from adults during the first year by their undeveloped crests and their duller and browner coloring everywhere; they become indistinguishable from the adults during the second winter.

Adult males have a semi-eclipse plumage in summer, in which the head and neck become largely mottled with brownish and the breast and flanks lose their brilliant colors and resemble those of the female. The double molt is probably not complete, though the whole plumage is changed at least once. The full plumage is assumed early in the fall, much earlier than in young birds, and is usually complete in October.

Food.—The hooded merganser lives and feeds almost exclusively on and in fresh water; I believe that some of its food is obtained on the surface, but it is an expert diver and finds much of its food on muddy or on stony bottoms. Its food is mostly animal, and consists largely of insects. Like other mergansers, it is expert at chasing and catching small fish, which probably constitute its chief supply; in muddy pools it finds frogs and tadpoles and snails, and other mullusks; on clear stony bottoms it obtains crawfish, caddis fly larvae, and dragon-fly nymphs; sand eels, small crustaceans, beetles, and various aquatic insects are also eaten. It is also known to eat some vegetable food, the roots of aquatic plants, seeds, and grain. Dr. F. Henry Yorke (1899) recognized among its vegetable food the following genera of water plants: *Limnobium, Myriophyllum, Callitriche,* and *Utricularia.*

Behavior.—Dr. D. G. Elliot (1898) writes of the flight of this species:

On the wing it is one of the swiftest ducks that fly, and it hurls itself through the air with almost the velocity of a bullet. Generally it proceeds in a direct line; but if it is alarmed at any object suddenly appearing before it, the course is changed with the swiftness of thought, and a detour made before again taking the first line of progression. Sometimes, without apparent reason, the course will be altered, and away it shoots at right angles to the first route; and again, it vacillates as though uncertain which way to take, or as if it was looking for a good feeding place. Usually five or six, but more frequently a pair, are seen flying together, and often, on dull days when the lookout in a blind is somewhat relaxed, and the sportsman is consoling himself for lack of birds with possibly a nap or the lunch basket, the first intimation of the presence of a hairy crown is given by one or more flashing close overhead with a startling whirr, and then as rapidly disappearing in the distance. It requires a steady hand and a correct eye to kill them on the wing, and the gunner must be ever mindful of the good old adage in duck shooting, "Hold well ahead." It rises from the water without any preliminary motions, and is on the wing at once, and in full flight, the pinions moving with a rapidity that almost creates a blur on either side of the body, the outline of the wing disappearing.

Dr. P. L. Hatch (1892) says:

Once in January, 1874, when the mercury had descended to 40° below zero while a north wind was blowing terrifically, I saw a flock of six of this species flying directly into the teeth of the blizzard at their ordinary velocity of not less than 90 miles an hour. The compactness of their flocks of half a dozen to 15 in their flight is characteristic, and their directness fully equal to that of the green-winged teal.

Mr. J. W. Preston (1892) observes:

A pleasing characteristic of the species is the manner of flying during nesting time. One may see them chasing round and round some wooded lake, speeding ever with a thrilling impetuosity; uttering a peculiar note as they glide along; then they have darted out into the forest, leaving the beholder pleased with the performance, and none the wiser as to the nest site. I timed one of this species, and it made its mile in less than one minute.

Audubon (1840) says: "When migrating, they fly at a great height, in small loose flocks, without any regard to order."

Of its swimming and diving habits, Doctor Elliot (1898) says:

The movements of this bird upon the water are quick and active, and it swims rapidly and dives with great celerity. It is a beautiful object, and few birds surpass the male in attractiveness as he swims lightly along, elevating and depressing his beautiful crest. If suspicious, this species will sink the body until the water is almost level with the back, and sometimes disappears beneath the surface, apparently without effort, as if some unseen hand was pulling it down. When wounded it is one of the most difficult birds to secure; and it dives with such quickness, remains under water so long, and skulks and hides with so much skill that it is very apt to make its escape, and always tries the patience of its pursuer, whether dog or man, to the utmost. Their progress under water is extremely rapid, and the wings as well as the feet are used as means of propulsion, perhaps more dependence being placed upon the wings, and they may be said to fly beneath the surface.

The same writer says of its voice: "It utters a hoarse croak like a small edition of the note of the red-breasted merganser."

Audubon (1840) writes:

Their notes consist of a kind of rough grunt, variously modulated, but by no means musical, and resembling the syllables "croo, croo, crooh." The female repeats it six or seven times in succession, when she sees her young in danger. The same noise is made by the male, either when courting on the water or as he passes on wing near the hole where the female is laying one of her eggs.

Fall.—A study of the migration records will show that the hooded merganser is not an early migrant in the fall, nor is it the very latest; the main flight comes along during the latter half of October and first half of November; the latest stragglers often linger until frozen out. Audubon (1840) gives an interesting account of its behavior at this season, which I quote, as follows:

At the approach of night, a person standing still on the banks of such a river as the Ohio first hears the well-known sound of wings whistling through the air, presently after, a different noise, as if produced by an eagle stooping on her prey, when gliding downward with the rapidity of an arrow, he dimly perceives the hooded mergansers sweeping past. Five or six, perhaps 10, there are; with quick beats of their pinions, they fly low over the waters in wide circles. Now they have spied the entrance of a creek; there they shoot into it, and in a few seconds you hear the rushing noise which they make as they alight on the bosom of the still pool. How often have I enjoyed such scenes, when enticed abroad by the clear light of the silvery moon, I have wan-- dered on the shores of la belle rivière to indulge in the contemplation of nature!

Up the creek the mergansers proceed, washing their bodies by short plunges, and splashing up the water about them. Then they plume themselves, and anoint their feathers, now and then emitting a low grunting note of pleasure. And now they dive in search of minnows, which they find in abundance, and which no doubt prove de- licious food to the hungry travelers. At length, having satisfied their appetite, they rise on wing, fly low over the creek with almost incredible velocity, return to the broad stream, rove along its margin until they meet with a clean sand beach, where they alight, and where, secure from danger, they repose until the return of day. A sly raccoon may, when in search of mussels, chance to meet with the sleeping birds, and surprise one of them; but this rarely happens, for they are as wary and vigilant as their enemy is cunning, and were the prowler to depend upon the hooded mergansers for food, he would be lean enough.

Game.—From the sportsman's standpoint this is not an important species. It is a difficult bird to hit on the wing, it is small and its flesh is not particularly attractive to eat; it is often very fat and when it has been feeding on grain or vegetable food its flavor is not bad. It is rather tame and unsuspicious, coming readily to decoys. It is known by a variety of names such as " wood sheldrake," "water pheasant," "hairy crown," etc.

Winter.—The hooded merganser is resident throughout the year over much of its range, wintering as far north as it can find open water in which it can obtain its food supply. Doctor Hatch (1892) says that, in Minnesota, " they stay as long as the ice will let them on the shores of the lakes, whence they go to open rapids, and late in November mostly drift more southward." Dr. Amos W. Butler (1897) writes, in regard to Indiana:

Throughout the State the hooded merganser may be found in winter, the more numerous the more open the winters, and always attracted to the open water, so that in the most severe winters they are most to be observed on the rapid streams of southern Indiana, where ripples and rapids are about the only places they can find at which to congregate.

Although a few migrate beyond our borders, the principal winter home of the species is in the States bordering on the Gulf of Mexico, where they frequent the inland waters, seldom if ever being seen on salt water.

DISTRIBUTION.

Breeding range.—Temperate North America, locally. East to western New Brunswick (St. Croix River), eastern New York (Adi- rondacks and Catskills), central Pennsylvania (Williamsport), and eastern South Carolina (Berkeley County). South to central Florida

(Titusville and Fort Myers), southern Tennessee (near Chattanooga), northeastern Arkansas (Big Lake), northern New Mexico, and north-western Nevada (Truckee River).   West to Oregon and northwestern Washington (near Tacoma).   North to southeastern Alaska (Stikine River), central British Columbia (Cariboo district), southern Mac-kenzie (Great Slave Lake), northern Manitoba (Churchill), eastern Ontario (Algonquin Park), and perhaps the interior of Labrador.

Winter range.—Mainly in the Southern States.   North to Massa-chusetts, Pennsylvania, Lake Michigan, Nebraska, Colorado, Utah, and southern British Columbia (Comox and Okanagan).   South to Cuba and southern Mexico (Orizaba and Jalapa).

Spring migration.—Early dates of arrival: Iowa, central, March 5; Minnesota, Heron Lake, March 20; Ontario, Ottawa, March 21. Average dates of arrival: New York, western, March 20; Ontario, Ottawa, April 18; Michigan, southern, March 19; Iowa, central, March 22; Minnesota, Heron Lake, April 5.

Fall migration.—Average dates of arrival: New York, western, October 15; Virginia, Alexandria, October 26.   Average dates of de-parture: Quebec, Montreal, October 29; Minnesota, southern, No-vember 10; Iowa, central, November 26.

Casual records.—Accidental in Bermuda (January 10, 1849, and December 23, 1850), Great Britain (Wales, winter 1830–31), Ireland (County Cork, December, 1878, and County Kerry, January, 1881), and Alaska (St. Michael, October, 1865).

Egg dates.—Michigan: Four records, April 22 to May 19.   Illinois: Three records, March 15, April 29, and May 5.   Iowa: One record, June 5.   Montana: Four records, April 27 to May 28.   Washington: One record, April 21.   Missouri: One record, June 13.

MERGELLUS ALBELLUS (LINNAEUS).

SMEW.

HABITS.

This is a Palaearctic species, of rare and doubtful occurrence as a straggler, on the North American continent.   It is included in our check list on the strength of a female in the British Museum pur-chased from the Hudson's Bay Company, said to have been taken in Canada, but with no data as to the exact locality; there is also a record of a female, which Audubon (1840) claims to have taken in Louisiana in 1819, which is open to doubt.   Evidently Wilson's (1832) references to the abundance of this species in New England were based on incorrect identifications.

On the Atlantic and Mediterranean coasts of Europe it is fairly common in winter and in the eastern Mediterranean quite abundant, whence it retreats northward in April to its breeding grounds in the northern portions of Europe and Asia.

Courtship.—Millais (1913) describes the courtship performance of the smew as follows:

The male swims slowly around the female, sometimes with the long scapulars slightly raised or expanded. The head and neck is often moved slowly forward in a pushing manner, and when about to make the act of display the neck is drawn back as far as possible between the scapulars. All this time the crest is raised and spread in a very peculiar fashion. It is separated into two parts, the front consisting of only a few feathers of the front of the crown. These stand quite clear away from the latter part of the crest, which is expanded above the nuchal patch of black. Often a single white feather stands out alone connecting the two sections of the crest. The nuchal patch lies flat, and the back of the head is not distended in any way.

The next act of show is to push farward the neck somewhat slowly and then back as far as possible, the crest on the crown being raised as already indicated, and the chin pointed upward, whilst there is a slight rise of the forepart of the body as it is lifted from the water. During this sedate movement the mating cry of the male is uttered. The throat is slightly swelled, and the note is a prolonged croak or grunt like the word "err-err-err-umph," the last sound being an exhalation to clear the lungs, and seeming to be an effort on the part of the bird. During this movement the bird is stationary, with the tail either lying under or on the surface of the water.

On the completion of the movement there is a quick forward dip of the head and bill, followed by a sudden rise of the forepart of the body out of the water, something like a "mallard and teal" show, but not nearly so upright. In fact, it is almost a forward movement. At the same time the feet are paddled vigorously to maintain equilibrium. The call is often made as the bird throws itself up and forward.

At the end of this movement the bird often drops to the water with neck out-stretched and parallel to the water, and when in full show often makes a little rush forward.

Nesting.—For a long time the breeding habits of the smew remained a mystery, until Mr. John Wolley received some authentic eggs and established the fact that the species bred in holes in trees in Swedish Lapland; he published a full account of it in The Ibis in 1859, from which Yarrell (1871) and others have quoted extensively. Probably all statements to the effect that the smew nests on the ground are erroneous, as it is known to nest only in hollow trees and in nesting boxes put up by the Laps for ducks to nest in. It is not known to nest north of the tree limit.

The down in the nest is described by Millais (1913) as "small and grayish white, freely intermixed with fine white feathers. Fragments of rotten wood and moss may also be found mixed with the down at the bottom of the nest hole or nesting box."

Eggs.—He says of the eggs:

Usually 6 to 9 in number, but 10 have been recorded. They are creamy in color and smooth in texture. Average size of 107 eggs, 52.42 by 37.46 millimeters; maximum, 58 by 40.5; minimum, 47.7 by 34. They are smaller on the average than widgeons'; decidedly shorter as a rule, and not quite so broad; but the measurements of the species overlap. Full clutches may be taken in northern Europe from the last week of May to the middle of June.

Plumages.—Dresser's description of the downy young, as given by Millais (1913), is as follows:

Upper parts, including the sides of the head below the eye, but only the back of the neck, dark blackish brown, darkest on the crown and the lower part of the back; at the base of the wing joint a white spot, and another close to it, but rather lower down the back, and on each side of the rump another white spot; below the eye a very small white spot; under parts white; breast and flanks pale grayish or sooty brown. One young bird, which can only be 2 or 3 days old, has the bill so slightly serrated that the serrations can only be seen when very closely looked into; but another, which is a few days older, has the serrations very distinct.

The sequence of plumages to maturity is thus outlined by Millais (1913):

The young male in first plumage very closely resembles the adult female and young female, and until December it is very difficult to tell the sexes apart except by dissection. At the end of five months, however, the young male begins to turn much darker. The nape is now often changed to new black feathers and the upper wing has a larger area of white; the lores, too, show many dark feathers. The tail is often complete by December. So the advent of the male plumage continues to advance on the upper parts until April when the usual halt takes place, until an eclipse plumage closely resembling that of the adult male is assumed. The wings, which are always the key to identification, are not the same as the adult male, and always have more or less brown or blackish edges on the upper coverts instead of being the pure white of the adult male. The immature male passes through the same stages as the other mergansers and assumes its first complete plumage in late November—that is, at 17 months.

The same writer describes the molts and plumages of the adult as follows:

The adult male assumes its eclipse plumage in June. As Naumann points out, it closely resembles that of the adult female, though I fancy that the bird from which his description was taken was not yet in full eclipse, as it differs somewhat from those I have seen.

In July the adult male has gained a very rich red-brown crest, somewhat fuller than the female, and it can always be distinguished from the female by the rich coloring of the wing, the white irides, and the black patch round the front of the eyes; also by its larger size black edges to outer white scapulars, and a few vermiculated feathers above the thighs on the flanks. In other respects the whole of the rest of the plumage is like the adult female, except the mantle, which is nearly black. Wings as in winter. The autumn molt proceeds in the usual manner, and the adult male regains its winter plumage by the end of November. Sometimes a few eclipse plumage feathers remain in the plumage until the new year, but this is unusual.

Food.—The food of the smew consists chiefly of small fish, crustaceans, small frogs, water insects, mollusks, and sand eels, which it obtains by diving. It is an expert diver and exceedingly swift in its pursuit of fish. Naumann, as quoted by Millais (1913), gives a pleasing account of a flock of smews fishing in a river full of ice:

To watch a flock of smews at their fishing unseen affords a pleasant amusement. At one moment all are swimming together, and then in a flash all have vanished from the surface, the water is stirred by their paddling in it, and finally one after another appears on the top again, but scattered, and where there is room, often 30 to 50 paces from the original place; they assemble again, dive yet again, and to the surprise of the observer they appear this time perhaps quite close to him on the surface. It is very wonderful how they obtain their means of food only by diving often from such small opening in the ice of only a few square feet; and they conduct their fish chase then under the ice roof, but they always come up again to the open places to breathe and rest for a few moments; and this is a proof that their sight under water must reach to a considerable distance. In places where the open water does not contain enough fish, or they have themselves caught or scared away a fish, they scour the bottom for insects or frogs taking their winter sleep in the mud, or for fish which have taken refuge and hidden there.

Behavior.—Millais (1913) says of the flight, swimming, and diving habits of the smew:

The flight is very rapid, and the neck held very stiff and straight. When going at full speed they swing from side to side, and often shoot down suddenly close to water. On alighting on the water they often dive at once as a precautionary measure and on rising to the surface stop, preen, and bathe. Like the other mergansers, they are constantly preening their feathers, whether on land or on the water. In winter they seldom come ashore, but in summer they often emerge from the water and lie for hours asleep amongst the stones or on some sand spit or island. Hennicke observed a flock of smews in Finland, in September, 1900, resting on tree trunks in the middle of the rapids of the Ulea River.

The swimming attitudes are the same as the red-breasted merganser, and they only "sink" the body in the water when alarmed. When on feed they swim lower, and the tail trails, or is sometimes a little lower than the line of the water.

They dive with swiftness, and apparently more vertically than the other mergansers—this may be due to their feeding on slower-moving fish—but they do not seem to range over the same extent of ground as the larger species. I have, however, seen a smew making long horizontal dives like a red-breasted merganser, and in this instance it was probably hunting for food or in pursuit of trout. Certainly the few smews I have seen on feed did not change their ground much, but came up again near to the spot they had dived, and it may be true, as some authors have asserted, that this is their general habit. I do not think that any of the mergansers use their wings under water as the eiders do. They all seem capable of swimming distances under water without coming to the surface to breathe. If the flock has separated, it soon swims together again before again diving.

Of the voice he says: "Smews very seldom make any cry, and in the winter only a harsh croaking note."

Winter.—The smew migrates to its winter home in southern Europe rather late in the fall, from the middle of October to the end of November, where it frequents the lakes and rivers until it is driven by the formation of ice in its favorite resorts to the estuaries, bays, and even the open seas, where it associates, on very intimate terms, with the golden eye. It is at all times shy and difficult to approach.

Naumann, as quoted by Millais (1913), speaks of its haunts, as follows:

The smew does not seem to like the open sea, and it is seen almost always near land, in summer in deep narrow gulfs running far inland, in estuaries, or in land lakes near and on other pieces of open water, less often on salt than on fresh. With us in the winter it most often keeps to the rivers and streams both in flat and hilly country, wooded or quite open country, and from there it visits other open places of the lakes, ponds, brooks, or even quite small springs. Should the cold become more severe, and should therefore fewer places remain free from ice, then they go the round from one to the other and betake themselves as soon as they are disturbed to the next place and continue thus, doing this daily for weeks, and repeat this series of changes though not at regular intervals, until the cold weather either forces them farther southwest or the approach of milder weather opens again larger places for them on the rivers and permits them to remain there. They can endure the most severe cold quite comfortably, and it is only the breaking up of the ice on the rivers which they hate, particularly if the so-called ground ice is driving hard; in that case they take refuge on the open places of quiet water in the neighborhood of the former and fly from one to another. In time of need they do not despise an occasional stay on the smallest springs and brooks, and in our neighborhood often appear at such times quite close to the villages.

DISTRIBUTION.

Breeding range.—Northern Europe and Asia. From northern Lapland and Finland eastward across northern Russia and Siberia to Bering sea.

Winter range.—On the eastern coasts from Norway to Morocco and inland as far south as the Swiss lakes, the Mediterranean, Black and Caspian Seas, Persia, Afghanistan, north India, China, Japan and the Commander Islands.

Casual records.—One reported taken by Audubon near New Orleans in winter of 1817. Specimen in the British Museum purchased from the Hudson's Bay Company, and one in the Tristam collection, both supposed to have come from North America.

Egg dates.—Northern Europe: Twelve records, May 23 to June 26; six records, May 28 to June 14.

ANAS PLATYRHYNCHA Linnaeus.

MALLARD.

HABITS.

Spring.—With the first signs of the breaking up of winter, when the February sun, mounting higher in the heavens, exerts its genial power on winter's accumulations of ice and snow, and when the warm rains soften the fetters that have bound the lakes and streams of the middle west, the hardy mallards, the leaders in the migrating hordes of wild fowl, leave their winter homes in the Southern States and push northward whenever they can find water, about the margins of the ponds, in open spring holes, and among the floating ice of rivers and streams, flushed with the spring torrents from melting snow banks. Because they follow so closely in the footsteps of retreating winter the earliest migrants have been termed "ice mallards" by the gunners. The spring migration starts in the Central Mississippi Valley soon after the middle of February and advances north as fast as conditions will permit. By the second week in March the advance guard has reached the Northern States, and large flocks may be seen circling about over the lakes in search of open water or dropping into sheltered pond holes to feed on the first tadpoles and other small fry thawed out by the warm rays of the advancing sun. It is usually three weeks or a month later before they penetrate into central Canada and they do not reach the northern limits of their breeding range in the Mackenzie region until the first week in May, or in Alaska until the middle of May.

Throughout all the central portions of its range, in the Great Plains region of the Northern States and central Canada it is one of the most abundant and most widely distributed of the ducks, as well as the best known and most important of our game birds; eastward of the Prairie States it diminishes in abundance and is almost wholly replaced by its near relative, the black duck.

Courtship.—The plumage of the mallard drakes is at its highest stage of perfection before the end of winter, and the first warm days stimulate these vigorous birds to migrate to their northern homes. Many of them are already mated when they arrive and the flocks of mated birds soon break up into pairs and fly about in search of suitable nesting sites. Others are busy with their courtships, which are conducted largely on the wing. I have seen as many as three males in ardent pursuit of one female flying about, high in the air, circling over the marshes in rapid flight and quacking loudly; finally the duck flies up to the drake of her choice, touches him with her bill nd the two fly off together, leaving the unlucky suitors to seek other mates.

Dr. Charles W. Townsend (1916) describes the courtship of this species as follows:

When the mallard drake courts, he swims restlessly about following or sidling up to a duck. She may lead him quite a chase before she vouchsafes to acknowledge his presence, although he is continually bowing to her, bobbing his head up and down in nervous jerks so that the yellow bill dips into the water for a quarter of its length and comes up dripping. He also rears himself up in the water and from time to time displays his breast. She occasionally turns her head to one side and carelessly dabbles her bill in the water, but sooner or later, if all goes well, she begins to bow also, less vigorously at first—not touching the water at all—and to the empty space in front of her. Suddenly she turns and the pair bow to each other in the same energetic nervous jerks, and, unless a rival appears to spoil the situation, the drake has won his suit.

Mr. H. Wormald (1910) has given a detailed account of the courtship of the mallard, illustrated with excellent drawings, to which I refer the reader. He says:

The performance usually begins by four or five drakes swimming round a duck with their heads sunk, and their necks drawn back, and in this attitude they have the appearance of being most unconcerned. This I will call action No. 1. After swimming round in this fashion for some little time, the mallards will suddenly lower their bills so that the tips of them are under the surface, and as they do so they stand up in the water and then rapidly pass their bills up their breasts. This motion is performed with somewhat of a jerk, and if one observes very closely, a tiny jet of water will be seen to be thrown out in front by the bill being jerked from the water; this is interesting, as one also finds this jet of water in the spring "show" of the golden-eye, but in this case it is made by the drake kicking out a small jet of water with his foot while he quickly throws back his head.

The mallard while performing action No. 2 as I will designate it, utters a low note rather difficult to describe, but I think it may be said to be a low whistle with a suspicion of a groan in it, as though it caused the bird an effort to utter. Following this, the mallards lower their breasts and raise their tails two or three times in quick succession; and this, which we may call action No. 3 is often followed by a repetition of actions Nos. 1 and 2. A quick "throw up" of head and tail, with the feathers of the head puffed out, is action No. 4, and this is followed quickly by action No. 5 in which the drakes stretch out their necks with their throats just over the water and swim rapidly about in different directions, when, apparently by common consent, they all come back to action No. 1, and go through the whole performance over again.

Nesting.—In North Dakota I found the mallard breeding quite commonly, in 1909, about the lakes and sloughs of Nelson and Steele Counties, although it was outnumbered by at least three other species, the blue-winged teal, the pintail, and the shoveller. It begins laying in that region early in May, though fresh eggs were found as late as May 31. The locality there chosen for its nest is generally on or near the edges of a slough or lake, either among dry, dead flags where the ground is dry or only slightly marshy, or upon the higher land not far from the water and among thick dead reeds. It also nests on the open prairies and often at long distances from any water. Two of the nests we found were on an island in Stump Lake in the middle of a patch of tall, dry, reedlike grass, locally called "queen of the prairie" (*Phragmites*) which grows higher than a man's head. The nest is usually well hidden and consists of a hollow in the ground, well lined with broken dead reeds or flags, apparently picked up in the immediate vicinity, mixed with dark gray down and a few feathers from the bird's breast; the down is thickest around the edges of the nest and increases in quantity as incubation advances.

During my two seasons spent in southwestern Saskatchewan, 1905 and 1906, mallards were frequently seen flying about in pairs up to the middle of June, indicating that they had not all finished laying at that time. They were not as common as several other species of ducks, but were seen on many of the lakes and nearly all of the creeks. Only seven nests were found during the two seasons and five of these found on the great duck island in Crane Lake on June 17, 1905; these five nests contained 1, 2, 6, 8, and 9 eggs, respectively, showing that they breed later in that region than farther south, although these may have been exceptional cases.

While my own personal experiences with the nesting habits of the mallard undoubtedly illustrate its normal habits, certain departures from its customary manners of nesting are worth mentioning. Dr. Morris Gibbs (1885) mentions finding a nest " placed in a hollow stub, similiar to the wood duck's nest."

Mr. L. E. Wyman says of its nesting habits in the vicinity of Nampa, Idaho:

Breeds in the tules and swampy creek bottoms, and to some extent around the reservoir, where lack of that sort of vegetation essential to its breeding operations has led to its nesting in some cases in alfalfa fields a quarter of a mile away, a well-beaten path connecting nesting site and water.

Mr. J. Hooper Bowles (1909) describes the nesting habits of the mallard in Washington as follows:

West of the Cascades the nest is often built at a considerable distance from water, a nest found near Spanaway Lake serving for an example. It was situated 150 yards from the lake under a pile of brush on a bushy hillside. The duck, when flushed, tumbled along the ground, feigning a broken wing, but she soon flew quacking to the lake, where she was very shortly joined by the drake. Other nests are built in the heavy fir timber, being placed at the base of a giant tree in exactly the same manner as nests of the sooty grouse.

Mr. Robert B. Rockwell (1911) thus describes a rather unusual nest which he found in the Barr Lake region of Colorado:

On May 11, 1907, while wading out from shore through a sparse, burn'd-over growth of cat-tails, skirting a small lake, a female mallard flushed noisily from a large musk-rat house and revealed a beautiful set of 11 eggs deposited in a hollow, scraped in the dead cat-tails and débris forming the house, and well lined with down. The house was very conspicuous, standing over 2 feet above the surface of the water surrounding it, and the nest was an open one, as can plainly be seen from the accompanying illustration. There was no apparent attempt at concealment. The female flushed when we were fully 30 yards from the nest, and the male swam about well out of gunshot. A week later (on the 18th) we succeeded in approaching to within 10 feet of the brooding female, who was in plain sight even from a considerable distance. The nest was in much the same condition as on the preceding visit, but the downy lining was much less in evidence. On the 24th we found that the muskrats had been adding to the house, with the result that the mother bird, in order to keep her treasures from being buried, had been forced to move her nest over toward the edge of the pile. In fact four of the eggs were missing on this date, and we surmised that they had been pushed off into the water during the moving process. A week later (May 31) the house had been built up much higher, and the nest was on the ragged edge of the pile, with the eggs apparently far advanced in incubation. On June 8 the eggs had been hatched, and in our examination of the nest we were surprised to find the four missing eggs deeply buried in the débris at almost the exact spot where the nest was located when first found. A fascinating bit of the family history would have undoubtly been revealed had we been enabled to observe the attitude of the busy muskrats toward the brooding mother bird, and the process of moving the nest.

Mr. J. Hooper Bowles has sent me the following interesting notes:

In the vicinity of Takoma, Washington, the mallards have an extremely wide range of variation in their nesting habits, both as to date and locations for nesting sites. Many of them are paired by the middle of January, and the first eggs are usually deposited during the last week in February. Fresh eggs may be found from this date up to the middle of June, but the great majority are hatched by the latter part of April.

The early nests are nearly always, in my experience, placed either in trees or far back in the dense fir timber on the ground, in the latter case usually at the base of some huge fir, or under a fallen log among dense brush, often a quarter of a mile from the water. It has always been a mystery to me how a bird of the open water, like a mallard, can find its way back to the nest through timber and brush so thick that it requires all the ardor of the oologist. I well remember my first sight of mallards under these conditions, when I was hunting horned owls in some very heavy timber. It was a pair flying about ten feet from the ground and passing only a few feet from where I stood motionless. They were evidently hunting a favorable nesting site and they threaded their way swiftly, but surely, among the tree trunks, seeming as much at home as any grouse.

When building in trees the nests are never in those of the large tree-nesting birds, but are usually built in the fork of some large tree where an abundant growth of moss and tree ferns make the site both secure and well concealed. I have found such nests as high as 25 feet above the ground, the great majority of my observations being made on the estate of Dr. G. D. Shaver, who has made an especial study of these ducks and showed me all the nests that he could find.

As the season advances more open situations are often selected, sometimes at the base of a small oak on the dry prairie, at others among the rushes in a marsh over several feet of water, and again on a floating log in some small woodland pond. The mallards seem to lose much of their habitual shyness when the nesting season approaches, having little hesitation in building close to human habitations. However, they are very artful when leaving and returning to the nest, being experts at crawling and hiding, so that few people have any idea that there is such a thing as a duck within a mile of them.

The male is the most attentive to the female during the nesting season of any of our ducks, being seldom far from the nest at any time. I once saw a drake guiding a brood of downy young through a very brushy swamp, and was fortunate enough to have them pass directly under me as I was standing on a low, rustic bridge. The female was nowhere to be seen, which is so unusual under the circumstances that I believe she must have met with some accident.

The mallard occasionally lays its eggs in the nest of other ducks. I have found what were apparently mallard's eggs in nests with canvasbacks and redheads. Nearly all ducks when nesting in close proximity are more or less addicted to this habit though the mallard is less often guilty of it than several other species.

The nest of the mallard is generally well lined with large fluffy down, "bister" or "sepia" in color, with conspicuous white centers and faintly indicated whitish or light brown tips. Distinctly marked breast or flank feathers, with central brown streaks, or broadly banded with dusky and tipped with brown, are usually found in the nest, together with more or less rubbish. The nest and eggs somewhat resemble those of the pintail, but both the down and the eggs are larger and the feathers are distinctive.

Eggs.—Only one set of eggs is normally laid by the mallard which usually consists of from 8 to 12 eggs, sometimes 6 eggs constitute a full set and sometimes as many as 15 are laid. The eggs of the mallard might easily be mistaken for those of the pintail, but they average slightly larger, a little lighter in color and are not quite so much elongated. The female mallard when flushed may be readily distinguished from the pintail by its larger size, shorter neck, and by its blue speculum with conspicuous white borders. The eggs are elliptical ovate in shape and vary in color from a light greenish buff to a light grayish buff, or nearly white, with very little luster. The measurements of 93 eggs in various collections average 57.8 by 41.6 millimeters; the eggs showing the four extremes measure **64** by 41.5, 63.5 by **45**, **52.5** by 39.5 and 53 by **38.5** millimeters.

Young.—Incubation, which is performed wholly by the female, lasts from 23 to 29 days, usually 26; it does not begin until after the last egg is laid, so that they all hatch out about the same time. As soon as the young have dried their downy coats and are strong enough to walk, they are led by their mother to the nearest water which is often a long distance away. The watchful mother is ever on the alert and at the approach of danger gives her note of alarm which sends the little ones scattering in all directions to hide in the underbrush or thick grass, while she diverts the attention of the intruder. She is very courageous in the defense of her young; I once surprised a female with her brood in a little pond hole in the timber; although the young were well hidden in the surrounding grass and bushes, the old bird was flapping about, within a few feet of me, splashing and quacking loudly, frequently rising and circling about me, then dropping into the pond again and showing every symptom of anxiety, totally regardless of her own safety; the young were too well concealed for me to find them and I left the anxious mother in peace. The drakes usually take no interest in family cares, after the eggs are laid, but gather in small flocks by themselves, molt into eclipse plumage and hide among the rushes in the sloughs where they spend the summer in seclusion. The female, according to Audubon (1840), cares for and rears the brood alone.

She leads them along the shallow edges of grassy ponds, and teaches them to seize the small insects that abound there, the flies, the mosquitoes, the giddy beetles that skim along the surface in circles and serpentine lines. At the sight of danger they run as it were on the water, make directly for the shore, or dive and disappear. In about six weeks those that have escaped from the ravenous fishes and turtles have attained a goodly size; the quills appear on their wings; their bodies are incased with feathers; but as yet none are able to fly. They now procure their food by partial immersions of the head and neck in the manner of the old bird.

Dr. Harold C. Bryant (1914) has noted that—

when diving to escape capture they would often cling to the weeds beneath the surface, and when finally forced to come to the top for air would expose to view the top of the bill only. They tried to escape by simply diving and clinging motionless to weeds more often than they attempted to swim long distances under water.

As soon as the young birds have acquired their first plumage, in September, they gather into flocks, old and young together, and feed in the grain fields, where they become very fat.

Plumages.—The downy young mallard, when first hatched, is richly colored; the upper parts, the crown and back, are "sepia" or "clove brown," darkest on the crown; the under parts, including the sides of the head and a broad superciliary stripe, are "napthalene yellow" more or less clouded, especially on the cheeks with "honey yellow" or intermediate shades; there is a loral and postocular stripe and an auricular spot of "clove brown"; four yellowish spots, two on the scapulars and two on the rump, relieve the color of the back.   As the young birds increase in size the colors of the upper parts become duller and lighter and the yellows of the under parts fade out and are replaced by more buffy shades.

The juvenal plumage comes in first on the scapulars and flanks at an age of about 3 weeks, then a week later, on the rump and breast and finally on the head and neck, when the bird is nearly 2 months old; the tail begins to appear with the first plumage, but the last of the down has disappeared from the neck before the wings are even started; these are not completed until after the young bird is fully grown or about 10 weeks old.   In this juvenal plumage the sexes are practically indistinguishable, though the male is slightly larger and has a larger bill.   In this plumage the young birds resemble the adult female to a certain extent, but they are darker and more brownish, especially on the chest and back; the latter is "hazel" or even as bright as "burnt sienna" in young birds.

From this time on the sexes differentiate rapidly in their steady progress towards maturity; this is accomplished during the next two months by a continuous molt which is, perhaps, accompanied by some sympathetic change of color in the growing feather.   The result is that the young birds have assumed by December, or when about 6 months old, a plumage which is practically the same as that of the adults, though the highest development of the plumage is not acquired until the following year.

The annual molts and plumages of the adult consist of a double molt of all the contour feathers, into the eclipse plumage in the summer and out of it again in the fall; the flight feathers are molted but once, while the drake is in the eclipse plumage, in August.   Thus instead of a nuptial plumage, worn in the spring and summer, and a winter plumage, worn in the fall and winter, we have a full plumage, worn in the winter and spring, and an eclipse, or a concealing, plumage, worn for only a month in the summer, but with much time consumed in the two transitional molts.   The same thing takes place, to a greater or a lesser extent, with nearly all of the ducks; the eclipse plumage is much more complete in the surface-feeding ducks than in the others, and it is more strikingly illustrated in the mallard than in any other species.   It seems remarkable, indeed, that such a brilliant and conspicuous plumage, as that of the mallard drake, should disappear entirely and be completely replaced with an entirely different plumage, which only an expert can tell from that of the somber, mottled female; but such is the case; the wings and the larger scapulars, which are molted only once, are all that remain to distinguish the male. I have seen males molting into the eclipse plumage as early as May 10, but usually the molt does not begin until the latter part of that month. I have seen drakes in full eclipse plumage as early as July 20, but usually it is not complete until August. It is worn for about a month, the earliest birds beginning to molt out of it in August. Some birds regain their full plumage in October, but some not until November or even later. Mr. John G. Millais (1902), one of the greatest living authorities on ducks, has made a very thorough and exhaustive study of this subject and has written a particularly full and detailed account of the plumage changes of the mallard. Although we may not wholly agree with all of his interesting conclusions, regarding color changes without molt and control of the molt, we must accept them as probably correct until they are proven erroneous.

The tendency of several species of ducks to hybridize is well known and many interesting hybrids have been described. The mallard seems to be more inclined to hybridism than any other species, particularly with its near relative, the black duck. Numerous specimens of hybrids between these two species have been collected, showing various grades of mixed blood; they freely interbreed in captivity and their offspring are perfectly fertile. Specimens have been described showing first crosses of mallard blood with the muscovy duck, the green-winged teal, the baldpate, and the pintail. In connection with plumages it may be worth mentioning that many sportsmen throughout the West recognize two varieties of mallards, the yellow-legged variety, which is the earlier migrant in the spring and the later in the fall, and the red-legged variety, which is more of a warm-weather bird; the former is supposed to breed farther north and to frequent the prairies exclusively whereas the latter is more often found in the timbered swamps and streams. Probably the differences in the two varieties are due to age rather than geographical variation.

Food.—Mallards are essentially fresh-water ducks and find their principal feeding grounds in the sloughs, ponds, lakes, streams, and swamps of the interior, where their food is picked up on or above the surface or obtained by partial immersion in shallow water. In Alaska and on the Pacific coast they feed largely on dead salmon and salmon eggs, which they obtain in the pools in the rivers. On or near their breeding grounds in the prairie regions they feed largely on wheat, barley, and corn which they glean from the stubble fields. On their migrations in the central valleys they frequent the timbered ponds, everglades, and wooded swamps, alighting among the trees to feed on beechnuts and acorns or to pick up an occasional slug, snail, frog, or lizard. In the South they resort to the rice fields and savannas in large numbers, feeding both by day and night if not disturbed; where they are hunted persistently they become more nocturnal in their feeding habits.

Mr. W. L. McAtee (1918) has published an exhaustive report, based on the examination of 1,578 gizzards of the mallard by the Biological Survey, from which I quote as follows:

Approximately nine-tenths of the entire contents of the 1,578 mallard stomachs examined was derived from the vegetable kingdom. The largest proportion of the food drawn from any single family of plants came from the sedges and amounted to 21.62 per cent of the total. Grasses rank next in importance, supplying 13.39 per cent; then follow smartweeds, 9.83; pondweeds, 8.23; duckweeds, 6.01; coontail, 5.97; wild celery, and its allies, 4.26; water elm and hackberries, 4.11; wapato and its allies, 3.54; and acorns, 2.34 per cent. Numerous minor items make up the remainder. Some of the stomachs of the mallards were interesting on account of the large numbers of individual objects they contained. For instance, one collected at Hamburg, La., in February, revealed about 28,760 seeds of a bullrush, 8,700 of another sedge, 35,840 of primrose willow, and about 2,560 duckweeds as the principal items, a total of more than 75,200.

The animal food of the mallard duck though extremely varied may be classed in five main groups: Insects, which constitute 2.67 per cent of the total diet; crustaceans, 0.35; mollusks, 5.73; fishes, 0.47; and miscellaneous, 0.25 per cent.

Dr. Thomas S. Roberts (1919) has published the following interesting note, showing the useful work done by mallards in destroying mosquitoes.

The late Dr. Samuel G. Dixon, while health commissioner of Pennsylvania, published an article in the Journal of the American Medical Association for October 3, 1914, detailing results of experiments made by him along this line. Two dams were constructed on a stream so that the ponds would present exactly the same conditions. One was stocked with gold fish and in the other 20 mallard ducks were allowed to feed. After several months the duck pond was entirely free from mosquitoes while the fish pond "was swarming with young insects in different cycles of life." Ten well-fed mallards were then admitted to the infested pond. At first they were attracted by the tadpoles but "soon recognized the presence of larvae and pupae of the mosquito and immediately turned their attention to these, ravenously devouring them in preference to any other food present. At the end of 24 hours no pupae were to be found and in 48 hours only a few small larvae survived.

Mr. Edward H. Forbush (1909) says:

It sometimes attacks sprouting or ripened grain but like most fresh-water fowl it is undoubtedly of service in destroying such insects as the locusts and army worms which sometimes become serious pests. Professor Aughey found in the stomachs of ten mallards taken in Nebraska 244 locusts and 260 other insects, besides mollusks and other aquatic food.

Mr. J. H. Bowles (1908) records an interesting case of lead poisoning among mallards which had been feeding in a marsh that for many years had been a favorite shooting resort. The ground must have been thoroughly sprinkled with shot for the stomachs of the dead ducks were well filled with the pellets which had probably been picked up by mistake for gravel.

One stomach contained 19 shot, one 22, and the other 27. The large intestine was heavily leaded and seemed contracted, while the lining of the stomach could be easily scaled off in quite large crisp pieces. The gastric juices had evidently worked on the shot to some extent, as most of them were considerably worn and had taken various shapes.

Dr. Alexander Wetmore (1919) has published an interesting paper on this subject, based on investigations made near the mouth of Bear River, Utah, in 1915 and 1916, in which he shows that lead poisoning is a real cause of mortality among this and other species of ducks, where these birds have been feeding on grounds which have been shot over for many years. No practicable remedy has been suggested.

Doctor Wetmore (1915 and 1918) has published two other papers, based on his extensive investigations in Utah, from which it appears that the great mortality among waterfowl around Great Salt Lake is due largely, if not wholly to alkaline poisoning. Countless thousands of ducks and other waterfowl have perished within recent years in this and other similar localities, apparently from disease. He explains the cause very well as follows:

After June 15, as the spring waters in Bear River recede, great expanses of mud flat are laid bare in the sun. Surface evaporation and capillary attraction rapidly draw the salts held in solution in the mud to the surface and there concentrate them. As the mud becomes drier these concentrates are visible as a white deposit or scale (efflorescence). This in many cases is exposed only an inch or so above the surrounding water level. In the large bays strong winds bank up the water and blow it across these drying flats. As it advances it takes rapidly into solution the soluble salts, largely sodium chloride, but containing calcium and magnesium chloride also. This inflow of water carries with it quantities of seeds and myriads of beetles, bugs, and spiders, washed out of crevices and holes in the dried and cracking soil. The ducks come in eagerly to feed on this easily secured food and work rapidly along at the front of the advancing water, each bird hurrying to get his fill. Many individuals in this way secure a sufficient quantity of these poisons to render them helpless. As the water recedes again small pools are left in shallow depressions, and other ducks and shore birds feeding in these are affected.

The only remedy suggested is to supply the birds with a sufficient quantity of fresh water, under which treatment they recover.

Behavior.—The wild mallard is an active, wary bird, well worthy of the prominent place it holds among the game birds of the world. It springs from the water, at a single bound, straight up into the air for several yards and, when clear of all surrounding reeds, bushes or trees, flies directly away in a swift, strong and well-sustained flight. Several loud quacks are usually uttered as the bird springs into the air. The mallard, especially the female, is a noisy bird on its feeding grounds, the loud quacking notes, suggesting familiar barnyard sounds, give timely warning to the ardent hunter, as he seeks his quarry among the reedy sloughs. The mallard is not a diving duck and ordinarily does not go below the surface of the water; when wounded, however, it is skillful in avoiding capture by swimming under water or hiding among the rushes, with only its bill protruding; it has even been known to hide under a lily pad, lifting the leaf above the surface to enable it to breathe. Dr. Wilfred H. Osgood (1904) relates the following interesting incident illustrating the hiding ability of the mallard.

One foggy morning as we were slipping down the current of one of the narrow side channels a brace of mallards flew across a small peninsula to our left and alighted in a little cove, whence they hauled out on the muddy bank. Thinking to secure a good fat duck for dinner, we quickly swung the canoe into an eddy and paddled upstream toward the little cove. One of the birds flew while out of range, and at about the same time the other somehow disappeared, although there was but a small patch of grass for concealment. Expecting the bird to rise at any moment, we paddled on but were beginning to feel baffled, when just before the canoe touched the bank, we found our game giving a very pretty exhibition of its confidence in protective coloration. It was a female mallard, and lay on the brown mud bank, strewn with dead grass and decaying matter, which blended perfectly with the markings of its back. It was not merely crouching, but lay prostrated to the last degree, its wings closely folded, its neck stretched straight out in front of it with throat and under mandible laid out straight, and even its short tail pressed flatly into the mud. The only sign of life came from its bright little eyes, which nervously looked at us in a half hopeful, half desperate manner. When a paddle was lifted, with which it could almost be reached, the bird started up and was allowed to escape with its well-earned life.

Game.—Local fall flights of mallards begin before the end of summer, late in August or early in September, soon after the young birds are able to fly, but these are mainly wandering, drifting flights from their breeding grounds or summer hiding places in the sloughs, to favorite feeding grounds in the vicinity, where wild rice is ripening or where grain stubble offers a tempting food supply. The real fall migration does not begin in earnest until late in September, when the first early frosts, the brilliant hues of ripening leaves and the falling crop of acorns and beechnuts remind them of advancing autumn. But the waning of the harvest moon and the crisp, clear nights of early October also remind the hunters of the glorious sport of duck shooting; in the stillness of the night they push their flat skiffs out through the watery lanes among the acres of reeds and buckbrush to the shallow ponds, overgrown with smartweed and wild rice, where the ducks are wont to feed, their wooden decoys are anchored in some conspicuous open space and their skiffs are carefully concealed in blinds of thick reeds and grasses, where they patiently await the coming daylight, listen for the quacking notes of the awakening ducks and watch for the passing flocks on the way to their feeding grounds. If they have not been shot at too much mallards come readily to the decoys, but they become wary with experience; artificial duck calls are used to imitate their notes, which are quite effective when skill-fully operated; line decoys, as they are called, are fastened to long lines run through fixed pulley blocks, so that they can be made to swim in towards the blind or out again, by pulling on the lines, to attract the attention of passing flocks. Large numbers of mallards are still killed in this way all through their main routes of migration, but they have decreased greatly in numbers owing to persistent shooting in both spring and fall and owing to the settlement and cul-tivation of their main breeding grounds in the northern prairie re-gions. The mallard is a splendid game bird and has always held the leading place among our wild fowl on account of its abundance, its wide distribution, and its excellent qualities as a table bird; in my estimation there is no duck quite equal to a fat, grain-fed mallard, not even the far-famed canvasback; unquestionably the mallard has always been our most important market duck and certainly more mallards have come into our markets than any other one species.

Winter.—The mallard is a hardy bird and its winter range is a wide one, reaching as far north as it can find open water. Hagerup (1891) found it "common the whole year round, but most numerous in win-ter, when they keep in small flocks along the shore," in southern Greenland. In Alaska the mallard winters at several places at the outlets of lakes, in open streams near the seacoast and about the Aleutian Islands. Although essentially a fresh-water duck through-out its general range, the mallard is forced by circumstances in Green-land, Alaska, the northern Pacific Coast, New England, and other northern portions of its scattering winter range to resort to the mouths of rivers and bays where it can find open water. The main winter range, however, is in the lower half of the Mississippi Valley, south of the line of frozen ponds, and in the Gulf States from Texas to Flor-ida. Here it lives and flourishes, mainly in fresh-water ponds, swamps, streams, everglades, and rice fields, fattening on the abun-dance of good food but still harassed by gunners and killed by market hunters and sportsmen in enormous numbers. "Big Lake, Arkansas, was and still is one of the favorite resorts, and during the winter of 1893–94 a single gunner sold 8,000 mallards, while the total number sent to market from this one place amounted to 120,000," writes Doc-tor Cooke (1906).

Mr. E. H. Forbush (1909) says:

In 1900 I visited a gunning preserve in Florida where northern sportsmen were shooting ducks by the hundred and giving them away to their friends and to settlers.

One of these gentlemen armed with repeating guns and supplied with a man to load and others to drive the birds to his decoys is said to have killed on a wager over 100 ducks in less than two hours. Even within the last two years reports of reliable observers on the Gulf coast aver that market hunters there have been killing 100 birds each per day.

The Houston (Texas) Post of January 29, 1908, asserted that during the previous week five citizens while hunting came upon a small lake into which the fowls were flocking in great numbers. Using their repeating guns and acting by a prearranged signal they flushed the game, emptied their guns, and gathered 107 killed not counting the wounded and missing. The birds were mainly mallards.

The foregoing quotation will serve to indicate the enormous slaughter which has been going on among our game birds, of which the mallard is merely a fair sample. This was due mainly to the increasing numbers of gunners and the improved effectiveness of firearms.

Owing to the prohibition of market hunting, the curtailing of the shooting season, and the establishment of breeding reservations and fall and winter sanctuaries, this rapid extermination has been checked and the birds are now holding their own and are even increasing in some places. The big reservations on the coast of Louisiana show the beneficial effect of protection. Here the mallards and other ducks gather in great numbers in the winter, feeding in the ponds or patches of open water in the marsh, or rising, when disturbed, in immense flocks with a mighty roar of thousands of wings.

DISTRIBUTION.

Breeding range.—Northern portions of the Northern Hemisphere. In North America mainly west of Hudson Bay and the Great Lakes. A few breed in Greenland, but these are considered subspecifically distinct *Anas platyrhyncha conboschas* Brehm, and are apparently resident. East sparingly to eastern Ontario, central New York (Cayuga County), northwestern Pennsylvania (Erie and formerly Williamsport), and central New Jersey (Passaic County and Burlington County). South to northern Virginia (upper James River) southern Ohio, southwestern Indiana (Knox County), southeastern Illinois (Wabash County), central Missouri (Johnson County), eastern Kansas (Johnson County), southern New Mexico, and northern Lower California (San Pedro Martir Mountains). West to the Pacific coasts of the United States, Canada, and Alaska, and west in the Aleutian Islands to Tanaga Island and probably farther. North to northern Alaska (Kotzebue Sound), the Arctic coast of Mackenzie (Mackenzie Delta and Anderson River), and the coast of Hudson Bay. In the Eastern Hemisphere it breeds in Iceland, throughout Europe (south of the Arctic Circle), in the Azores, and in northern Africa; in Asia from Turkestan to China, Japan, Chosen, the Kurile Islands, Kamchatka, and the Commander Islands.

Winter range.—Practically all of North America, south of Canada. East to the Atlantic coast, the Bahama Islands, and rarely to the Lesser Antilles (St. Vincent and Grenada). South to central Florida (Cape Canaveral), the Gulf coasts of Louisiana and Texas, and to southern Mexico (Jalapa and Colima). North along the Pacific coast to the Aleutian Islands; in the interior to central Montana (Fergus County), southern Wisconsin, and the Great Lakes; and on the Atlantic coast regularly to Virginia, irregularly to New York and New England, and casually to Nova Scotia.   In the Eastern Hemisphere it visits the Azores, Madeira, and the Canary Islands, and ranges south in Africa to the Tropic of Cancer and to India and Burma.

Spring migration.—Early dates of arrival: Manitoba, Aweme, March 24; Saskatchewan, Qu' Appelle, March 26; Mackenzie, Fort Resolution, May 7; Fort Providence, April 27; and Fort Simpson. May 3; Alaska, Kowak River, May 17.   Average dates of arrival: Pennsylvania, Erie, March 5; New York, central, March 23; Ontario, southern, March 24; Ottawa, March 27; Indiana, Frankfort, February 21; Missouri, central, February 26; Illinois, Chicago, March 19; Iowa, Keokuk, February 24, and Spirit Lake, March 10; Minnesota, Heron Lake, March 11; South Dakota, central, March 16; North Dakota, Larimore, March 28; Manitoba, Aweme, April 3; Saskatchewan, Qu' Appelle, April 10.   Late dates of departure: North Carolina, Raleigh, April 7; Mississippi, Shellmound, April 5; Missouri, central, March 28; Texas, northern, May 6.

Fall migration.—Average dates of arrival: Virginia, September 21; Illinois, Chicago, September 27; Iowa, Grinnell, September 17; Texas, northern, October 11: Panama, Miraflores, November 26. Average dates of departure: Quebec, Montreal, October 26; New Brunswick, Scotch Lake, November 7; Ontario, Ottawa, November 5; Manitoba, Aweme, November 12; Illinois, Chicago, November 13; Minnesota, southern, November 22; Iowa, central, November 15; Nebraska, central, November 18.   Late dates of departure: Quebec, Montreal, November 13; Ontario, Ottawa, November 14; Manitoba, Aweme, November 23; Minnesota, southern, December 11; Iowa, central, November 27; Nebraska, central, November 26.

Casual records.—Accidental in the Bahamas, Cuba, Jamaica, and Grenada; and the Hawaiian Islands.

Egg dates.—North Dakota, Minnesota, and Wisconsin: Thirty-one records, April 29 to July 6; sixteen records, May 17 to June 1. California and Utah: Twenty-nine records, March 25 to 27; fifteen records, April 27 to May 20.   Alberta and Saskatchewan: Thirteen records, May 15 to July 4; seven records, June 5 to 17.   Oregon and Washington: Twelve records, March 17 to July 2; six records, May 7 to 24.   Northern Alaska: Two records, June 9 and 19.

ANAS NOVIMEXICANA Huber.

NEW MEXICAN DUCK.

HABITS.

Contributed by Wharton Huber.

In the valley of the Rio Grande River from El Paso, Texas, north to Albuquerque, New Mexico, this northern form of the *diazi* group makes its home. Whether on the mud flats in the river, the numerous alkali ponds, or cat-tail swamps through the valley this duck is ever watchful and wary of man.

I have observed several New Mexican ducks about 3 miles north of the city of El Paso, Texas, hence they probably range down the river possibly as far as the Big Bend country in Texas. In June, 1915, I saw five individuals at Belen and two at Albuquerque, New Mexico, on the mud flats in the Rio Grande River.

Courtship.—During the months of April and early May, 1920, I watched the courtship of several pairs of these ducks along the Rio Grande River west of Las Cruces, New Mexico. In April, two, three, and sometimes five New Mexican ducks could be seen on the mud flats in the middle of the river, as often with flocks of mallards as alone. When with a flock of mallards they would stay together and not mix with the former. The male could be seen bowing to the female and occasionally pecking and pulling at her wing feathers. When in the water the male would swim close to the female he had chosen, generally behind her, swim close up and pull at her feathers quacking all the while. If another (presumably a female) came too close he would swim rapidly at the intruder until she was driven to a safe distance. Returning to his prospective mate he would bob his head up and down a number of times quacking contentedly. Early in May these ducks were evidently mated as they were always seen in pairs or single birds.

On May 7, 1920, while watching a pair of the ducks on a mud flat in the middle of the Rio Grande River west of Las Cruces, New Mexico, I witnessed a very interesting performance. Both ducks took flight simultaneously, rising in the air at an angle of about 30°. They were flying slowly, their wings seeming to raise higher than in ordinary flight, both quacking incessantly. They passed the point where I was concealed about 400 feet away and about 300 feet high, the male (as I afterward learned) directly above the female. Making a large circuit over the land the male all the while keeping his position directly above the female, they swung again over the river coming head up into the light wind, whereupon they set their wings and descended to the water, the female slightly in the lead. Immediately upon alighting copulation occurred.

Nesting.—The nest and eggs of the New Mexico duck, so far as I know still remain to be described. Although I hunted almost daily during the last half of May and the first half of June, I was not successful in locating a single nest of this species.

Young.—On July 20, 1920, in a cat-tail (*Typha latifolia*) swamp of about 7 acres extent, 4 miles southwest of Las Cruces, New Mexico, I flushed a female that evidently had young. She flew over the cat-tails in circles while I spent over an hour wading the swamp looking for the young, apparently not at all afraid of me, as she passed time and time again within a few feet of my head. I did not, however, find the young. On July 27, 1920, a young fully feathered male and an adult female were collected from a flock of 12 individuals that were feeding where the overflow from an irrigation ditch ran into the Rio Grande 5 miles southwest of Las Cruces, New Mexico. From this date on flocks of from 10 to 25 young accompanied by old females could be seen feeding along the river bank. Even at this age they were extremely wary, and one could get within range only by the greatest stealth and stalking.

Food.—The feeding habits of this species are similar to the mallard. They feed along the river banks, in the drainage canals, ponds, and cat-tail swamps. In the spring the flooded alfalfa fields are favorite grounds for food. Ever extremely wary, they pass much of the daytime on the mud flats in the middle of the river. At dusk they seek their favorite feeding grounds, a cat-tail swamp or flooded alfalfa field even though it lie close to a ranch house or small settlement. The food I found to consist of green shoots of alfalfa and cat-tail, grass roots, corn, wheat, and numerous small fresh-water shells together with the larger seeds of weeds and grasses.

Behavior.—The flight of the New Mexican duck is similiar to that of the mallard, but it is a stronger and somewhat faster flyer. It was during the very heavy wind storms lasting two or three days that occur in March in southwestern New Mexico that I noticed the greater strength of flight of this species over the mallard. One could easily distinguish an individual of this species in a flock of mallards by its darker color and conspicuous pyrite yellow bill.

While feeding in the ponds and flooded alfalfa fields it keeps a short distance away from the other ducks. Food is obtained in the deeper water by thrusting the head straight down and keeping the body nearly submerged by the use of the feet, the tail only standing straight up above water. I have never seen this duck dive even when wounded. Several times while hunting with decoys on some of the larger ponds, combined flocks of mallards and New Mexican ducks would alight a hundred or more yards beyond the decoys. Ever wary and suspicious the New Mexican ducks would feed by themselves and never venture nearer the decoys, while the unsuspecting mallards would soon be swimming in and out amongst the wooden ducks.

DISTRIBUTION.

Contributed by Wharton Huber.

Although little is yet known of the range of this species, we do know that it is most plentiful along the Rio Grande River from Albuquerque, New Mexico, south to El Paso, Texas. Wetmore in his paper on Birds of Lake Burford, New Mexico, 1920, speaks of seeing—

on May 25th (1918), a large very dark-colored duck in company with a mated pair of mallards. It had white bars on either side of the speculum and was much darker in color than the female mallard, resembling a black duck markedly. It is possible that this was a female mallard, but it seemed to have a clear olive green bill and was larger, thus resembling a male of the black duck group (possibly *A. diazi*).

This evidently was an individual of the present species in the northwestern part of the State of New Mexico (Rio Arriba County). Probably the lakes and streams of Chihuahua, Mexico, will shortly be included in the known range of this duck.

[Author's note: A female, now in the Conover collection in the Field Museum in Chicago, was taken in Cherry County, Nebraska, on October 17, 1921.]

ANAS RUBRIPES TRISTIS Brewster.

BLACK DUCK.

HABITS.

The black duck, or dusky duck, as it was formerly and more properly called, for it is far from black, fills an even more important place among the wild fowl of eastern North America than does the far-famed mallard of the interior and western part of our continent. The black duck, by which name it is universally known among gunners, is decidedly *the* duck of the Eastern States, where it far outnumbers all other species of fresh water ducks. The West has many other species to divide the honors with the mallard, but in the East the black duck stands practically alone. It is the black duck more than any other that is suggested to my mind by those classic lines of William Cullen Bryant:

> Vainly the fowler's eye
>   Might mark thy distant flight to do thee wrong,
> As, darkly lined upon the crimson sky,
>   Thy figure floats along.

Spring.—When the gentle breath of spring calls him from his winter home on the New England seacoast the black duck seems to know in some mysterious way that the ice is going out of the lakes and streams far inland, perhaps he is impelled by the increasing warmth of the sunshine or by some impulse of returning love to leave behind him his fruitful feeding grounds in the salt marshes and the tidal estuaries; rising high in the air he sets his course toward his summer home in far-distant lakes and swamps; the sea knows him no more until the following autumn or winter. He is not the earliest of the migrants, nor yet the latest, but the latter part of March generally sees him well on his way. Black ducks migrate in pairs or in small flocks in the spring, usually at a great height; they are always extremely cautious, and before alighting in a woodland pond for the night, they usually circle around it several times and then scale down on stiff decurrent wings to alight in the open water far out from any dangerous cover; here they rest in peace and safety, but at the first blush of dawn their silver-lined wings are flashing over the tree tops and they are off for another day's journey.

Courtship.—While afield in early April in search of hawks' nests and other springtime treasures, I have often seen the spirited nuptial flight of this species. Near some woodland reservoir I have heard the loud quacking notes and looking up, have seen a pair, or perhaps, three, of these ducks flying over the tree tops at full speed; the courtship chase seems to be a test of speed and energy, a sort of aerial game of tag, as they sweep around again and again in a large circle or back and forth over the pond or swamp which they will probably choose for a summer home; finally the bride yields to the suitor of her choice and they fly off together or drop down into the the water. Mr. Edmund J. Sawyer (1909) describes a spectacular courtship performance of this species which I have never seen; he writes:

Most interesting were the actions of one pair that, from the time the flock came, constantly raced from end to end of the pond, one bird closely pursuing the other. Now and again the chase became too hot, and the leading bird in a thrilling swirl of water rose several feet into the air, followed immediately by the other. Toward the farther end of the pond they would splash into the water, soon to take wing again in the opposite direction. As, with necks stretched far out and downward, the pair flew half the length of the pond—2 or 3 rods—while the other ducks looked quietly on or went indifferently about their feeding, making the water dance till it seemed alive with ducks, the scene was really spectacular. Again and again I heard the pair of ducks break from the water, and the splash, splash, as they dropped into the pond again. Each time they rose, it seemed as though they must discover me, for at such times I was in open view, had they glanced in my direction.

Nesting.—The black duck nests in a variety of situations and does not seem to show any preference for any particular kind of surroundings provided it can find sufficient concealment. This makes it one of the hardest ducks' nests to find, for one never knows where or how to look for it and can only happen upon it by chance; I have spent many hours hunting for it in vain, around the edges of swamps or grassy meadows, along the shores of ponds, in thickets of underbrush or even in the borders of the woods near such places. The nest is generally placed in dry ground, but usually not very far from the water. The first nest I found was in the Magdalen Islands on June 21, 1904. It was in the center of a little islet or "nubble" in a small pond hole in the East Point marshes, a favorite breeding place of this and other ducks. The nest was prettily located and well concealed in a thick clump of tall dead grass; the hollow in the ground was lined with dry grass and only a little down, as the eight eggs that it contained were fresh; as with other ducks, more down would be added as incubation advanced. On another similar "nubble," two days later, we noticed a pathway leading through the grass to a clump of low bayberry bushes, and on investigation we unearthed a black duck's nest with four eggs completely buried under the dry bayberry leaves and rubbish; the eggs were perfectly concealed under natural surroundings and there was nothing to indicate a nest except the obscure pathway.

Audubon (1840) found nests in Labrador—

embedded in the deep moss, at the distance of a few feet or yards from the water. They were composed of a great quantity of dry grass and other vegetable substances; and the eggs were always placed directly on this bed without the intervention of the down and feathers, which, however, surrounded them, and which the bird always uses to cover them when she is about to leave the nest for a time.

Mr. M. A. Frazar (1887) reported the nests that he found, in the same region, as "being generally placed upon the outreaching branches of stunted spruces." Mr. John Macoun (1909) publishes an account of a nest, found by Rev. C. J. Young near Brockville, Ontario, on May 24, 1897, on the edge of a floating bog; he writes:

The place where the nest was made was not exactly wet, as there was a matted foundation of dry weeds among which it was well concealed, composed of dry grass and well lined with the down of the bird. Incubation had commenced about a week, which would make the time of commencing to lay about the first week in May in this case.

I have never succeeded in finding a nest in New England, but Mr. E. A. Samuels (1883) gives a satisfactory account of the nesting habits here, as follows:

The nest is built about the last week in April or the first in May. It is placed in a secluded locality in a tussock of grass, or beneath a thicket of briers or weeds; usually in a meadow, near a pond or stream, but sometimes in a swamp in which a small brook is the only water for miles around. This species sometimes follows these small brooks up to their sources; and I once found one with a nest on a low stump that overhung a small spring on the side of a hill, a mile from any other water. The nest of this species is constructed of pieces of grass and weeds, which are neatly arranged into a structure 18 inches in diameter on the outside, and 3 or 4 in depth. This is hollowed for perhaps an inch and a half or 2 inches, and lined with the down and feathers from the breast of the parent bird.

LIFE HISTORIES OF NORTH AMERICAN WILD FOWL.   53

Mr. Robert T. Moore (1908) found a nest of this species in New Jersey on May 22, 1908, in a wooded point surrounded by meadows; he describes it very attractively, as follows:

The body of the nest filled the space between the roots of a large maple. Dark-green lichens spotted the tree forming a beautiful background, while light green was the color of the huckleberry bushes branching above and grouping on the left. The front and right were screened by a bunch of soft brown grasses, which converged above with the huckleberry bushes and made it impossible to thrust in a hand without breaking the grasses. The nest proper concaved about a depression 8 inches in diameter. It was filled with pine spills, bracken, and leaves of oak and maple, no down having yet been inserted. The eggs were packed closely, the leaves sticking up between them. In color they were cream buff, some of them having a slight greenish tinge. The whole interior of the nest was soft brown, leaves, spills, and eggs lending various shades, but all molding into each other. These browns harmonized with the greens above, and made a most attractive home. Four days later the nest contained 12 eggs, so full as to have the appearance of convexity. Three eggs had been laid in four days. Down was now present, having been inserted in little bunches over the inside of the nest, adding a touch of warmth.

The black duck has been known to nest in old deserted crow's and hawk's nests in trees. Mr. Edwin Beaupré (1906) records two such cases in Ontario, as follows:

The first instance occurred June 10, 1904, when, on a small island in the St. Lawrence River, a pair of these ducks had taken possession of an old crow's nest, and on the date of discovery had laid 10 eggs. The nest was saddled on a limb of a large elm 45 feet from the ground. With the exception of a liberal supply of down furnished by the bird the nest was in its original condition and so completely was it concealed by the foliage that the presence of the duck in her snug retreat would never have been suspected had she not been accidently observed flying to the tree. The difficulty I experienced in photographing the nest adds to the value of the excellent negative I secured.

April 29, 1905, I located the second nest; in this case, owing to the bareness of the trees, concealment was impossible. The duck had laid 10 eggs in a last year's nest of the red-shouldered hawk in a basswood tree 50 feet up, and the appearance of this large bird sitting on her nest among the naked branches was truly most unique.

The down in the black duck's nest much resembles that in the mallard's; it is large and fluffy in form, and "bister" or "olive brown" in color, with whitish centers, which are not so conspicuous as in the nest of the mallard. Mixed with the down a few characteristic breast or flank feathers are usually to be found; these dusky feathers with a central buffy streak or buffy edgings will serve to identify the nest.

Eggs.—The black duck lays from 6 to 12 eggs in a set, usually 8 or 10; it has been said to lay as many as 15, but anything beyond 12 is unusual. The eggs closely resemble those of the mallard and can not with certainty be distinguished. In shape they vary from elliptical ovate to nearly oval. The shell is smooth, but has very little luster. The color varies from dull white or creamy white to various pale greenish buffy shades, such as "pale olive buff" or "Marguerite yellow." The measurements of 82 eggs in various collections average 59.4 by 43.2 millimeters; the eggs showing the four extremes measure **64** by 43, 61.2 by **44.8**, **55** by 42.5 and 58.5 by **41** millimeters.

Young.—Incubation lasts for 26 or 28 days and is performed by the female alone. The males usually desert the females as soon as the eggs are laid and flock by themselves, leaving their mates to hatch the eggs and care for the young. In this the duck proves to be a faithful and devoted mother; she sits close upon the eggs, particularly as the hatching time approaches. The process of hatching is described by Mr. Charles S. Allen (1893) in the following interesting detail:

The exact method adopted by the bird in freeing itself from the shell proved interesting. I will describe the procedure as it occurred in an egg that I took from the nest before the first crack had appeared. While examining it there was evidence of a strong muscular effort on the part of the bird inside, and a small disk of shell was chipped out and raised above the surface at about one-third of the distance from the end; then came a second or two of rest, followed by what felt like a scramble inside; then a second of quiet and the horny little knob on the end of the bill was driven through the shell one-eighth of an inch to the right of the first puncture. This routine was repeated over and over until some 25 or 30 punctures had been made, completely encircling one end of the egg, each being about one-eighth of an inch to the right of the preceding one. The efforts seemed stronger as it started around the same circle again, and the cap of the shell would be lifted a little each time, showing that it was attached by little more than the tough membrane beneath the shell. Before the second circle was half completed, it tore the cap loose so that it could be raised like the lid of a box, with 1 inch of the membrane acting as a hinge. In freeing itself from the shell the neck was stretched out and the little one breathed for the first time. Then the shoulders were pushed out into my hand, free of the shell, one wing after the other being freed, while the bird lay gasping and gaping widely with its bill. In half a minute more it was entirely free from the shell and lay weak and helpless in the sun, its wet, slimy skin absolutely bare, save here and there small dark hairs widely separated. As it began to dry it gained in strength and made feeble efforts to stand, resting on the whole length of the tarsus. In drying, the hairs no longer adhered to the skin. Soon each little pointed hair began to crack and split open, and from this protective casing there came a light fluff of down nearly as large as the end of one's finger. It was more surprising than the bursting of a grain of pop corn, though far less rapid. It took comparatively few of these yellow and brown fluffs to convert the naked weakling into a beautiful downy duckling that stood up boldly in my hand and began to notice what was going on about it, especially the calls of the parent bird close by. Each went through the same procedure, invariably breaking the shell from left to right. They showed no fear and would cuddle under one's hand very confidingly.

The young remain in the nest for a few hours after hatching sometimes not over an hour, until they have gained a little strength, become dry and freed the fluffy down from its wet sheaths. Then the careful mother leads them forth to introduce them to the world and its many dangers, teaching them how to escape from their various enemies, how to hide in the thickest grass, under the leaves or any other object that will cover them, how to crowd and " freeze " in any little protecting hollow, how to swim and dive in the water or scurry away to the nearest cover when they hear her warning call to hide. It is surprising how soon they learn the art of concealment, how quickly they obey the call, and how suddenly they vanish so completely that it is about useless to hunt for them. In the meantime the devoted mother, utterly regardless of self, uses every art known to her fertile brain to attract attention to herself and away from her young, flopping along the ground or water, as if hopelessly crippled, within a few yards, or even feet, of her giant enemy, returning again and again to throw herself at his feet beseechingly. Once I surprised a mother duck, far from land, swimming across a bay with a brood of little ones close at her heels; she would not desert them and my thoughtless boatman fired at her; fortunately he missed her and fortunately for him he did not shoot again or he would have measured his length overboard. The old duck sometimes makes long journeys over land on foot with her little brood. Judge J. N. Clark (1882) relates the following incident:

One of my neighbors, sitting by a window, had his attention called to a brood of young ducks running across the street. It was an old black duck and her young. He saw them enter a cow yard, and in one corner she called her brood under her wings and covered them. As he went near she flew some 15 rods and watched his movements, quacking her displeasure as he proceeded to capture her young ones. He secured 10 of them, all the brood but 2. After he had examined all he cared to he set them at liberty, and together they started on a run through Main street, continuing for 40 rods before they turned aside, a distance which they accomplished inside of five minutes; for the little things could run like squirrels.

I once found a little one that had come to grief on its overland journey; with one foot hopelessly entangled in some vines, it had fallen into a wagon rut; it was still alive, but had been deserted by its mother. I can imagine the consternation of the poor mother, who, after exerting every effort to free it from its predicament, was finally obliged to abandon it to save the rest of the brood.

Plumages.—The downy young black duck resembles the young mallard, but the color of the upper parts averages darker, much darker in many cases, and it extends farther down on the sides of the breast and flanks, often invading the belly; the under parts are less yellowish; and the dusky stripes on the head are darker, more pronounced, and more extensive. There is much individual variation in the latter character in the 17 specimens in my collection; in all there is a dusky stripe, of greater or less intensity, from the bill to the eye and from the eye to the occiput; in most of them there is a dusky rictal spot, and a dusky auricular spot, though in some the former is lacking; in some these spots are joined in a stripe; and in one very swarthy individual, in which even the lower parts are largely dusky, these two stripes are very broad and coalesce on the cheek.

The colors of the upper parts, including the crown down to the bill and the head stripes, vary from dark "mummy brown" to "bister" or "sepia." The sides of the head, including a broad superciliary stripe, are "buckthorn brown" or "honey yellow" becoming paler on the throat to "light buff" or "cream buff." Similar but duller shades appear on the under parts, from "Naples yellow" to "cream color," shading off to dull grayish white on the belly. The inner edge of the wing and the scapular and rump spots are pale yellowish buff.

The development of the young bird to maturity is practically the same as in the mallard, the sexes being indistinguishable and the wings being acquired last. The growth and development of the flight feathers in the young goes on simultaneously with the molt of the adults, so that both reach the flight stage together in September. During the first fall and perhaps for some time after that, I do not know how long, young black ducks can be distinguished from old birds, but during the first year, perhaps during the fall and early winter, they are rapidly becoming adult in appearance. Young birds during their first fall and winter may be recognized by the more striped appearance of the under parts, due to the fact that the feathers of the breast and belly are centrally black quite to the tip and broadly edged on the sides only with brown or buff; whereas in adults these feathers are very broadly dusky and only narrowly margined with buff, giving the under parts a much darker appearance; the lighter color of the neck is not so sharply separated from the dark colors of the body in the young as in the adult; and young birds have more conspicuous light edgings above and a partially immaculate chin and throat.

The age and seasonal changes in this species are not well marked or conspicuous, and one can not discuss them very far without becoming hopelessly involved in the much argued, sad case of the red-legged black duck, which has never been positively or even convincingly proven or disproven.

I have often been asked if the black duck has an eclipse plumage, with the double molt common to all the surface-feeding ducks. The eclipse plumage, if it had one would not be conspicuous and the double molt could be detected only by dissection or close inspection. It begins to molt very early in the summer and is in more or less continual molt for three months or more, but, as there is no necessity of an eclipse plumage for concealment, I doubt if there is an actual double molt. Lord William Percy, the British expert on ducks, tells me that none of the ducks in which the sexes are alike have an eclipse plumage; probably he is correct in this statement. The black duck then has, probably, only one annual molt, the postnuptial, which is prolonged and complete; the remiges are all molted at about the same time, so that the bird becomes practically flightless for a while.

The black duck crosses freely with the mallard and the two species are so closely related that the hybrids are fertile. A number of cases of first crosses have been recorded and specimens showing signs of mixed blood are rather common.

Food.—Black ducks are surface feeders or dabblers in shallow waters, where they can reach bottom by tipping up their tails and probe in the mud with their bills. In the shallow, muddy ponds and swamps where they spend the summer they feed largely on aquatic insects and their larvae, salamanders, tadpoles and small frogs, leeches, various worms, and small mollusks; many varieties of snails are found on the stems of sedges and grasses; small toads are not despised and even small mammals are eaten occasionally. With all this variety of animal food they mix a fair proportion of vegetable diet; seeds of aquatic and land plants are picked up and the succulent roots of many water plants are pulled up and eagerly devoured. Dr. F. Henry Yorke (1899) records the following genera of plants as recognized in the food of the black duck: *Limnobium, Zizania, Elymus, Danthonia, Piper, Myriophyllum, Callitriche,* and *Utricularia.*

Dr. Leonard C. Sanford (1903) writes:

In localities where blueberries grow near the water they are a favorite food. On the Magdalen Islands the writer has frequently seen black duck feeding high up on the hills among the blueberry bushes, in company with Hudsonian curlew.

In the fall, when the grains are ripened, they resort to the grain fields and feast on wheat, barley, buckwheat, and Indian corn. Later in the season they visit the timber where acorns and beechnuts are to be found in the vicinity of woodland ponds. The rice fields of the South are fruitful feeding grounds in winter where they grow fat and rich in flavor. On the seacoast in winter they resort mainly to the salt marshes to feed at night, returning to the open sea or to large bodies of water during the day; in the marshes and meadows they feed mainly on snails, bivalves and other small mollusks, crustaceans, and perhaps some vegetable food.

Dr. J. C. Phillips (1911) sent a lot of stomachs of ducks and geese, shot in Massachusetts in the fall of 1909, to Mr. W. L. McAtee for analysis; he quotes from Mr. McAtee's report as follows:

The contents of the black ducks' stomachs (29 in all, 4 empty) was 88.4 per cent vegetable, the principal items being seeds of bur reed (*Spanganium*), pondweed (*Potamogeton*), bullrush (*Scirpus*), eelgrass (*Zostera*) and mermaid weed (*Proserpinaca*), and buds, rootstocks, etc., of wild celery. The animal matter, amounting to 11.6 per cent, included, in the order of importance, snails, ants, chironomid larvae, bivalves, crustacea, and insects. The percentage of mineral matter of the gross contents was 36.5.

Mr. Ora W. Knight (1908) says:

I have known individuals to so gorge themselves with huckleberries in late August that they would go to sleep under the bushes near the water, and one which I started from under my feet in this condition when I too was after huckleberries was unable to fly, it was so gorged, but it managed to scramble into the water and swim away, disgorging itself until finally able to rise and fly away, all the time quacking incessantly.

He also speaks of two birds—

killed in winter on the Penobscot River [which were] literally crammed [with the fruit of] *Lepaigyraea canadensis* Nuttall, another shrub not known from this particular region.

Mr. Elon H. Eaton (1910) writes:

I shot a black duck from a flock of 75 birds, which were returning to Canandaigua Lake from a flooded cornfield. From its gullet and gizzard I took 23,704 weed seeds, which, together with a few pebbles, snail shells, and chaff, were the sole contents of its stomach. Of these seeds, 13,240 were pigweeds (*Chenopodium* and *Amaranthus*), 7,264 were knot grass (*Polygonum*), 2,624 were ragweed (*Ambrosia*), and 576 were dock (*Rumex*).

Behavior.—The black duck starts into flight, from land or water by a powerful upward spring, rising perpendicularly 8 to 10 feet into the air before it starts away in its swift and direct flight. When once under way its flight is strong and swift, usually high in the air, unless forced by strong adverse winds to fly low; its long neck is outstretched and its wings vibrate rapidly, the white underside of the wings flashing in the light and serving as a good field mark at a long distance. When descending from a height to alight in a pond the pointed wings are curved downward and rigidly held, as the smooth body glides through the air, tipping slightly from side to side, gradually dropping in a circle until near enough to check its momentum with a few vigorous flaps and drop into the water, feet first, with a gentle, gliding splash.

On land the black duck walks with ease and grace, running rapidly, if necessary, and holding its head high. It is ever on the alert and can seldom be surprised. It swims lightly and gracefully and with some speed. It does not ordinarily dive, but it can do so, if necessary, as every gunner knows who has wounded one and chased it. I have read that this duck can detect the presence of danger by the sense of smell, but I doubt it; it would not come so readily to well-concealed duck stands, where human beings are living constantly, if its nostrils were very keen. I should think it more likely that it depends on its sight and hearing, both of which are very acute and highly developed.

Dr. D. G. Elliot (1898) very aptly says:

Its note is so like the mallard's that it is difficult to distinguish them apart, and every few moments the quacks are shot forth in abrupt vociferations, as if the bird had just reached the limit of its power for suppressing them, and the voice had gained strength and sonorousness by long confinement.

The drake has only a low reedy quack, whereas the duck's note is a loud and resonant quack, in which she indulges freely; most of the noise is made by her. These ducks are generally very noisy while feeding and, as they are very alert and wary, their loud notes serve as timely warnings to other species on the approach of danger.

Fall.—Black ducks usually flock by themselves mainly because, in the regions where they are most numerous, there are comparatively few of the other surface-feeding ducks of similar habits. The early flights in September are often associated with blue-winged teal and later flights with a few of the other western ducks. The earliest flights consist mostly of young birds, often not fully feathered, and probably they are made up of family parties. The flocks are usually small, often less than 10 birds and seldom more than a dozen; large flocks are very rare except when congregated in winter quarters.

Mr. Walter H. Rich (1907) writes of the migration in Maine:

When the summer wanes and the young birds have become strong enough to journey, straggling ducks begin to make their appearance in the salt marshes, then in small bunches a few at a time; as cold weather approaches they gather at the sea into flocks ranging from 20 to 200 birds. Near my home they gather winter after winter at the mouth of a fresh-water river in a body of, at times, as many as 5,000 birds, coming in at night and spending their days on the salt water, except in bad weather, when they huddle on the ice at a safe distance from the shore. From the 1st of September such of their number as are not inclined to brave the rigors of a New England winter begin their longer journey to southern waters, and up to the middle of December the migrant birds continue to pass.

Dr. J. C. Philips says in his notes, published by Doctor Townsend (1905), referring to the fall migration at Wenham Lake, Massachusetts:

It has always seemed to me that there were three more or less distinct flights of black ducks observed here at the pond. The outside dates for these flights are about as follows: September 14 to October 5; October 1 to October 31; November 1 to November 20. These dates vary, of course, according to the season.

The first of the ducks are hastened along by an early frost or cool northwest weather, and their approach can be predicted almost to a certainty by a flight of ospreys, which precedes and accompanies them. The ospreys begin to come by in some numbers two or three days before the ducks arrive, and their flight seems to be at its height during the first day or two of the duck flight. Black ducks on this flight are very often accompanied by pintail and blue-winged teal.

The first and second flights sometimes merge into one another, but are commonly separated by an interval of some days to a week, or more, during which time few birds are observed. The second flight is scattered over a longer period and is accompanied by various other varieties of ducks. Widgeon and mallard are often seen with black duck at this time and sometimes pintail. The red-legged subspecies is common during the flight but rare among the early ducks. The second flight is much more pronounced during certain weather. Thus on the end of a stormy northwest wind or during brisk southwest weather, more birds are noted than at other times. At night, there are apparently many bunches which alight in the pond for a very short period of rest, and which leave of their own accord. These night flights are seen almost entirely during southwest winds and probably occur as often on dark as on moonlight nights.

The last flight is a more scattering and irregular affair and consists mostly of the red-legged variety. Some of these birds probably winter not far away. The red-legs average heavier and are a much wilder bird. They take to the larger ponds only and nearly always approach decoys with caution.

Prof. Lynds Jones (1909) reports this as "the commonest of the larger ducks, if, indeed, it is not the commonest of all ducks" in Ohio. He says:

Gunners report "millions" in the height of the gunning season. Such an estimate appears less extravagant when one realizes that the birds, almost crazed by the constant rattle of the guns, are flying back and forth and up and down, the same individuals reappearing many times in the course of an hour. I have seen many hundreds in a single day, but I doubt if more than a few thousands are even present on any day.

Game.—Whereas, this is only one of the many birds which interest ornithologists and bird protectionists, it is the bird of all others which interests the wild-fowl gunners of the Eastern States; it is the most important object of their pursuit, the most desirable as a game bird, one of the shyest, most sagacious, and most wary of ducks and the one on which their best efforts are centered. Therefore, I have always thought that it ought to be considered and treated more from a sportsman's standpoint than from any other and that any legislation for its protection ought to give due consideration to the rights of the sportsmen in the pursuit of such a noble game bird. To prohibit shooting it during January has always seemed to me unfair to many sportsmen on our seacoast, to whom it is not available during earlier months. The black duck has shown marked success in the struggle for existence; it is so sagacious, so wary, and so alert that it is one of the best equipped species to survive, even in a thickly settled region where it is constantly beset by hunters, but where, fortunately for its welfare, numerous safe refuges have been established. For these reasons it is hardly in need, as yet, of very stringent protective laws; therefore, I see no reason why sportsmen should not be allowed a reasonable amount of sport at its expense.

The methods employed for shooting black ducks are many and varied, but they all depend on the strategy and skill of the hunter in outwitting one of the keenest of game birds. They will not ordinarily come to wooden decoys, for their keen eyes readily detect the deception, but on the islands off the coast of Maine I have had fair sport over wooden decoys anchored just off the rocks where we lay concealed; the birds came in singly or in twos or threes, circling wide at first and then coming in to inspect the decoys; on discovering their mistake they would mount into the air and swerve off, but sometimes too late to escape the shot. In the winter when the ponds were frozen over, we used to find good shooting, without any decoys at all, where springs or small streams emptied into the salt water bays; here the ducks came in to drink or bathe in the only available fresh water under the cover of darkness; it is almost useless to attempt this kind of shooting except on moonlight nights and even then it is difficult and unsatisfactory.

Mr. Rich (1907) describes the method used in Maine, as follows:

Probably the most of these birds which fall a prey to the gunner's wiles are shot from "sinkboxes" and "blinds" in the reed-grown corners of fresh-water ponds, using live decoys to lure the birds on to their destruction. The successful duck shooter must be up betimes and be ready to endure much discomfort, for he must be at his position before daylight in order to get the cream of the shooting, and, where gunners are as numerous as in my section, a late comer is apt to find every stand occupied.

The decoys are placed before the blind, anchored, as a rule, so that one old drake is somewhat separated from the rest, and being dissatisfied and lonesome, he keeps up a continual remonstrant conversation with the rest of his flock. If a bunch of birds is passing, never fear but he will see them and find means to let the strangers know of his presence and whereabouts, and they, with a sudden turn from their course, with necks outstretched and wings stiffly set, come in at full speed. Now they turn away, careering around the pond two or three times because the foxy old fellow who leads them is not just suited with the appearance of things—some small matter of suspicion in his mind—but next time around a bird or two in the tail of the flock, more hungry than wise, drop out with slanting flight, then another and yet more, until finally the main body comes in like a flight of arrows. Splash! Splash! They have settled just outside the line of decoys and begin to swim in toward them. Now the gunner waits until they are bunched at a little distance from his "tolers," which, if old hands at the business, at once swim away from their visitors, and when his feathered assistants are surely safe the gunner pulls trigger where there is the greatest number of heads. The encore when the survivors rise like the scattered fragments of a bursting shell will hardly account for more than a pair, but usually the "pot shot" with the first barrel has done grand service toward thinning the game supply, and it is no common occurrence for one gun in experienced hands to gather in nearly all of the flock.

A modification of this method, more highly developed and modernized, is practiced in Massachusetts. On the shore of a pond frequented by migrating waterfowl, or on an island in it, a permanent camp is built, known as a " duck stand," at which one or more of the gunners live constantly all through the shooting season. This consists of a small house or shanty equipped with sleeping bunks for severa men, a stove for cooking and for heating it and shutters to prevent the lights showing through the windows at night. Along the shore is built a fence or stockade just high enough so that a man can shoot over it; there are portholes cut in the fence so that several men can shoot through it without being seen. The house and the fence are completely covered with branches of freshly cut pine and oak with the leaves on them, which renders the whole structure practically invisible from the lake. The stand is built where there is a beach or a point in front of it, or where a sandy beach can be artifically made. Various sets of wooden decoys or "blocks," as they are called, are anchored at some distance out in the lake. A large supply of live decoys, semidomesticated black ducks, mallards and Canada geese, are kept in pens, inside or behind the enclosure, and a few are tethered on the beach, anchored in the water near it or allowed to roam about. Sometimes a few are kept in elevated pens back of the stand, so arranged that the pens can be opened by pulling a cord and allowing the ducks or geese to fly out and meet the wild ones. With all this elaborate equipment ready for action the gunners, I can hardly call them sportsmen, spend their time inside the house, smoking, talking, playing cards, or perhaps drinking, while one man remains outside on the watch for ducks. Should a flock of wild ducks alight in the pond, he calls the others and they all take their places at the portholes, with heavy guns, ready for the slaughter. The quacking of the decoys gradually tolls the wild birds in toward the beach or perhaps the fliers are liberated at the critical moment. Each gunner knows which section of the flock he is to shoot at and waits in anticipation until the birds are near enough and properly bunched, when the signal is given to fire. If the affair has been well managed most of the flock have been killed or disabled on the water, but, as the frightened survivors rise in hurried confusion, a second volley is poured into them and only a few escape. The wounded birds are then chased with a boat and shot. There is no method of duck shooting which is more effective and deadly; with gunners constantly on the watch and decoys always ready to call a passing flock, very few ducks get by without an attempt being made on their lives, and often these attempts are only too successful. Probably before many years this form of duck shooting will be prohibited by law, as too destructive, and the more sportsmanlike method of shooting flying birds from open blinds will give the ducks some chance for their lives.

Winter.—When the swamps, ponds, and lakes of the interior are closed with ice the black ducks are driven to the seacoast to spend the winter. They linger in the lakes, even after they are partially frozen over, as long as an open water hole remains, resorting to the spring holes and open streams, visiting the grain fields and marshes or other places where they can find food and resting during the day in large flocks on the ice, where they sleep for hours while some of their number act as sentinels. On the coast their daily routine is to spend the day at sea or on large open bays and to fly into the marshes, meadows and mud flats to feed at night. At the first approach of daylight, long before the rosy tints of sunrise have painted the sky, black ducks may be seen, singly or in small scattered parties, winging their way out to sea, high in the air, their dark forms barely discernible against the first glow of daylight. At a safe distance from land they rest on the tranquil bosom of the sea or sleep with their bills tucked under their scapulars. It must be half-conscious sleep, or perhaps their feet work automatically, for they never seem to drift much. When the open sea is too rough their resting places are in the lee of ledges in little coves or in the bays.   Often times they rest and sleep in large numbers on drifting ice, on sand bars or even on unfrequented beaches.   In very stormy weather they are often driven into the bays and harbors in enormous numbers.   During that memorable storm of November 27, 1898, I was out duck shooting all the morning at Plymouth, Massachusetts; the gale was so fierce that we could hardly walk against it and the driving snow and sleet was almost blinding; black ducks were driven inland in large numbers, the little pond holes in the woods were full of them and they could hardly fly against the storm; they even sought shelter in the orchards among the houses, and large numbers were killed.

DISTRIBUTION.

Breeding range.—Northeastern North America.   East to the Atlantic coast, from New Jersey northward.   South to eastern North Carolina (Pamlico and Currituck Sounds), northern Ohio (Lake County), northern Indiana (Lake County), northern Illinois (Calumet marshes), and northwestern Iowa (Spirit Lake).   West to central Minnesota (Kandiyohi County), eastern Manitoba (Lake Manitoba), and the west coast of Hudson Bay (Churchill).   Seen in summer at Fort Anderson. North perhaps to the latter point, to Ungava Bay, and northern Labrador (Okak).   Northern breeding birds are supposed to be the red-legged subspecies.

Winter range.—Eastern United States, mainly coastwise.   South to central Florida (Gainesville), the Gulf coasts of Alabama and Louisiana, and to south central Texas (San Antonio and Corpus Christi). West to eastern Nebraska (Lincoln), but rare west of the Mississippi Valley.   North to southern Wisconsin (Delavan), northern Ohio (coast of Lake Erie), northwestern Pennsylvania (Erie), sometimes the lake regions of central New York, on the coast of New England, and as far east as Nova Scotia (Chignecto Bay).   Casual farther west or north of above-named points and comparatively rare west of the Alleghenies.

Spring migration.—Early dates of arrival: Maine, southern, March 19; Quebec, Montreal, March 27; Quebec, Quebec, April 6;   Prince Edward Island, April 5; Ontario, southern, March 16; Ontario, Ottawa, March 21.   Average dates of arrival: Maine, southern, April 7; Quebec, Montreal, April 14; Quebec, Quebec, April 18; Quebec, Godbout, April 21; Prince Edward Island, April 23; Ontario, southern, April 7; Ontario, Ottawa, April 14.   Late dates of departure: North Carolina, Raleigh, April 11;   Florida, Wakulla County, May 2.

Fall migration.—Average dates of arrival: Virginia, Alexandria, September 30; South Carolina, Mount Pleasant, October 22 (earliest); Florida, Wakulla County, November 16.   Average dates of departure: Ontario, Ottawa, November 7; Quebec, Montreal, November 6; Prince Edward Island, November 13. Late dates of departure: Ontario, Ottawa, November 21; Quebec, Montreal, November 14; Prince Edward Island, December 8.

Casual records.—Accidental as far west as Saskatchewan (Davidson) and California (Willows, Glen County, February 1, 1911). Said to have occurred in Bermuda, Cuba, and Jamaica.

Egg dates.—Ontario, Quebec, and Nova Scotia: Thirty-six records, April 30 to June 28; eighteen records, May 20 to June 6. Massachusetts and Rhode Island: Seven records, April 23 to June 2. New York: Six records, April 18 to June 19. New Jersey: Eight records, April 25 to July 3. Virginia and Maryland: Several records, April 20 to May 10.

ANAS RUBRIPES RUBRIPES Brewster.

RED-LEGGED BLACK DUCK.

HABITS.

When our late lamented friend, William Brewster (1902) described the above subspecies, he started a controversy which has led to endless discussion and which has never yet been satisfactorily settled. A still further complication arose when the old, well-established name, *obscura*, was shown to be untenable; for this necessitated adopting Mr. Brewster's name, *rubripes*, for the species, to which he (1909) tacked on still another new name, *tristis*. The incident was sad enough to warrant the name, but our old friend was hardly recognizable after all the changes; fortunately we can still call him by the old familiar name, the black duck. Let us be thankful for the much-needed stability in the English names.

Sportsmen and others have long recognized the existence of two kinds of black ducks, the smaller birds with olive or brownish legs and olive colored bills, which appear early in the fall, and the larger birds with reddish or orange-colored legs and yellowish bills, which come later in the season and presumably from farther north. But whether these differences represent two geographical races and should be recognized in nomenclature, is another question.

It seems to me that the characters on which Mr. Brewster (1902) based his new form, *Anas obscura rubripes*, are the characters of the adult, while those which he leaves for *Anas obscura obscura* are those of immature birds. If we may reason by analogy from what takes place in the closely related mallard, we might expect to find in young black ducks a rapid approach toward maturity during the first winter, producing a plumage in the following spring which is practically, but not quite, adult. Then, if this theory is correct, the first winter plumage would be characterized by the olive bill, the dark pileum, the imperfectly spotted chin and throat, and the brown legs. The birds might be expected to breed in this plumage, as the mallards do. At the first postnuptial molt, which is complete in August, it would then assume a plumage indistinguishable from adults, or nearly so, characterized by the yellow bill, the feathers of the pileum edged with grayish or fulvous, the throat and chin wholly spotted with blackish, and the red legs and toes. All of these characters probably become more pronounced in very old birds and perhaps the many puzzling intermediates are birds of the second year. There is another character which seems to be more pronounced in old birds, of the red-legged type, and that is the white tips of the greater wing coverts, forming a narrow white border of the speculum, which is conspicuous in older birds and either lacking or inconspicuous in birds of the first year.

As to the evidence in the case let us consider briefly a few salient points. The strongest claim that the red-legged black duck has to recognition as a distinct subspecies is based on the well-established fact that nearly, if not quite, all the early migrants are brown-legged birds and that very few, if any, of the large, red-legged birds are seen or shot much before the 1st of October, the heavy flight coming after the middle of that month and presumably from farther north. This claim is somewhat weakened when we consider that the great bulk of the species nest far north of the points where observations have been made and records kept and that undoubtedly the younger and more tender birds migrate first and the older and hardier birds later, as is the case with some other species of ducks. In this connection Dr. Charles W. Townsend (1905) writes:

Assuming, for the sake of argument, that *rubripes* is merely the adult male of *obscura*, it is interesting to note the similarity in seasonal distribution, between these two forms and the adult male red-breasted merganser as compared with the very differently plumaged females and immature. In both cases the small, obscurely dressed birds come first during the early autumn, while the large showy birds come in late September and in October. In both, these large birds are abundant in the winter, and the smaller ones are less common, while in both, the two forms appear again in the spring. The remark of Doctor Phillips that "the first flight of black ducks consists mostly of young and often imperfectly feathered birds" is interesting in this connection.

In order to have any standing a subspecies must be shown to have a distinct breeding range, which has not been demonstrated in this case. Mr. Brewster (1902) was able to find only four specimens of breeding ducks which he could unhesitatingly refer to the red-legged race, one from Ungava, northern Labrador, one from Moose Factory on James Bay, one from Cape Hope, Severn River, and one from Fort Churchill. One of these seems doubtful, the Ungava bird, which the collector, Mr. Lucien M. Turner, describes in his original notes, as follows: "In the specimen procured by me the bill is of a dusky olive color; the nail black; the tarsus and toes deep orange red; webs, as well as under surface of toes and posterior portion of tarsus, blackish." The form *rubripes* is supposed to have a yellow bill. We have specimens of the old form, *tristis*, from southern and northern Labrador, Okak, and from Newfoundland; and I doubt if Mr. Brewster felt confident that all of the birds from the Hudson Bay localities, mentioned above, were *rubripes*. On the other hand we have some evidence to indicate that red-legged birds breed farther south. Dr. Jonathan Dwight (1909) had "a number of freshly killed birds" sent to him from Long Island, New York, "that scarcely needed dissection to prove them to be breeding birds. They were shot at various dates in April and all had red legs." Moreover, an adult male, killed on Long Island June 11, 1909, came fresh into his hands, which had the red legs and other characters supposed to belong to the northern race; it was "in full postnuptial molt, and evidently was recently mated." The only summer specimen I ever shot on Cape Cod had red legs. Mr. Horace W. Wright (1911) says of the birds seen by him, which were probably breeding near Jefferson, New Hampshire:

Most of those birds which have been seen on the ponds in the summer, near enough to distinguish whether they were of the type *rubripes* or *rubripes tristis*, have been of the latter type. Perhaps only two have been distinctly seen which were of the former type, namely, on July 30, 1908. These took wing so near to us on our approach that the red legs were clearly seen.

Mr. Edwin Beaupré writes to me from Kingston, Ontario, as follows:

Owing to the great number of black ducks in this vicinity this season (1920) the time was considered opportune for looking into the question of the subspecies. Between September 1 and November 4, 1920, 20 specimens were available for determination; of these five had red legs and were much larger than the brown-legged birds. The presence of these red-legged ducks in this locality September 1 is a reasonable indication of their having bred here.

Mr. P. A. Taverner has recently sent me some colored drawings of the bills and feet of black ducks, from which it appears that the breeding ducks of the Ottawa River region and of the Gaspé region are red-legged; also that what are evidently young birds show, at least a tendency toward red-legs.

It will be seen from the above remarks that the known facts regarding the distribution of the two forms are not conclusive either one way or the other, so we must turn to what little other evidence we have. Doctor Dwight (1909) says that the differences between the two forms—

are exactly the ones that distinguish old birds from young whether they occur in the United States or Canada. My evidence on this point is conclusive for I have skinned and dissected fully 50 specimens representing many localities, north and south, besides examining dozens of others shot by friends or found hanging in the markets.

We must admit that Doctor Dwight is an experienced expert in such matters and that his opinion ought to carry weight. Doctor

Townsend (1912) adds to our knowledge of the subject by recording the results of his observations on some black ducks hatched in Massachusetts and reared in confinement; he writes:

When 4 months old one of the females had a pure buffy throat, while the other female's throat had a few scattered spots on it. All three males had more or less fine spotting on a buffy ground. The bills of the females were dark greenish black, their tarsi brownish, while the bills of the males tended more to greenish yellow and their legs to orange. The next spring the bills of the males were slightly lighter in color, but by no means yellow, and their tarsi were possibly a little brighter orange. A study of the plumage showed, however, no suggestion of either an eclipse or a nuptial dress. In the third spring the appearance was essentially the same. The surviving male had a dark crown and nape, a buffy throat, fairly well, but not thickly spotted, a greenish yellow bill and orange feet—not by any means the coral red feet of *rubripes*. The female had a dark olive-green bill, dirty-yellow tarsi and an unspotted buffy throat. Their size was that of the smaller race.

This certainly proves that the bills of young black ducks grow yellower and the feet grow redder, as the birds grow older. We do not know how long it takes a black duck to acquire these evidences of age. We do know, however, that we have a similar case in the mallard, in which the hunters recognize a red-legged variety which migrates early in the fall and late in the spring, probably the younger birds, and a yellow-legged variety, which is the last to come in the fall and the first to appear in the spring; the latter is known as the "ice mallard" and is probably the very old bird. Mr. Fred H. Kennard (1913) thinks that he has settled the controversy by the discovery of a young bird with red legs. He says:

While at Monomoy Island, Massachusetts, during the last two weeks of October 1912, with a couple of friends, we shot a number of black ducks of the red-legged kind (there were no green legs), among which were several that were apparently young birds; and on October 25, there fell to one of our guns a female, which from its size, plumage, and general characteristics, was so evidently young that there could be no possible doubt about it. I personally skinned and sexed this specimen, which showed its immaturity in all those ways familiar to those who handle birds. It must have been one of a very late brood, for its upper mandible was a steel gray, and had not yet begun to show those shades of light olive green of the adult bird, and the "nail" at the end of the upper mandible was hardly darker than the rest of the bill, and nothing like the dark and glossy black of the adult bird. The lower mandible was pinkish and still quite soft and pliable, as in the case of very young ducks, and *the bird had red legs*.

Dr. John C. Phillips (1920) recognizes a distinct difference in habits between the two forms which he sums up as follows:

The habits that characterize the two forms as they appear in autumn in New England may be thus summed up: *Anas rubripes tristis*—Breeding locally and often migrating as early as, or before, mid-September, or at least "shifting ground" from inland nesting grounds to better feeding grounds near coast. Feeding in both ponds and salt meadows, but if in salt meadows, resorting to fresh water once or twice a day. Much less nocturnal in feeding habits than *rubripes*, because less shy, and much less inclined to spend day on open ocean. Prefers good fresh water and brackish water food, but spends the winters on the coast of New England in small numbers along with *rubripes*. Reaches great size at times. Largest male 3 pounds 10 ounces; largest female 2 pounds 15 ounces (Squibnocket, 1919). More difference in size between sexes than in *rubripes*! Comes readily to live decoys, no matter how extreme the voice may be (too high or too low); and is more loquacious than the red-legged form.

A. rubripes rubripes.—Late migrant never becomes localized except near sea, and where marine food in the form of small mollusca is abundant. Very seldom resorts to small ponds or bogs, but likes large open sheets of fresh water near ocean, to which it often makes daily trips to drink and rest but not to feed. Is better able to sit off-shore in rough seas; and in general appears a more rugged bird with heavier feathering and superior resistance to extreme cold. In winter, it does not depend on ponds for fresh water, but obtains a sufficient supply in small springs about salt meadows at low tide.

This is a much more wary bird, is more silent itself, and comes less easily to live decoys, toward which it manifests an instinctive fear, especially if they be loud or shrill callers. In the salt meadows the best gunners prefer seaweed bunches or canvas sacks, and find the live decoys useless, especially late in the season.

When a flock of *rubripes* alights on a pond near a shooting stand, they nearly always keep at a safe distance until perfectly satisfied of their surroundings. Then, more often than not, they will swim away from the stand and its live decoys. If they approach the stand, which they do with the utmost caution, and with necks erect, they are not apt to keep closely together as *tristis* does.

Extreme weights not much above that of *tristis*. Heaviest male noted by myself, 3 pounds 12 ounces. Average is a good deal heavier than *tristis*, females perhaps more nearly size of males than in *tristis*, but no figures at hand to bear out this point.

To sum up the evidence it does not seem to have been proven that a northern race, with a known breeding range, exists; but it does seem to have proven that the characters ascribed to it, are to be accounted for, at least partially if not wholly, by age variations. I am still prepared to believe that a northern race exists, but we need more evidence to prove it.

ANAS FULVIGULA FULVIGULA Ridgway.

FLORIDA DUCK.

HABITS.

Up to about 1874, when Mr. Ridgway described this species, the dusky ducks of eastern North America, from Texas to Labrador, were all regarded as one species. This well-marked southern species, characterized by its smaller size, lighter color, and particularly by its immaculate buffy throat, inhabits Florida and the other Gulf States. It is not known to intergrade with the northern black duck, and there is a considerable hiatus between the breeding ranges of the two species. The southern species has since been split into two subspecies, the Florida duck, restricted to Florida, and the mottled duck found in Louisiana and Texas; whether these two forms inter-grade in the intervening States, or where they meet, does not seem to have been determined. Should a hiatus be found to occur between their breeding ranges it might be proper to regard them as distinct species, though all three forms are closely related and probably the intergrades have only recently disappeared.

In the central and southern portions of Florida this duck is an abundant resident bird. I have met with it frequently in the various portions of Florida that I have visited. On the islands in Indian River, where there were muddy ponds surrounded by marshes, we usually found a pair of these ducks, which were probably breeding there but had their nests too well concealed in the luxuriant growth of tall, thick grass for us to find them. We saw them occasionally in the inland lakes of southern Florida, but we found them most abundant in the extensive marshes of the upper St. Johns River; here they found ample feeding grounds and playgrounds among the dense tangles of vegetation, pond lilies, bonnets, water hyacinths, water lettuce, and other aquatic plants; the dense clumps of taller growth and the impenetrable saw-grass sloughs offered them concealment from their enemies; and they found safe sleeping and resting places in the centers of the larger bodies of water.

Nesting.—As I have not been fortunate enough to find one of their nests, I shall have to quote from the observations of others. Dr. D. G. Elliott (1898) says:

It breeds in April, and the nest, formed of grass and similar materials and lined with down and feathers, is placed upon the ground in the midst of matted grass, or under a palmetto, or some sheltering bush, near water.

The following account is published by Baird, Brewer, and Ridgway (1884) based on the excellent field notes of Mr. N. B. Moore:

This duck hatches in Florida from the first to the last of April, only one set of eggs being laid in a season, unless it fails in raising its first brood. The nest is always placed on the ground, and the number of eggs is usually 9 or 10. In one instance a nest was discovered which was nearly 300 yards from water, and other nests were met with still farther from water. The one first referred to was cautiously concealed in a thick mass of dead grass held upright by green palmettoes, about 2 feet high. Mr. Moore once noticed a pair of ducks fly from a pond, near which he was seated, and pass over the pine barrens. One of them dropped among the grass; the other returned to the water. Suspecting that the birds might have a nest, he visited the locality the next day, when the birds behaved as before. He soon made his way to the spot where the female alighted, and found her in a somewhat open space. On her return to the pond he soon discovered her nest. It was carefully screened from view on all sides, and so canopied by the standing grass that the eggs were not visible from above. There was a rim of soft down, from the mother's breast, around the eggs, partly covering those in the outer circle. On viewing the nest the next day this down was found to have been drawn over all the eggs. Mr. Moore took them and placed them under a hen; and six days after they were hatched. This was early in April. It would appear, therefore, that the statement that the male forsakes his mate during incubation is not well founded; for in this instance the male bird, about the twenty-fourth day of incubation, still kept in the vicinity of the nest. It is, however, the universal belief that he does not assist in rearing the young.

Mr. C. J. Maynard (1896) "found them breeding on Indian River, the nests being placed on the drier portions of the marshes, in grass which was about 18 inches high."

Eggs.—The Florida duck lays about 8 or 10 eggs which are similar to those of the black duck, but slightly smaller or shorter and rounder. In shape they are elliptical oval to oval. The shell is smooth and in some specimens slightly glossy. The color is creamy white or greenish white. The measurements of 52 eggs in various collections average 57 by 44.3 millimeters; the eggs showing the four extremes measure **62** by 46, **49.8** by **49,** and 55 by **40.5** millimeters.

Young.—The period of incubation is probably the same as with the black duck, 26 to 28 days. It is performed wholly by the female, although the male does not entirely desert her. Mr. C. J. Maynard (1896) writes of the behavior of the mother and young:

The eggs were deposited during the first and second weeks of April; then about the 1st of May, I would frequently see flocks of little downy ducklings following the female, but unless I took care to conceal myself, I did not enjoy watching these little families long, for as soon as the parent became aware of my presence, she would emit a chuckling note, when away they would scamper, helter-skelter, into the nearest grass, where it was impossible, upon the most careful search, to discover a single young. I once surprised a brood, when they were some distance from any place of shelter, for they had ventured out upon the mud of a creek, at low tide, and I chanced to come out of the high grass, just in front of them. The old duck appeared to comprehend the situation at once, for she came directly toward me, driving her brood before her, hoping to engage my attention by a display of bravery, while the young escaped into the sheltering vegetation behind me; but placing my gun on the ground, I stooped down and grasped two of the little fellows, as they were running past. The diminutive ducklings uttered shrill cries when they were captured, which drove their parent nearly frantic, for regardless of possible consequences, she dashed about in front of me, with ruffled feathers and half-closed wings, often coming within a foot of me, at the same time, quacking loudly. This outcry attracted the attention of the drake, but he did not approach very near, merely circling about, some 50 yards distant, quacking softly. Leaving the old female to care for the remainder of the brood, I carried my captives into camp and placed them in a box, the sides of which were about a foot and a half high, but young as they were, they managed to escape.

Plumages.—As I have seen but few downy young of the Florida duck, as the series of immature birds available for study is very scanty and as the two subspecies are so much alike in these respects, I prefer to refer the reader to what I have written about the plumages of the mottled duck, which will probably fit this subspecies equally well.

Food.—According to Mr. W. L. McAtee (1918) the southern black ducks eat a larger proportion of animal food than their northern relatives. Based on the examination of 48 stomachs by the Biological Survey he found that 40.5 per cent of their food consisted of animal matter. Mollusks compose five-eighths of the animal food and snails as large as 1 inch in diameter are eaten. Insect food consists of dragon-fly nymphs (rarely adults), water bugs, caddis larvae, and a variety of beetles and flies, including horsefly larvae. Crawfish and small fishes are eaten in small quantities. Of the vegetable food he says:

Grasses are the most important element of the vegetable food of the southern black duck, forming almost half of it. Frequently the rootstocks are dug up and devoured, and some stems and leaves are eaten. Of the grass seeds consumed, cultivated rice is most important. Most of that found in the stomachs was waste, being taken in winter, and as it included red rice, some good was done by eating it. However, as the southern black duck spends the summer in the country where much rice is grown, it has the opportunity of feeding upon the crop in the younger and more appetizing stages. It is said to do this sometimes to a destructive extent. However, the game value of the duck makes it undesirable to take aggressive measures against it on behalf of the rice crop. A toll large enough, if not too large, is taken of the birds during the hunting season.

Next to grasses the seeds of smartweeds are preferred. They form almost a tenth (9.54 per cent) of the total diet. No fewer than 800 seeds of prickly smartweed (*Polygonum sagittatum*) were taken from a single stomach. The seeds and tubers of sedges compose the next largest item, namely, 6.34 per cent. Seeds of water lilies and coon tail make up 3.11 per cent, and seeds, stems, and foliage of pondweeds and widgeon grass, 1.6 per cent. Other items of vegetable food worth mentioning are bayberries and seeds of buttonbush.

Behavior.—I can not find much published on the habits of the Florida duck, in all of which it undoubtedly closely resembles the black duck. It is, of course, a surface feeder, but that it can dive, if hard pressed, I have learned to my sorrow in attempting to chase wounded birds; I have seen one dive and swim for several yards under water until it could find concealment among aquatic vegetation, where it remained hidden, probably with its bill protruding, and was never seen again. In flight, appearance, and behavior it is much like the black duck, the white lining of its wings being very conspicuous, but it is not nearly so shy as the northern species, perhaps because it is less hunted.

Game.—It is not an important factor as a game bird, because it is not migratory. It inhabits chiefly the less frequented and most inaccessible places in Florida, seldom visited by sportsmen. What few sportsmen visit its haunts usually come in the winter, when this species is widely scattered, or in the spring, when it is mated and breeding, and their time is usually fully occupied with hunting other, more numerous, species which offer better return for their trouble. For these reasons I have never heard of the Florida duck being systematically hunted and I doubt if its numbers are being seriously reduced except where its haunts are becoming thickly settled, cultivated, or drained. It is the same with them as with many of the western ducks, civilization and agriculture are killing them off faster than gunpowder.

Fall.—Mr. Moore says, in the notes referred to above—

that in August, September, and the first part of October parties of from 5 to 20 of
this species leave the fresh ponds and fly across the bay to sand bars on the inner
sides of the Keys, where they spend the night in the pools or coves near the mangroves and return at sunrise the next morning. Those at this time were all males;
but in January, February, and March mated birds, flying in pairs, spend their nights
in the same places.

DISTRIBUTION.

Breeding range.—Florida, mainly in the southern half. Said to be
absent from northeastern Florida, but breeds on the eastern coast
at least as far north as northern Brevard County and Orange County
(Banana River and St. Johns River) and probably farther. Breeds
along the northwestern coast of Florida and probably intergrades
with *maculosa* between Florida and Louisiana.

Winter range.—Apparently the same as the breeding range, but
perhaps some of the West Indies may be included.

Egg dates.—Florida: Fifteen records, February 28 to May 22; eight
records, April 9 to 25.

ANAS FULVIGULA MACULOSA Sennett.

MOTTLED DUCK.

HABITS.

Mr. George B. Sennett (1889) first called attention to the characters which separated the ducks of the species *Anas fulvigula* which
inhabit Louisiana and Texas from those found in Florida. He described the Texas bird as a new species and, as the two forms have
not, apparently, been shown to intergrade, perhaps he was justified
in doing so. In his description he sums up the characters, as follows:

The most marked differences between *A. maculosa* and *A. fulvigula* are that the
cheeks of the former are streaked with brown, while those of the latter are plain buff;
the speculum is purple instead of green; the general effect of the coloration, especially on the under sides, is mottled instead of streaked; the light color everywhere is
a pale buff or isabella color instead of a rich, deep buff; and the tail markings also
are different, as indicated.

Dr. D. G. Elliot (1898), in commenting on these characters, says:

The streaked cheeks are to be seen among some individuals of the Florida dusky
duck, and the color of the speculum is at times merely a question of light, purple and
green in metallic hues, being often interchangeable. An ornithologist might readily
recognize to which form most of his specimens belonged, but the ordinary observer
would probably have difficulty in distinguishing them.

Dr. John C. Phillips (1916), who has made a careful study of these
ducks, has this to say on the subject:

In January, 1914, while paying a visit to Mr. E. A. McIlhenny, at Avery Island,
Louisiana, I was able to collect a series of seven of the mottled ducks from the Vermilion Bay region. There are six adult males and one female. Taken as a whole,

this Louisiana series is even darker than the Texas series; the breasts of the males are very dark, glossy chestnut, and the ground color of the cheeks and chin is distinctly more rufous than in the Texas series or in the Florida series. The cheeks are also quite heavily streaked, and this streaking extends in all cases far below the superciliary stripe; in the Florida ducks the streaking of the cheeks is finer and does not extend so far ventrally on to the chin, while the lores are plain buff and the chin itself is paler in all cases. The pileum of the mottled ducks from Louisiana is more solid black and less streaked black than is the case with the Florida birds; if anything it is darker than the Texas birds. On the upper surface of the Louisiana series and the Texas series the light edges of all the feathers (back, scapulars, rump, and tail) are darker and richer brown, but especially is this so in the Louisiana birds. The speculum character noticed by Sennett does not seem to me to hold good. It was said to be more green and less purple in *fulvigula* than in *maculosa*.

To sum up, I should say that the only character which seems important in distinguishing *A. f. maculosa* and *A. f. fulvigula*, aside from the generally darker tone of the former, is the coarser and more consistently striped head and neck of *A. maculosa*. In all cases the feathers bordering the sides of the culmen, the lores, are dotted with black in *maculosa* and plain buff in *fulvigula*. I believe the richer and more ruddy ground color of the head and neck of *A. f. maculosa* from Louisiana is partly due to the color of the water and mud in the Vermilion Bay region. These Vermilion Bay ducks are certainly more highly colored than ducks from the Brownsville region of Texas. The form *A. fulvigula maculosa*, therefore, will probably remain as a valid race.

The characters are slight, but fairly constant, and the new form, whether species or subspecies, seems to be distinct.

Nesting.—Audubon (1840) was the first to describe the nesting habits of this duck, although at the time he did not consider it as anything but a common black duck. He writes:

On the 30th of April, 1837, my son discovered a nest on Galveston Island, in Texas. It was formed of grass and feathers, the eggs eight in number, lying on the former, surrounded with the down and some feathers of the bird, to the height of about 3 inches. The internal diameter of the nest was about 6 inches, and its walls were nearly 3 in thickness. The female was sitting, but flew off in silence as he approached. The situation selected was a clump of tall slender grass, on a rather sandy ridge, more than a hundred yards from the nearest water, but surrounded by partially dried salt marshes.

Mr. George F. Simmons (1915) thus describes a nest found in a prairie pond near Houston, Texas:

As is the case with all ponds in this section of prairie, the whole with the exception of a small spot near the center was thickly covered with tall grass, rushes, water plants of various sorts, and sprinkled with a few bushes or reeds, locally known as "coffee-bean" or "senna."

The nest itself was placed about 8 inches up in thick marsh grass and rushes, over water 4 inches deep, and was neatly hidden by the tops of the grasses and rushes being drawn together over the nest. It was but 2 or 3 inches thick, a slightly concave saucer of dead, buffy rushes and marsh grass, supported by the thick grasses and by two small "coffee-bean" reeds. The lining was of smaller sections and fragments of the rushes and marsh grass, and a small quantity of cotton; and the 11 eggs were well, though not thickly surrounded by down and soft feathers evidently from the breast of the parent.

Mr. George B. Benners (1887) found three nests near Corpus Christi, Texas; "the nests were built on the edge of the river's bank and were so carefully concealed that if the birds had not flown up we would never have noticed them." Mr. James J. Carroll (1900) says that in Refugio County, Texas, it "breeds along the mainland near the beach and on the islands in April."

Eggs.—The eggs of the mottled duck are indistinguishable from those of the Florida duck, except that they seem to average a little smaller. The measurements of 75 eggs in various collections average 54.9 by 40.5 millimeters; the eggs showing the four extremes measure 60 by 40.2, 56.5 by **43, 51** by 41 and 54.5 by **38** millimeters.

Plumages.—The downy young of the mottled duck is similar to that of the black duck, but it is somewhat lighter colored and the dark markings on the head are much more restricted and paler. The upper parts are "mummy brown" varying to "Dresden brown," and to lighter on the forehead and flanks; the sides of the head, including a broad superciliary stripe, are "Isabella color" or "honey yellow" paling to "cream buff" or "cartridge buff" on the chin and throat; a dusky stripe extends from the bill to the eye and from the eye nearly, or quite, to the occiput; the under parts are "cream buff" or "cartridge buff"; the color of the back is relieved by scapular and rump spots of "cream buff" and the edge of the wing is the same color. The colors become paler with age. The progress toward maturity is apparently the same as in the black duck, the changes are not conspicuous and not easily traced beyond the earlier stages. The juvenal body plumage is worn for only a short time during the first fall; in this the broad edgings of the back and scapulars are "wood brown" or "avellaneous" and those of the lower parts are paler, varying from "avellaneous" to "vinaceous buff," the dusky markings on the breast are more longitudinal, less rounded, than in adults. The juvenal wing, characterized by its duller colors and by its incomplete speculum, is worn all through the first year until it is molted at the first complete postnuptial molt; the speculum in the young male is not only much duller in its metallic purple, but the color is much more restricted, occupying less area, and the black borders are narrower and lacking in velvety luster. During the late autumn and winter the progress toward maturity is rapid, until by spring the body plumage is practically adult and only the wings remain to distinguish the young bird. After the first postnuptial molt, when the bird is a little over a year old, the young bird is practically indistinguishable from the adult. The broad edgings are much more richly colored than in the young bird, varying from "tawny" on the back to "hazel" on the scapulars and from "hazel" to "amber brown" on the breast; the blackish markings on the under parts are more rounded and blacker; the dark colors of the breast are more sharply separated from the buff of the neck; and the speculum is now complete with its brilliant metallic purple, bordered with broad stripes of velvet black.   There is probably no recognizable eclipse plumage, but a complete annual molt in summer.

Behavior.—I have never seen this species in life and find practically nothing published on its habits, but there is no reason to suppose that it differs materially in its behavior from the closely related Florida duck.   It seems to be practically resident throughout its range.   In Louisiana it is called *canard noir d'été,* summer black duck, as the northern black duck is found there only in winter and is called *canard noir d'hiver.*   Messrs. Beyer, Allison, and Kopman (1907) say, of the status of *Anas fulvigula* (now restricted to *maculosa*) in Louisiana:

A regular resident on the coast, and especially on the islands, whence its local name, *canard des isles.* Its numbers are greatly increased during the winter, and at that season it may be found on open lakes, even in the northern part of the State.

DISTRIBUTION.

Breeding range.—Mainly on the coasts of Louisiana and Texas, less common inland, up the Mississippi valley in Louisiana, and westward nearly to central Texas.   South to the mouth of the Rio Grande and perhaps into northern Mexico.

Winter range.—Approximately the same as the breeding range.

Casual records.—Has wandered to Colorado (near Loveland, March 15, 1889, and November 6, 1907); Kansas (Neosho Falls, March 11, 1876).

Egg dates.—Texas: Twelve records, April 18 to August 20; six records, May 3 to 16.   Louisiana: Four records, April 23 to June 1.

EUNETTA FALCATA (Georgi).

FALCATED TEAL.

HABITS.

This beautiful duck is a resident of eastern Asia, breeding abundantly in the southern half of eastern Siberia and migrating to its winter home in southeastern Asia.   It has occurred several times as a rare straggler in Europe, but has been recorded only once in North American territory.   Our reason for including it in our next check list is that Dr. G. Dallas Hanna has recently (1920) recorded the capture of a specimen in the Pribilof Islands, as follows:

A male of this beautiful crested teal was secured on St. George Island, April 18, 1917. Its gorgeous coloration was admired by all who saw it. The native hunters there do not readily distinguish the several species of ducks, and this was called by them "mallard," which name is applied to at least eight separate kinds.

As I knew nothing whatever about this duck and had no references to it in my index, I appealed to Dr. John C. Phillips, who is preparing an extensive monograph on the ducks of the world, and he very generously has sent to me his references and his unpublished manuscript on this species. With deep gratitude to him for his unselfish courtesy, I shall quote freely from his manuscript, merely selecting and arranging such parts as it seems best to use.

Spring.—Very little information and few dates are available for a study of the migration of these ducks. According to Dybowski (1868) they arrive in Dauria in April. Prjevalsky (1878) states that they appear at Lake Hanka from the middle of March to the middle of April, at which latter date Radde (1863) saw them arrive on the middle Amur. They appeared at Utskoi-Ostrog on May 3 (Middendorff, 1853) and at Nikolaievsk on May 18 (Schrenck, 1860). On the upper Amur they did not appear until early June.

The time of breeding, so far as one can judge, is not particularly early. They are said to nest early in June in Transbaikalia, and in east Siberia they begin to nest in late May (Taczanowski, 1873) continuing through June and perhaps into July (Baker, 1908).

Courtship.—The display as observed in captivity was first described by Finn (1915). He describes it as essentially like that of typical ducks, but the erection of the long crest made the head look enormous. He continues—"There was the same rear up, with the head bent down, followed by an upjerk of the hind parts; the long sickle-shaped tertials, so noticeable in this species, seemed little if at all expanded, and were not so prominent in the display as one would have expected from their abnormal character. But what especially attracted my attention, as I had noted the display of the male of this duck some time before, was that the females displayed simultaneously with the males, and with the same gestures."

Nesting.—The nest, so far as known, is always on the ground, built in swamps and along the low-lying banks of the larger rivers. It is rather well built, of leaves, grass, or rushes, compactly put together and lined with a very heavy complement of down. It is said to be not particularly well hidden, but hard to get at on account of the treacherous nature of the ground (Baker, 1908).

Eggs.—The eggs are six to nine in number, probably averaging eight. They are smaller than those of the mallard and are colored like those of the gadwall, although the yellow tinge is somewhat more pronounced (Taczanowski, 1873). The average of 21 eggs measured by Jourdain was 56.2 by 39.65 mm., the maximum being 58.5 by 39 and 55 by 41.5, the minimum 53 by 41 and 57 by 38 mm. (Hartert, 1920). The length of the incubation period is not known. Baker (1908) seems to think that the drake assists, at least occasionally, in the duties of incubation, but I hesitate in attributing such habits to the males of any Palaearctic ducks. Baker also says that the male is seldom found far from the nest.

Food.—There are no detailed notes available, but the food seems to be of a vegetable nature (Stejneger, 1885; Radde, 1863).

Behavior.—The falcated teal lacks much of the elegance of the true teal or the mallard. It appears short, chunky, and large-headed for a surface-feeding duck. The long sickle-shaped tertials and short tail give the body a very stumpy appearance. Heinroth (1911) says that a male in the Berlin Gardens always kept his head and neck well drawn in, so that the mane lay on the upper part of his back. The writer never saw these crest feathers lifted, and the impression created was more like that of a diving duck.

There are no recorded observations as to the flight, except that it is said to be swift and teallike, which probably means that it is more erratic than that of the mallard. In Assam it appears singly or in pairs, more rarely in small parties. But in northern China, Prjevalsky (1878) speaks of their arriving on spring migration in large numbers and associating very commonly with other species of ducks.

The voice of the male is a short low trilling whistle (Walton, 1903), or, according to Prjevalsky (1878), a "tolerably loud and piercing whistle." Although I have never heard the note myself I gather from Heinroth's (1911) account that it is decidedly teallike. The note of the female is the typical mallardlike *quack*, said to be five times repeated (Finn, 1915).

Game.—Even to this day these birds are probably little disturbed over the greater part of the breeding area. But along the coasts of Japan and southern China they are undoubtedly hunted on an increasingly large scale. Great numbers were killed in the Pekin region over 40 years ago, especially in spring (David and Oustalet, 1877), and more recently a great many have been shipped from Hankow to the markets of Europe (Ghidini, 1911).

Fall.—In autumn they leave early, disappearing from the Amur region evidently in late September and early October. According to Dybowski (1868) some stay in Dauria till late December. It is an interesting fact that of the specimens taken in Burma and India a great majority are females (Finn, 1909).

DISTRIBUTION.

Breeding range.—Eastern Siberia. East to Kamchatka, probably the Commander Islands (Bering Island), the Kurile Islands and northern Japan (Yezzo). South on the mainland to the vicinity of Vladivostok and to approximately the northern border of Mongolia. West nearly, if not quite, to the upper Yenesei River. North to about 65° N., not quite to the Arctic Circle.

Winter range.—Southeastern Asia. East to Japan. South to the Japanese Archipelago, Formosa, southern China, Burma, and north central India (Delhi). The western and northern limits seem to be not well determined.

Casual records.—Rare straggler to Europe; specimens have been taken in Sweden, 1853, Hungary, 1839, and Bohemia. One taken on St. George Island, Bering Sea, on April 18, 1917.

CHAULELASMUS STREPERUS (Linnaeus).

GADWALL.

HABITS.

The arrival of the ducks on their breeding grounds in the great wildfowl nurseries of northwest Canada is a spectacular performance. I shall never forget the sights I saw, one cold, rainy day, June 13, 1905, as I walked down toward the great sloughs at the head of Crane Lake, Saskatchewan; hundreds of ducks arose from the wet meadows, from the sloughs, and from an island in the lake, flying around in great loose flocks; a great cloud of them rose, like a swarm of mosquitoes, from the mouth of Bear Creek; most of them were gadwalls, but there were also large numbers of canvasbacks, redheads, shovellers, and blue-winged teal, as well as lesser numbers of lesser scaups, mallards, baldpates, and ruddies, with a few Canada geese; the air seemed to be full of ducks, flying in all directions in bewildering clouds; I have never seen so many ducks before nor since. This was the center of their abundance in one of the greatest duck-breeding resorts I have ever seen. Probably all of the ducks had arrived on their breeding grounds at that time, but evidently many of them had not mated and others had not finished laying.

On June 17, 1905, Mr. H. K. Job and I made a careful census of the ducks breeding on the island, referred to above, by dragging a long rope over it as thoroughly as we could and by noting and recording the nests found by flushing the birds. The island was about 300 or 400 yards in length by about 100 yards in width, fairly high at one end and everywhere covered with a thick growth of grass, through which were scattered on the higher portion numerous small clumps and in some places large patches of rose bushes, offering ideal conditions as a breeding ground for ducks. There were several small ponds near the center of it lined with fringes of cat-tails and bullrushes. On the lower portion of the island the grass was shorter, and where it extended out into a point the ground was bare. A colony of common terns occupied this point, which was also the favorite resort of a flock of white pelicans, which may have bred here later in the season. Marbled godwits, Wilson phalaropes, and spotted sandpipers were breeding here, as well as western savanna sparrows.

A pair of crows had a nest in the only tree on the island, a small willow, and they must have fared sumptuously on stolen duck's eggs. A pair of short-eared owls had a nest on the island containing young in various stages of growth. We were unable to drag the whole island, as the rose bushes were too thick in many places, but in the course of two hours' work we recorded 61 nests, as follows: Mallard, 5 nests; gadwall, 23 nests; baldpate, 3 nests; green-winged teal, 2 nests; blue-winged teal, 10 nests; shoveller, 7 nests; pintail, 8 nests; and lesser scaup duck, 3 nests. The ducks were identified to the best of our ability by eyesight; the female gadwalls and baldpates were very difficult to distinguish and there may have been more of the latter than we supposed, but certainly both the species were nesting there, as we saw a number of males in the small pondholes; the green-winged teals' nests were identified by seeing the female join a male of that species. We started a number of ducks, mostly pintails, where we failed to find nests, which probably meant broods of young and which were not counted. Most of the sets were incomplete or fresh, indicating that the ducks were only just beginning to lay; we therefore must have overlooked a great many nests, where the eggs were covered and no ducks flushed, as we found a number of such nests by accident. Considering these facts, making allowance for the unexplored parts of the island and judging from the immense numbers of ducks that were flying about or bedded out on the lake, I considered it fair to assume that at least 150 pairs of ducks were breeding or preparing to breed on this one island. In addition to the species above recorded, we saw on the island several American mergansers, a white-winged scoter and one cinnamon teal, making a total of 14 species of ducks which were probably breeding on the island or in sloughs around it. As may be imagined, it was with considerable interest and pleasant anticipation that I revisited this island in 1906, but I was most keenly disappointed to find it practically deserted. Instead of the immense flocks of ducks which I had seen rise from the sloughs like clouds of mosquitoes, only a few scattered flocks were seen. As we walked across the island expecting to see ducks flying up all about us, hardly a duck arose, and in place of the 60 odd nests that we expected to find only 3 nests were found. The mystery was soon solved by finding a nestful of broken eggs and bunches of yellowish hair clinging to the rose bushes. A coyote had been living on the island and had cleaned out all of the nests, and driven the ducks away. The destruction of the bird population of the island had been still further carried on by a family of minks and the entrance to their burrow was strewn with feathers. Whether the ducks will ever return to this island or not is an open question, but probably they have moved to some safer spot.

The prevailing impression which seems to exist in the minds of many writers that the gadwall is nowhere an abundant species should be dispelled by the foregoing account of its abundance in Saskatchewan, where it was at that time, and probably still is, the most abundant of all the ducks. It is undoubtedly steadily decreasing, as all the other ducks are, for advancing civilization and the demands of agriculture are usurping its breeding grounds. If a few such places, as I have just described, which are not particularly valuable for agricultural purposes, could be set apart as breeding reservations for waterfowl, this and many other species might be saved from extermination, which otherwise seems inevitable. Since the above was written I have learned from Mr. Hoyes Lloyd that much of the land around Crane Lake has been secured by the Canadian Government and that the locality described above will be included in an extensive bird reservation.

Courtship.—Dr. Alexander Wetmore (1920) has given us the following interesting account of the courtship of the gadwall:

The mating flight of the gadwall is always interesting and is seen constantly when the birds are on their breeding grounds. Here at Lake Burford opportunities for observing it were excellent. The flight was usually performed by two males and one female. In the beginning two males approached a female in the water, calling and bowing. She usually rose at once and flew with a slow flapping flight, mounting in the air with the males in pursuit, calling and whistling constantly. First one and then the other of the males swung in front of her, set his wings, inclined his body upward to show his handsome markings, and, after a few seconds, dropped back again to his former position. Late in the season there was always one of the males who was favored and who displayed more often than the other, flying close to the female, so that in passing his wings often struck hers, making a rattling noise. After a short time the second male often left the pair and returned to the water. The birds frequently mounted until they were 300 yards or more in the air, and darted quickly from side to side, flying now rapidly and now slowly. When the flight was over the birds descended swiftly to the water again. I was never able to ascertain whether there were some extra males about or not, as, though, there were usually two with the female in this flight I found them at other times always in pairs.

The female gadwall, like the mallards, also came out in the short grass of the shore and walked about with head down, quacking loudly, an action that I took for part of the mating display.

When the birds were in the shelter of the rushes they went through other mating actions of interest. The male swam toward the female bowing by extending his neck until the head was erect and then retracting it, bringing his bill down onto his breast. He then approached pressing his breast against the sides of the female and shoving her easily, first on one side and then on the other, biting her back and rump gently as he did so. After a few seconds she lowered her body in the water and copulation took place with the female entirely submerged save for the crown of her head while half of the body of the male was under water. As the female emerged the male turned immediately to face her and bowed deeply, giving a deep reedy call as he did so.

Nesting.—In North Dakota, in 1901 we found the gadwall breeding quite commonly on the islands in the larger lakes, particularly on the islands in Stump Lake which are now set apart as a reservation and protected. Baldpates and lesser scaup ducks were breeding abundantly on the same islands, far outnumbering the gadwalls; there was also a breeding colony of double-crested cormorants on one island and colonies of ring-billed gulls and common terns on two of them. The gadwall's nests were usually well concealed in thick rank grass, tall reeds, dense clumps of wild rye, or patches of coarse weeds; they were always on dry ground and never very near the water. The nests consisted of hollows scooped in the ground and well lined with strips or pieces of reeds, bits of dry grass, and weed stems, or whatever material could be most easily gathered in the vicinity, mixed with the down from the bird's breast; with incomplete sets or fresh eggs very little down is found, but as incubation advances the down is added until the eggs are surrounded and sometimes entirely covered with a profusion of dark gray down, which is usually mixed with bits of grass or straw.

Although the gadwall seems to prefer to nest on islands we found a number of nests in Saskatchewan in meadows or on the open prairie at long distances from water, where we flushed the birds from their nests as we drove along; such nests were well concealed in thick grass, which was often arched over them, or were hidden under small sage or rose bushes. I have always found the gadwall a close sitter, flushed only when closely approached, but Mr. W. L. Dawson (1909) has noted variations in this respect, as follows:

The bird's behavior when surprised depends altogether upon the stage of incubation reached. In general, the bird sits close until discovered; after that, if the eggs are fresh, the duck may flee upon sighting her enemy a hundred yards away; but if the eggs are near hatching, she will endeavor to lead the investigator astray by painfully dragging herself through the grass. If too much harassed, however, she will desert her eggs outright rather than wait for what she regards as an inevitable doom, and the same remark will apply to almost any of the nesting ducks.

The down in the nest of the gadwall is smaller than that of the mallard, darker colored, and otherwise different in appearance. In color it is dark "hair brown," almost "fuscous," with whitish centers and grayish tips. The breast feathers mixed with the down are characteristic of the species, small, light colored, and with variable patterns of dusky markings in the center, but with light tips. It is difficult to distinguish, in the field, the nest of the gadwall from that of the baldpate, as the eggs of the two are indistinguishable, and the females are much alike. But the gadwall has more white in the speculum, the bill is yellower, and the breast is spotted, all difficult points to see as the female flies from the nest.

Eggs.—The gadwall lays from 7 to 13 eggs, but the usual set consists of 10, 11, or 12 eggs. I have occasionally seen one or two eggs of the lesser scaup duck in a gadwall's nest, and Mr. William Spreadborough, "on June 29, 1894, at Crane Lake, Saskatchewan, took a nest of this species containing 13 eggs, 7 of which were of the lesser scaup," according to Macoun (1909). Undoubtedly the baldpate occasionally lays in the gadwall's nest, as the two species are often intimately associated, but the eggs are nearly indistinguishable.

The eggs of the gadwall are nearly oval in shape and are usually shorter and more rounded than those of the baldpate, but there is much individual variation in both species. Their color is a dull creamy white, somewhat whiter than the baldpate's on the average. The measurements of 100 eggs in various collections average 55.3 by 39.7 millimeters; the eggs showing the four extremes measure **59.5** by 39.5, 57.5 by **43.5, 49.5** by 38 and 51 by **34.5** millimeters. The period of incubation, which is performed wholly by the female, is about 28 days.

Plumage.—The downy young of the gadwall is very much like that of the mallard, except that it is decidedly paler and less richly colored; the pale yellow of the under parts is more extensive on the sides and head extending nearly around the neck where it is separated by a narrow dark stripe on the nape, the light superciliary stripe is broader; the dark loral and postocular stripe is narrower and the uricular spot is hardly noticeable. The upper parts are "bister," deepening on the crown to "bone brown"; the under parts are "cartridge buff," paler on the belly and deepening to "cream buff" or "Naples yellow" on the neck and sides of the head; the light patches on the scapulars and sides of the rump are buffy white.

The young birds become paler as they grow older. The plumage develops in the same sequence as in the mallard. Young birds gradually develop, during the first fall and winter, a plumage closely resembling that of the adult; by the month of March this first winter, or first nuptial, plumage is generally complete; there are usually a few spotted feathers scattered over the under parts, as signs of immaturity and the wings are much duller than those of adults, with little if any chestnut in the coverts. The molt into the eclipse plumage begins in June and during the transitions of this first double molt, young birds become indistinguishable from adults.

Adult males begin to molt into the eclipse plumage about the last of May or first of June and by the end of June many dark, brown-edged feathers, like those of the female are scattered through the breast and flanks. I have seen males in full eclipse as early as August 10 and as late as September 8; this plumage is an almost complete reproduction of the female's, or young male's, excepting, of course, the wings, which are molted only once in August. The molt out of the eclipse plumage consumes about two months; I have seen birds in this molt from September 8 to November 23, but I am inclined to think that old drakes usually attain their full plumage by November 1st or earlier. There is no winter or spring molt in old birds. Hybrids occur occasionally; Mr. William G. Smith (1887) mentions a beautiful male hybrid in which the color and size were about equally divided between the gadwall and the baldpate; he also "killed two specimens of gadwall with a distinct black ring about the neck. They were male and female and were together."

Food.—Like the mallard, the gadwall is a clean feeder, which makes its flesh desirable for the table. It consumes a great variety of food, most of which is obtained by tipping or dabbling about the edges of marshy ponds, sloughs, or grassy, sluggish streams; it can dive well for its food, however, when necessary. Its vegetable food consists of tender grasses, the blades, buds, seeds, leaves, and roots of various aquatic plants, nuts, and acorns; it visits the grain fields to some extent, where it picks up wheat, barley, buckwheat, and corn.

According to Mr. Douglas C. Mabbott (1920):

In habits the gadwall resembles the mallard, feeding either on dry land or in shallow water near the edges of ponds, lakes, and streams, where it gets its food by "tilting" or standing on its head in the water. The food of both the gadwall and the baldpate, however, is quite different in some respects from that of the mallard. These two feed to a very large extent upon the leaves and stems of water plants, paying less attention to the seeds, while the mallard feeds indiscriminately on both or even shows some preference for the seeds. In fact, in respect to the quantity of foliage taken, the gadwall and the baldpate are different from all other ducks thus far examined by the Biological Survey. They are also more purely vegetarian, their diet including a smaller percentage of animal matter than that of any of the other ducks.

As computed from the contents of 362 stomachs collected during the six months from September to March, 97.85 per cent of the food of the gadwall consists of vegetable matter. This is made up as follows: Pondweeds, 42.33 per cent; sedges, 19.91; algae, 10.41; coon tail, 7.82; grasses, 7.59; arrowheads, 3.25; rice and other cultivated grain, 1.31; duckweeds, 0.61; smartweeds, 0.59; wild celery and water weed, 0.53; water lilies, 0.52; madder family, 0.37; and miscellaneous, 2.61 per cent.

Considerably more animal food is taken in summer than in winter, owing, of course, to the fact that more is available at that time of the year. The percentage of animal food for the summer months is higher also because there are included in the averages analyses of numerous stomach contents of ducklings, which feed to a great extent upon insects. All of the 11 stomachs collected during the month of July (9 from North Dakota and 2 from Utah) were of young ducklings. A computation of the average contents of this series produced the following results: Water bugs, 56.18 per cent; beetles, 7.09; flies and their larvae, 2; nymphs of dragon flies and damsel flies, 0.27; other insects, 2; total animal food, 67.54 per cent; pondweeds, 12.55 per cent; grasses, 5.09; sedges, 2; water milfoils, 0.55; smartweeds, 0.09; miscellaneous, 12.18; total vegetable food, 32.46 per cent.

The animal food of adults includes small fishes, crustaceans, tadpoles, leeches, small mollusks, water beetles and other insects, larvae, and worms.

Behavior.—The gadwall can walk well on land, where it forages for oak mast in the woods and for grain in the open fields, often a long distance from water. It takes flight readily from either land or water, springing into the air and flying swiftly away in a straight line. When migrating, it flies in small flocks of about a dozen birds; in appearance and manner of flight it greatly resembles the baldpate, but the male can usually be distinguished from the latter by the white speculum and the brown wing coverts; a similar difference exists between the females, but only to a slight degree; practiced gunners claim to recognize other field marks, but they have proven too subtle for my eyes, and I have frequently mistaken one species for the other. The gadwall ought not to be mistaken for any other species, except the baldpate or the European widgeon, but it frequently is confused, by ignorant gunners, with the young males and females of the pintail, though its flight and general appearance are entirely different; the name "gray duck" has been applied to both the gadwall and the pintail, which has led to much confusion of records and to erroneous impressions as to the former abundance of the gadwall in New England, where, I believe, it has always been a rare bird.

Doctor Wetmore (1920) describes the notes of the gadwall as follows:

The call note of the female is a loud quack that is similar to that of the female mallard but is pitched slightly higher and is not quite so loud and raucous. Considerable experience is required, however, to distinguish with certainty the calls of the two birds. The male has a loud call like *kack kack*, a deep reedlike note resembling the syllable *whack*, and a shrill whistled call.

The gadwall associates freely with other species of similar habits and tastes, particularly with the baldpate and pintail, with which it seems to be on good terms.

Game.—There seems to be a difference of opinion among sportsmen as to the food value of the gadwall; some consider it a close second to the mallard and others say it is hardly fit to eat; probably this is due to the different kinds of food that it lives on in various localities.

Mr. Dwight W. Huntington (1903) writes of his experience in shooting it:

I found it fairly abundant in North Dakota and usually shot a few gadwalls with the other ducks. One day when shooting on a little pond quite near the Devils Lake, I shot a large number of ducks, and nearly all of them were gadwalls. They came quite rapidly toward evening, and standing in the tall rushes without much effort at concealment, I had some very rapid shooting. Far out on the lake the swans and geese were trumpeting and honking. Large flocks of snowgeese, or white brant, as they call them in Dakota, were always in the air; the mallard, sprigtails, teal, and all the ducks were flying everywhere; but the gadwalls were the only ducks which came to me in any numbers. Had I put out only gadwall decoys, there might have been a reason for this, but I had no decoys that day at all. In fact the ducks were always so abundant that I could kill far more than I could carry, without decoys, and an ambulance from the garrison came out to carry in the game.

Winter.—As the gadwall is one of the later migrants northward in the spring, not appearing usually until the ice is all out of the ponds, so it is also one of the earlier ducks to leave in the fall and start on its short flight to its winter home in the Southern States, principally in the lower Mississippi valley, and in Mexico. The gadwall is primarily a fresh water duck, breeding far in the interior and wintering principally in the inland ponds, marshy lakes, sloughs, and swamps, where it can find mild weather and plenty of food; but it frequents to some extent the brackish pools and estuaries along the coasts of Louisiana and Texas, where it is very common. Messrs. Beyer, Allison, and Kopman (1906) say of its winter movements in Louisiana:

As in the case of the mallard, the first came by the early or the middle part of October, and continue to increase decidedly until the middle of December, then remaining *in statu quo* or showing something of a decrease, according to the nature of the winter, until the middle of January. A strong northward movement begins at that time, and while it consists largely of individuals that have wintered in Louisiana, it is doubtless augmented also by the first passage of transients. This later movement continues more or less freely until about March 15, after which date, duck migration is restricted almost entirely to a few species, among which the gadwall is seldom if ever found.

DISTRIBUTION.

Breeding range.—Temperate regions of the Northern Hemisphere. In North America, east to Hudson Bay (Churchill), southeastern Manitoba (Shoal Lake), southern Wisconsin (Lake Koshkonong, formerly), and formerly in Ohio. South to central Minnesota (Becker County), northern Iowa (Kossuth County), southwestern Kansas (Meade County), southern Colorado (La Plata County and San Luis Valley), northwestern New Mexico (Lake Burford), southern California (San Jacinto Lake, Riverside County). West to the interior valleys of California (San Joaquin and Sacramento valleys), Oregon (Camp Harney), and Washington (Brook Lake). North to central Alberta (Lesser Slave Lake), and northern Saskatchewan (Athabasca Lake). Has been found breeding casually on Anticosti Island, Gulf of St. Lawrence. In the Eastern Hemisphere, Iceland and the temperate regions of Europe and Asia, from the British Isles, Denmark, Sweden, and Holland to Kamchatka; also southern Spain and northern Algeria.

Winter range.—Southern States and Mexico, east to the Atlantic coast from Maryland (Chesapeake Bay) to southern Florida. South to Jamaica, rarely, south central Mexico (Guadalajara), and southern Lower California (San José del Cabo). West to the Pacific coast. North rarely to British Columbia (Chilliwack) more commonly to the coasts of Washington (Puget Sound region), and Oregon, Utah, northeastern Colorado (Barr Lake), northern Arkansas (Big Lake), and southern Illinois (Mount Carmel). In the Eastern Hemisphere, the British Isles, the Mediterranean basin, northern Africa (to the Sudan and Abyssinia), northern India, China, and Japan.

Spring migration.—Early dates of arrival: Iowa, southern, March 10; Minnesota, Heron Lake, March 17; Montana, Terry, April 1; Manitoba, Aweme, April 23; Saskatchewan, Indian Head, April 18; Alberta, Edmonton, May 5. Late dates of departure: Lower California, Colnett, April 8; Oklahoma, Caddo, April 2.

Fall migration.—Early dates of arrival: Lower California, southern, September 27; Massachusetts, Essex County, October 2; Rhode Island, October 8. Late dates of departure: Ontario, Ottawa, October 29; Rhode Island, Point Judith, November 11; Long Island, Oakdale, December 13.

Casual records.—Accidental in Bermuda (December, 1849) and Alaska (St. Paul Island, November 13, 1911).

Egg dates.—North Dakota: Twenty-seven records, May 18 to July 16; fourteen records, June 7 to 19. Manitoba and Saskatchewan: Twenty records, June 6 to July 3; ten records, June 13 to 27. California and Utah: Sixteen records, April 16 to July 20; eight records, May 15 to June 17.

MARECA PENELOPE (Linnaeus).

EUROPEAN WIDGEON.

HABITS.

This is an old-world species which has occurred frequently as a straggler on both coasts of North Amercia, as well as in the interior. The Atlantic coast records are nearly all fall and winter records, but in the interior its occurrence seems to be wholly in the spring and on the Pacific coast in the winter. As I know nothing about the habits of this foreign species from personal experience with it and as comparatively little has been written about it in American bird books, I am quoting freely from Mr. J. G. Millais (1902) who has given a very satisfactory life history of the widgeon.

Spring.—As the spring approaches we see on fine days the flocks of widgeon splitting up into smaller parties and engaged in pairing. By the end of March many widgeon have paired, and proceed to their breeding grounds together; but in most cases the northern movement is undertaken in a series of small flocks, which gradually detach themselves from the main bodies. These small parties of from 25 to 30 birds follow one another in their migration, often stopping for a few days at some halting place, like the Shetlands or the Norwegian fiords, till, by the middle of April, none are left on our coast except a few stragglers.

Courtship.—The actual courtship of the widgeon differs somewhat from that of other surface feeders, and the display of the male bird is an interesting one. A female having shown herself desirous of selecting a mate, five or six males crowd closely round, hemming her in on every side and persecuting her with their attentions. If she swims away, they follow her in a close phalanx, every male raising his crest, stretching out his neck close over the water, and erecting the beautiful long feathers of the scapulars to show them off. He also depresses the shoulder joints downward, so as to elevate the primaries in the air. All the time the amorous males keep up a perfect babble of loud "Whee-ous," and they are by far the noisiest of ducks in their courtship. Occasionally the cock birds fight and drive each other off, but ducks are not, broadly speaking, pugnacious birds, and success in winning the admiration of the female is rather a matter of persistent and active attention than physical force.

Nesting.—The nest of the female widgeon is generally placed at from 10 to 20 yards from the nearest water, and generally in coarse grass or heather. Sometimes, like the mallard, she will wander far in the tundra, and one of the only two nests I have found in Scotland I stumbled on by accident right in the middle of a grouse moor, and far from the lake near which I had been searching the whole morning. Generally from 7 to 10 cream-colored eggs are laid.

He also refers in a footnote to a nest he found at Scampston, Yorks, which was placed in nettles; I have found the nest of the American widgeon in a similar situation.

Eggs.—The eggs of the European widgeon are indistinguishable from those of the American widgeon or baldpate. The set usually consists of 7 or 8 eggs, but sometimes as many as 9 or 10 are laid. The measurements of 117 eggs, as given in Witherby's Handbook (1920), average 54.7 by 38.7; the eggs showing the four extremes measure 59.5 by 38.5, 58 by 41, and 49.9 by 35.2 millimeters.

Plumages.—The European bird is so closely related to our own, its downy young and its sequence of plumages are so similar, that I prefer to refer the reader to some of the leading British manuals rather than attempt to quote from them. Mr. Millais (1902) has treated the subject very fully and A Practical Handbook of British Birds, by H. F. Witherby and others, describes the plumages of this and other birds on the British list most exhaustively and satisfactorily.

Food.—Mr. Millais gives the following interesting account of the feeding habits of the widgeon:

In a regular feeding ground, generally some long open stretch of mud covered with *Zostera marina*, it is interesting to see the careful manner in which widgeon approach it. The first little pack will come flying up against the wind and alight on the water, at about two or three hundred yards from the shore, after having previously swung round once or twice to ascertain that no enemy is approaching. This generally takes place when the tide is half ebbed. Out on the water they remain packed close together and very quiet till the first green fronds of their favorite food are observed floating on the surface away inshore. Then the whole gathering begins slowly going shorewards, till at last one bird bolder than the rest swims in and commences picking at the floating weed. Even then they are subject to sudden fears, and, when about to follow their leader, will often suddenly put up their necks and swim rapidly out, the cocks whistling loudly. Once, however, they have reached the food, their taste for more generally asserts itself, and precautions against surprise are somewhat relaxed, as they one and all move in to still shallower water and commence to turn upside down so as to pull up the *Zostera* and eat the root, by far the most succulent part.

Sometimes widgeon, which are both conservative as to their beats and modes of life, will pay little attention to a vegetable diet, but live almost exclusively on animal food. Such I find to be the case with the birds living on the sandy coast near the town of Dornoch in Scotland, where all conditions are purely marine. The widgeon here feed by day and live entirely on small cockles. This renders their flesh poor, bitter, and quite uneatable. I have shot a good few of them there and found all to be the same, whilst birds from the other side of the same firth, and living on the *Zostera* beds to the west of Tain, were fat and as good as widgeon generally are. In spring widgeon are great grass eaters, and later on, like teal and garganey, they devour an enormous quantity of flies. One day in Iceland I observed with a telescope a small party of male widgeon whose wives were engaged in domestic affairs, paddling along the edge of a small lake near Myvatn, and picking the flies off the stones in hundreds. This particular insect, a sort of stinging house fly, is very nutritive and tastes like a piece of sugar. As you are obliged to eat plenty of them yourself, for they are always getting into your mouth, you soon get used to them, and swallow them with equanimity, and it is a common sight to see the Icelandic children of the Myvatn district picking these natural lollipops off their faces and eating them by dozens.

In certain northern firths, where widgeon and brent geese frequent the same ground, it is no uncommon sight to see widgeon in small parties of half a dozen "jackaling" the food which has been torn up by the large birds. The brent can reach far below the surface and tear up the *Zostera* and they themselves only eat the root and allow the fronds to drift away. These are eagerly devoured by the widgeon when they are hungry.

Behavior.—Macgillivray (1852) says of the behavior of widgeons:

They are frequently seen in very large flocks, but usually in small bodies, seldom intermingling with other species. They swim with great ease, and have a rapid direct flight, taking wing easily from the water, and producing a whistling sound as they fly. They are much addicted to garrulity, and at night especially emit a whistling cry, on account of which they have obtained the name of "whew-ducks."

Evidently the field marks of the widgeon are the same as those of our baldpate, but the two species can be readily distinguished by the reddish head of the adult male and the general ruddy tinge of plumage in the female and young male of the European bird; also the axillars in the baldpate are pure white in both sexes, whereas in the European bird they are freckled or clouded with gray.

Mr. Millais (1902) says:

The call of the male widgeon is a loud "whee-ou," a note both wild and musical, and dear to the heart of every gunner that has wandered on the coasts; it also makes a very peculiar "cheeping" note (rather like the call of the twite) when frightened. The female also has two calls, both somewhat similar yet quite distinct, both a sort of throaty croak one being used to attract the attention of others of her species; the other, somewhat harsher, is emitted in moments of fear.

Fall.—By the middle of August the old females and young begin to join together, and are generally the first to commence the southern migration. These are then followed, in September, by the stragglers and males with brown shoulders molting into white, and adult males still in nearly complete eclipse but showing the first signs of winter plumage in the upper scapulars. When all the males arrive they mix indiscriminately with other widgeon, and so the addition to the ranks is swelled gradually until November, when the large winter packs are formed.

Widgeon are more or less marine in their habits, and after arriving on our coasts, in September, they increase in numbers until December, when great packs are sometimes formed in estuaries suited to their tastes. They are the mainstay of the professional punt gunner, being numerous and always a marketable commodity, and it is interesting to note the appreciable change in the habits of the birds due to this enemy. By nature the widgeon is not necessarily a purely nocturnal feeder. In his summer home, where he is subject to little molestation, he feeds regularly in the early morning and late evening, resting only during the warm hours in the middle of the day. Now notice what happens when he arrives on the British coasts. At first the small packs continue to feed in daylight, as during summer, but a couple of raking shots in their midst, carrying death and destruction, tell them that this is too dangerous, so they become purely nocturnal feeders for the remainder of the season, and rest or fly about by day well out in the firths or open sea, according to the proportion of harassment. Where widgeon have been kept continually on the move; that is, after a series of gales sweeping over their resting grounds, as well as when several shots have been fired at them on the mud flats, they sometimes assemble in immense flocks, either on the principle of mutual protection or that "misery loves company". I have on more than one occasion seen the entire stock of widgeon frequenting a certain firth merged into one great gathering, which could not have contained less than five to seven thousand birds.

DISTRIBUTION.

Breeding range.—Northern parts of the Eastern Hemisphere. Iceland, the Faroes, Shetland and Orkney Islands, Scotland, and northern England. North in Europe and Asia to 70° N.; east to Kamchatka and perhaps to the Aleutian Islands.

Winter range.—The British Isles, the whole of southern Europe and northern Africa, south to Abyssinia; also southern Asia, to southern Japan.

Casual records.—North American records must be regarded as casuals until a definite breeding record for this continent can be established. Atlantic coast records are mostly in fall and winter, from October 20, 1899 (Halifax, Massachusetts) to March 25, 1899 (Keuka Lake, New York). Interior records are mostly in spring, from March 23, 1896 (English Lake, Indiana), to April 18, 1904 (Sandusky, Ohio). Greenland records fall between September 29, 1900, and December 17, 1900. The Pacific coast dates are mostly in December and February. The Alaska dates are, Unalaska, October 12, 1871, and Pribilof Islands, May 27, 1872 and April 30, 1911. Probably two migration routes reach the United States, one through Greenland to the Atlantic coast and one through the Aleutian Islands to the Pacific coast. Accidental in Spitsbergen, the Azores, Madeira, Canary and Marshall Islands.

Egg dates.—Iceland: Twenty records, May 12 to June 21; ten records, May 25 to June 15.

MARECA AMERICANA (Gmelin.)

BALDPATE.

HABITS.

I have always thought that the proper name for this species is the American widgeon, for it is certainly very closely related to and much resembles in many ways its European relative. The name widgeon is applied by gunners to various species of fresh-water ducks which they can not recognize, especially to the females; gadwalls, pintails, and the present species seem to be very confusing to sportsmen and are usually all lumped together as "widgeons." This name does well enough in the fall and winter, when associated with other species, but when seen in the spring, in the full glory of its nuptial plumage, with its glistening white crown, the name baldpate seems more appropriate; the name baldpate always suggests to my mind the mated pairs of the handsome ducks that I have so often seen swimming in these little ponds or streams of the western plains or springing into the air, if we drove too near, with a great display of their striking color patterns.

Spring.—The baldpate is not one of the earliest migrants; the ice has long since disappeared and spring is well under way before it starts, and many of the birds do not arrive in their breeding grounds in the Northern States until the latter part of May. Turner (1886) says of its arrival at St. Michael, Alaska:

It arrives about the 25th of May or even later. It is not at all gregarious, being found solitary or in pairs. It frequents the marshes, preferably those which are overflowed by the higher tides when it arrives. As soon as the season is advanced, the greater part of the snow is gone, and the little rivulets are full of muddy water, they resort to these places for food. They seem to delight in shoveling among the mud in search for their food. They plunge their heads at times completely under the soft soil to obtain a tender root or slug.

Mr. Aretas A. Saunders writes me, of its arrival in Montana, that it "arrives on its breeding grounds in the latter part of May. It is not mated before arrival, like the mallard and gadwall, but is seen more frequently in flocks of 5 to 15, during the first part of the spring migration."

Courtship.—Evidently the mating occurs after its arrival on its breeding grounds. I have never seen the courtship of this species, but it probably does not differ materially from that of the European widgeon, which is described under that species.

Dr. Alexander Wetmore (1920) describes the courtship flight of the baldpate, as follows:

The mating flight of this duck resembles that of the preceding species (gadwall), but is performed with more dash and speed. The birds fly swiftly and erratically. The males dart ahead of the female setting and decurving their wings and throwing their heads up, exhibiting their striking markings to the best advantage. The female calls *qua-awk, qua-awk* and the males whistle *whew whew* constantly during this performance. Occasionally as a pair swung in low over the water the male darted ahead and, with decurved wings and head thrown up, scaled down to the surface. Two males and a single female invariably took part in the display flight which began, as in the gadwall, by the males approaching the female, bowing and whistling and then following her as she rose in the air.

Nesting.—In North Dakota in 1901 we found the baldpate breeding abundantly, principally on the islands in the larger lakes. The baldpate is a late breeder, very few of the eggs being laid before June 1, and the majority of the sets are not completed until the second week in June or later. The greatest breeding grounds of this species were on the four small islands in the western end of Stump Lake, so graphically described by my companion Mr. Herbert K. Job (1898) under the appropriate title "The Enchanted Isles." These islands are now included in the Stump Lake Reservation. The islands were devoid of trees but supported a rank growth of grasses, tall coarse weeds, and various herbaceous plants, as well as several tall thick clumps of *Phragmites communis*, and patches of wild roses; they were high in the centers, sloping gradually down to gravelly beaches, with numerous loose rocks and bowlders scattered over them. Here, on June 15, 1901, we found no less than 15 nests of baldpates; probably there were more nests, which we did not find, as it was raining very hard when we explored the island where they were breeding most abundantly, so we made only a hurried search of about half an hour, finding 12 nests in this short time. We also found here numerous nests of other ducks, mallards, gadwalls, pintails, lesser scaup ducks, and white-winged scoters, besides a breeding colony of double-crested cormorants and large numbers of nesting ring-billed gulls and common terns. Though we were tramping around in a drenching downpour, the cloud of gulls and terns screaming overhead and the ducks flushing under our feet every few steps created enough excitement to make us forget our discomfort.

The nest of the baldpate is built on dry ground, often at a considerable distance from the water, in a slight hollow generally well lined with bits of dry grass and weed stems, with a plentiful supply of light-gray down surrounding the eggs, which increases in quantity as incubation advances. The bird frequently covers the eggs with the down when she leaves the nest, completely concealing them and making the nest almost invisible, even in an open situation, which is often selected. The nests which I found on the islands described above were located as follows: The first was well concealed in the center of a thick clump of goldenrod growing on the beach; it was lined with dried leaves and rubbish, with very little down around the eight fresh eggs. The second was in the center of a clump of nettles near the upper edge of a stony beach; it contained eight fresh eggs which were laid on the bare stones, one of them plainly visible in the center of the nest, and surrounded by a little down; it contained 10 eggs and a good supply of down two weeks later. The third nest was on higher ground, concealed in rather tall prairie grass; the 11 eggs in it were heavily incubated; it was profusely lined with down, mixed with bits of dry grass and weeds. The 12 nests found on June 15 were mostly under rose bushes, among the rocks, many of them in open situations; they contained from 9 to 11 eggs each. One of the nests contained a white-winged scoter's egg and one an egg of the lesser scaup duck, both of which were nesting on the island.

According to Baird, Brewer, and Ridgway (1884), Kennicott found several nests of the baldpate, on the Yukon, fully half a mile from the river:

He invariably found the nest among dry leaves, upon high, dry ground, either under large trees or in thick groves of small ones, frequently among thick spruces. The nest is rather small, simply a depression among the leaves, but thickly lined with down, with which, after incubation is begun, the eggs are covered when left by the parent. The nest is usually placed at the foot of a tree or bush, with generally no attempt at concealment. The female, when started from her nest, rises silently into the air, and usually flies to the nearest water, though sometimes she will alight on the ground a few rods distant.

The nest of the baldpate can only with difficulty be distinguished from that of the gadwall, as explained under the latter species. A careful study of the color patterns in the wings of the two females will help the collector to recognize the female as she flies from the nest. And the nest is distinguishable on careful comparison. The down in the baldpate's nest is lighter and smaller; it is "light drab" in color, with whitish centers and conspicuous whitish tips. The breast feathers in the nest are either pure white or with pale brownish or grayish centers.

Eggs.—From the writings of others I infer that the baldpate lays from 6 to 12 eggs, but, from my own experience, I should say that the usual full set consisted of from 9 to 11 eggs. The eggs are absolutely indistinguishable, with any degree of certainty, from those of the European widgeon or the gadwall, though, as a rule, the baldpate's eggs are slightly more elongated and of a purer, deeper cream color than those of the gadwall. They are creamy white in color, varying from deep cream to nearly white and are nearly elliptical ovate in shape. The shell is clear, smooth, rather thin and somewhat glossy, resembling in color and texture certain types of hen's eggs.

The measurements of 81 eggs in the United States National Museum and the author's collections average 53.9 by 38.3 millimeters; the eggs showing the four extremes measure **60 by 40, 58.2 by 40.2, 50.5 by 38 and 54 by 36** millimeters.

Young.—I can not find any definite data as to the period of incubation, but the European bird is said to incubate for 24 or 25 days, and our bird probably sits for the same period. This duty is performed by the female exclusively, though the male does not wholly desert her until the molting season arrives. The care of the young also rests with the female, and she guards them with jealous devotion. Nelson (1887) relates the following incident:

I once came suddenly upon a female widgeon, with her brood of 10 or a dozen little ducklings, in a small pond. As I approached the parent uttered several low, guttural notes and suddenly fluttered across the water and fell heavily at my feet, so close that I could almost touch her with my gun. Meanwhile the young swam to the opposite side of the pond and began to scramble out into the grass. Willing to observe the old bird's maneuvers, I continued to poke at her with the gun as she fluttered about my feet, but she always managed to elude my strokes until, just as the last of her brood climbed out of the water, she slyly edged away, and suddenly flew off to another pond some distance. I then ran as quickly as possible to the point where the ducks left the water, yet, though but a few moments had elapsed, the young had concealed themselves so thoroughly that, in spite of the fact that the grass was only 3 or 4 inches high and rather sparse, I spent half an hour in fruitless search.

By the last of August or the first of September the young are able to fly and are flocking with their parents for the autumnal flight. Baird, Brewer, and Ridgway (1884) quote Kennicott as saying:

The young, while unable to fly, are frequently found seeking the shelter of grassy lakes. As soon, however, as they can fly they return to their favorite river shores and open feeding places, where they obtain aquatic insects, a few small shells, and the seeds and roots of various plants. In the fall the broods often separate before leaving for the South; this they do about the middle of September.

Plumages.—Ridgway (1887) describes the downy young as follows:

Above, dark olive brown, relieved by a spot of greenish buff on posterior border of each wing, one on each side of back, and one on each side of rump; top of head and hind neck, dark olive, like back; rest of head and neck, with lower parts, pale olive buff or fulvous, the side of the head with a dusky streak, extending from bill, through eye, to occiput.

When about 4 or 5 weeks old, in August, the young baldpate assumes its first complete plumage, the wings being the last to reach full development. In this first mottled plumage the sexes are much alike, but in the male the gray feathers of the back begin to appear in September and the progress toward maturity proceeds rapidly; the brown mottled feathers of the back are replaced by the gray vermiculated feathers of the adult and the mottling in the breast disappears, leaving the clear vinaceous color of maturity; so that by December or January the most forward birds have acquired a plumage which closely resembles that of the old bird, except on the wings, which still show the gray mottling on the lesser wing coverts peculiar to young birds. In some precocious individuals the lesser wing coverts become nearly pure white before the first nuptial season, but in most cases the immature wing is retained until the first postnuptial molt, which is complete. With both old and young birds the molt into the eclipse plumage begins in June and the molt out of this into the adult winter dress is not completed until October or November. At this molt the white lesser wing coverts are assumed by the young, old and young birds becoming indistinguishable. The seasonal molts of the adult consist of the prolonged double molt of the body plumage, into the eclipse in June and July and out of the eclipse in September and October, and the single molt of the flight feathers in August. Old males in the eclipse plumage closely resemble females, except for the wings, which are always distinctive.

In the female the sequence of plumages is similar. During the first winter and spring, young birds make considerable progress toward maturity, but can be recognized by the immature wings. The fully adult plumage is acquired during the second fall and winter.

Food.—The baldpate feeds on or near the surface by dabbling in the mud or tipping up in shallow water. Where not disturbed it is liable to feed at any time during the day, though it is always more active in the early morning or toward night. But, as it seldom enjoys much security, it more often spends the day skulking in the reeds, dozing on some sunny bank or playing about on open water at a safe distance from land; then as dusk comes on it repairs in small flocks to its feeding grounds, where it can feed in safety during the greater part of the night. Its food is largely vegetable, consisting mainly of the seeds and roots of grasses and various water plants. Dr. John C. Phillips (1911) records the contents of stomachs of this species, taken in Massachusetts, as entirely vegetarian, consisting of "pond-weed, wild celery, water-lily seeds (*Brasenia*), burweed, and smart-weed seeds (*Polygonum hydropiper*), also "mineral matter 65 per cent." Dawson (1909) says that in Washington, "in late January and February, they confine their feeding largely to the water-soaked fields, digging up the young grass with their bills and eating roots and all."

Its well-known habit of robbing the canvasback on its feeding grounds in Chesapeake Bay has been often described; the baldpate, being a poor diver and yet extravagantly fond of the succulent roots of the, so-called, wild celery, has to be content with what small bits of this delicacy the canvasback lets drop or what it can steal from this expert diver on its return to the surface. In its winter home in the Southern States it feeds largely in the broken-down rice fields, where it finds an abundance of food and becomes very fat. Audubon (1840) says that it eats "beechnuts, small fry, and leeches." Warren (1890) found that two baldpates, taken in Pennsylvania, "had fed almost entirely on insects, chiefly beetles and crickets." Mr. F. C. Baker (1889), who dissected a large number, taken in Florida, found that their stomachs "contained shells of *Truncatella subcylindrica* (Say) and small seeds."

Mr. Douglas C. Mabbott (1920) says:

The vegetable food of the baldpate for the eight months from September to April averaged 93.23 per cent. This consisted of the following items in the order of their importance: Pondweeds, 42.82 per cent; grasses, 13.9; algae, 7.71; sedges, 7.41; wild celery and waterweed, 5.75; water milfoils, 3.48; duckweeds, 2.2; smartweeds, 1.47; arrow grass, 0.36; water lilies, 0.26; coontail, 0.24; and miscellaneous, 7.63 per cent.

Animal food amounted to 6.77 per cent of the contents of the 229 baldpate stomachs included in the computation. Even this figure is probably unduly large, because the greater part of the animal matter consisted of snails found in the gizzards of a series of ducks from southern Oregon, the only lot of birds found feeding almost exclusively upon such food. More than nine-tenths of the animal food (6.25 per cent of the total) consisted of mollusks, the remainder being made up of insects (0.42 per cent) and miscellaneous matter (0.1 per cent).

Behavior.—When alarmed the baldpate rises quickly from the water, almost perpendicularly, making a rattling sound with its wings, and flies rapidly away. Its flight is swift, strong, and direct; when migrating or when flying to or from its breeding grounds it flies in small flocks of irregular formation and at no great height from the ground or water. On its breeding grounds it is quite tame, but during the shooting season it becomes very shy. The drake is easily recognized by its striking color pattern, displaying so much white in the wings, the white crown, and the white belly. But the duck might easily be mistaken for the female gadwall, though it has a white unspotted breast and shows more white in the greater wing coverts, which is quite conspicuous as the duck flies away from its nest. It can readily be distinguished from the mallard duck by its smaller size and by the absence of the conspicuous white borders of the blue speculum. Doctor Townsend (1905) has called attention to the fact "that the under surface of the wings of the baldpate is gray, that of the mallard snowy white."

The ordinary call note of the male baldpate is a whistling *whew, whew, whew,* which is uttered on the wing or while feeding and swimming. Mr. J. H. Bowles (1909) has well described its vocal powers, as follows:

Their principal call is a lisping, throaty whistle, repeated three times in quick succession. It is surprisingly light in character for the size of the bird, and serves to confirm the bird's position on the list next to the teals. Although quite impossible to describe, the note is rather easily imitated when heard a few times, and frequently proves a valuable addition to the repertoire of the wild-fowl hunter. The only other note I have heard them utter is a low, short chattering, somewhat resembling that of the pintail, but greatly reduced in volume. Their quacks, or squawks, of alarm also express the limit of terror, but are still pathetically inadequate in comparion with those, say, of a hen mallard.

Doctor Yorke (1899) describes the cry as " a whistle like the last note of a Bartramian sandpiper." The female has a soft guttural note, which can hardly be called a "quack" and a louder cry, which Eaton (1910) says resemble "the syllables *kaow, kaow*." Mr. Aretas A. Saunders writes to me in regard to the notes of this species:

They also rarely quack like the mallard and gadwall, but this note is less nasal than the mallard's and not so loud and sonorous as the gadwall's. The whistle differs in pitch with different individuals, and one may frequently hear whistles of two or three different pitches coming from the same flock of birds.

On their breeding grounds baldpates are associated on friendly terms with various other species, as has already been shown above. In their winter feeding resorts they associate with canvasbacks, redheads, and scaup ducks, stealing from them what bits of food they can grab. Neltje Blanchan (1898) says of this performance:

Such piracy keeps the ducks in a state of restless excitement, which is further induced by the whistling of the widgeons' wings in their confused manner of flight in and around the feeding grounds. Here they wheel about in the air; splash and splutter the water; stand up in it and work their wings, half run, half fly along the surface; and in many disturbing ways make themselves a nuisance to the hunter in ambush.

Doctor Townsend (1905) describes similar behavior with the American coot, as follows:

I have seen a flock of five baldpates eagerly following half a dozen American coots that were frequently diving in a pond and bringing up weeds from the bottom. The baldpates gathered about the coots as soon as they emerged on the surface and helped themselves to the spoils, tipping up occasionally to catch some sinking weed. They seemed even to be able to perceive the coot coming up through the water, for they would begin to swim toward the spot just before the coot emerged. The coots appeared to take the pilfering as a matter of course; in fact they pilfered from each other, and continued to work for themselves and the poachers.

Such behavior has earned for the baldpate the local name of "poacher."

Fall.—On the fall migration the baldpate starts rather early in September, well in advance of the heavy frosts.

On its migration in Montana, according to Mr. Aretas A. Saunders, it—

associates with many other species of ducks, most frequently with the shoveller. Flocks of these two species, mixed, are quite common in the spring migration. I have observed, with this species, on a small pond in the spring migration, the following other species: Gadwall, shoveller, blue-winged teal, pintail, lesser scaup, goldeneye, and buffle-head. In the large flocks of ducks that gather on the larger alkali ponds in the fall migration, this is one of the commonest species, and is associated commonly with the shoveller and lesser scaup.

Game.—As a game bird it will not rank in importance with several others, though its vegetable diet, especially when it has been feeding on the Chesapeake with the canvasbacks, makes its flesh very palatable and desirable. It is a favorite too with many sportsmen on account of its swiftness, its boldness, and its readiness to come to decoys. Blanchan (1898) says:

The gentlemen hidden behind "blinds" on the "duck shores" of Maryland and the sloughs of the interior and with a flock of wooden decoys floating near by, or the nefarious market gunner in his "sink boat" and with a dazzling reflector behind the naptha lamp on the front of his scow, bag by fair means and foul immense numbers of baldpates every season, yet so prolific is the bird, and so widely distributed over this continent, that there still remain widgeons to shoot. That is the fact one must marvel at when one gazes on the results of a single night's slaughtering in the Chesapeake country. The pothunter who uses a reflector to fascinate the flocks of ducks that, bedded for the night, swim blindly up to the sides of the boat, moving silently among them, often kills from 20 to 30 at a shot.

Winter.—After loitering along its way for several weeks in a most leisurely manner, as if waiting for the young birds to fatten and grow strong the baldpate finally reaches its winter home before cold weather sets in, spreading out from its inland breeding range to winter largely on the coasts, as well as in the lower Mississippi Valley. Its winter habits in the Chesapeake Bay, which marks the northern limit of its winter abundance on the Atlantic coast, have been referred to above and have been well described by others. It is common on the coast of Louisiana associating with mallards, gadwalls, pintails, and lesser scaup ducks. On the Pacific coast it winters abundantly as far north as Puget Sound, though according to Bowles (1909) it is not so common there as formerly; he says:

During fall, winter, and spring it is most numerous of all ducks in Washington, save possibly the bluebills and scoters. Large numbers of them congregate upon the tide flats of Puget Sound, and the bird is abundant also on the interior waters.

Constant persecution, however, has greatly reduced their ranks, as is the case with the entire duck family, and possibly for this reason their migratory habits have undergone a marked change. Eight or ten years ago they used to appear in enormous flocks during the first week in October, at which period I have seen on the Nisqually Flats, near Tacoma, what was estimated at about 500,000, all in the air at one time. For the past two or three years, however, no widgeon to speak of have appeared before November or December, and then in such greatly reduced numbers as to give rise to serious fear, not only as to the abundance, but as to the existence of future generations.

DISTRIBUTION.

Breeding range.—Northwestern North America. East to Hudson Bay, southeastern Manitoba (Shoal Lake), and formerly southern Wisconsin (Lake Koshkonong and Horicon marsh). South to northern Indiana (English Lake, rarely), perhaps northern Illinois, northern Nebraska (Cherry County), northern Colorado (Boulder County), northern Utah (Bear River), northwestern Nevada (Truckee Valley), and northeastern California (Modoc County). Seen in summer and probably breeding in northwestern New Mexico (Lake Burford) and northern Arizona (Mogollon Mountains). West to the interior of Oregon (Camp Harney) and Washington (Tacoma), central British Columbia (Fraser Valley), and central Alaska (Yukon River). North to northern Alaska (Kotzebue Sound) and northern Mackenzie (Franklin Bay).

Winter range.—All of North America south of the Northern States. East to the Atlantic coast, rarely from southern New England (Boston), and regularly from Maryland (Chesapeake Bay) southward. South to the Lesser Antilles (St. Thomas, Trinidad, Guadeloupe, St. Croix, etc.) and Costa Rica. West to the Pacific coast of Central America, Mexico, the United States, and southern British Columbia (Vancouver Island). North in the interior to southern Nevada (Pahrump Valley), central Utah (Provo), northeastern Colorado (Barr Lake), and southern Illinois (Ohio Valley).

Spring migration.—Average dates of arrival: Rhode Island, Newport, March 19; Ontario, Ottawa, April 20; New York, western, March 23; Pennsylvania, Erie, March 24; Ohio, Oberlin, March 17; Michigan, southern, March 25; Colorado, Loveland, March 10; Nebraska, central, March 17; Iowa, Keokuk, March 15; Minnesota, Heron Lake, March 29; Manitoba, southern, April 20; Saskatchewan, Indian Head, April 24; Alberta, Edmonton, April 17; Mackenzie, Fort Simpson, April 28; Alaska, Knik River, May 10, and Kowak River, May 22. Late dates of departure: North Carolina, Raleigh, April 26; Lower California, La Paz, April 1, and Colnett, April 1.

Fall migration.—Early dates of arrival: Pennsylvania, Beaver, August 30; Massachusetts, Marthas Vineyard, August 31; Maine, Merrymeeting Bay, September 20; Connecticut, East Hartford, September 29; Rhode Island, Middletown, September 20. Late dates of departure: Alaska, Kowak River, September 20, and St. Michael, October 1; Alberta, Edmonton, November 6; Ontario, Ottawa, November 6; Nova Scotia, Sable Island, November 7; Iowa, Keokuk, November 18.

Casual records.—Accidental in Bermuda (October, 1854, and October, 1874), Cuba, Jamaica, Porto Rico, and St. Thomas. Rare on migrations in Labrador (Hamilton Inlet, Natashquan, and Old Fort Bay, November 27, 1880) and New Brunswick (St. John, January, 1880). Accidental in Aleutian, Commander, and Hawaiian Islands, in the Azores, British Isles (six or more records) in France, and in Japan.

Egg dates.—Arctic America: Twenty-two records, June 5 to July 4; eleven records, June 16 to June 25. North and South Dakota: Twenty-one records, May 25 to July 13; eleven records, June 2 to 23. Alberta and Saskatchewan: Fourteen records, June 1 to 25; seven records, June 13 to 17. Utah: Nine records, May 5 to June 17; five records, May 10 to June 3.

NETTION CRECCA (Linnaeus).

EUROPEAN TEAL.

HABITS.

This well-known and widely distributed Palaearctic bird has always appeared on our check list as an occasional visitor or straggler with its name enclosed in brackets. And such I always believed it to be until our expedition visited the Aleutian Islands in 1911 and definitely established it, as a regular summer resident at least in North American territory. I stated in my report on the results of this expedition (1912) that:

The European bird is supposed to occur only rarely, or as a straggler, in the Aleutian Islands and the American bird is recorded by nearly all of the writers on Aleutian ornithology as the common breeding teal of the region. Teal of one of these species were common on all of the islands; we saw them frequently and found them breeding in nearly all suitable places along the small water courses and about small ponds. Doctor Wetmore found a nest containing 10 fresh eggs on June 7, near Unalaska, and shot the female; unfortunately the male was not secured. We naturally assumed that these were American green-winged teal and, therefore, made no special effort to shoot males on any of the eastern islands, but I now sorely regret that we did not collect at least a few males as the females of the two species are nearly indistinguishable. Among the western and central islands we collected quite a series of both sexes and every male taken proved to be an European teal; not a single male green-winged teal was collected or identified anywhere. On my return to Washington I looked through the National Museum collection for specimens from the Aleutian Islands and found only two males, No. 85615, collected by Lucien M. Turner on Atka Island, June 28, 1879, and No. 192391, collected by Dr. J. Hobart Egbert on Kiska Island, July 14, 1904; both of these proved to be typical European teal. Therefore, failing to find any positive evidence to prove that the green-winged teal breeds on the Aleutian Islands, we must assume for the present, on the strength of what evidence we have, that the European teal is the common breeding species of this region, where it is fairly abundant, and that the green-winged teal, which is so abundant on the main land of Alaska, occurs on the islands rarely, if at all.

Courtship.—Mr. John G. Millais (1902) has given us a beautiful colored illustration of the courtship of the teal in which a number of handsome males are seen displaying their plumage, sitting upright in the water around the different females or venturing nearer to offer their attentions. I quote from his description of it as follows:

It is a pretty sight, this spring display of the teal, all the more so as many take part in it, and the positions of the male birds are curious and extravagant. As if by mutual consent, several drakes raise their bodies from the water, erect the tail, arch the neck and pass their bills down the chest, at the same time they give voice to the low double whistle. During this movement the female sometimes permits one or even two drakes to approach her closely, whilst all the others are disposed in a circle or semicircle near at hand; but if any male that has not found favor in her eyes seeks to approach she will drive him off at once—an ignominious position which he seems to accept without question. It is only after some days of this volatile flirtation that the female eventually goes off with one male and remains strictly monogamous for the rest of the season, for after the end of April one never sees amongst teal the *tertium quid* arrangement so common with other ducks.

Nesting.—The same writer says of its nesting habits:

The nest is to be found in almost any sort of sheltered position near the water, but the female evinces a marked partiality for placing it in heather. In Scotland I have usually noticed it in open heaths, sometimes far from the lake or bog, but generally near to a burn that leads to them. The eggs number from 8 to 15, are of a creamy-white color, sometimes with a faint tinge of green, which fades soon after their contents are extracted: size 1.8 by 1.2 inches. None of the ducks show such an affection for their young as the female teal; when flushed with her young brood she will display greater bravery in their defence and evince more solicitude for their welfare than almost any bird. Teal drakes, on occasion, like the mallard and the shoveler, will sometimes even betray a very distinct alarm when their wives and families are threatened, for I once disturbed a teal duck with young on an open moor at Cawdor. The drake was with her, and he, much to my surprise, was almost as anxious as the female to lead me away, resorting several times to the broken-leg feints of his distressed partner.

The nest that we found at Unalaska, referred to above, was made of down, feathers, and bits of grass; it was well concealed in a thick clump of tall, coarse, dead grass, not over 10 feet from the bank of a swift and shallow stream, which flowed in a winding course through a broad alluvial plain back of Iliuliuk village. The plain was more or less gravelly in places and was partially covered with coarse grasses and scattered clumps of small willow bushes; it was surrounded by steep grassy hills, rising beyond to snow-capped mountains and narrowed at one end into deep valleys and gorges through which two mountain streams came tumbling down over the rocks to form the little river which had evidently formed the plain.

Similar picturesque valleys and flat alluvial plains were found on many of the Aleutian Islands; these were the favorite resorts of the teal, which were usually common and sometimes abundant among the western and central islands. With them we sometimes found red-breasted mergansers and a few mallards and scaup ducks, which were flushed from the streams or seen swimming on the little ponds.

The down in the nest of the European teal is very small and very dark colored, dark "hair brown" or "clove brown" with large conspicuous white centers. The breast feathers in the nest are small, with dusky centers and buff tips.

Eggs.—This teal has been known to lay as many as 16 eggs, but the usual numbers run from 8 to 12. In shape and color they are indistinguishable from eggs of the green-winged teal. The measurements of 100 eggs, given in Witherby's Handbook (1920), average 44.6 by 32.6 millimeters; the eggs showing the four extremes measure 49.5 by 34, 47.6 by 35.2, 41 by 32.9 and 42.2 by 31.2 millimeters.

Young.—The period of incubation is short, 22 days, and is performed by the female alone. The young teal have many enemies to contend with during their early existence, among which certain individuals of the brown-headed gull, which seem to develop murderous instincts, are most destructive, as the following striking instance related by Mr. Millais (1902) will illustrate:

About the year 1884, the brown-headed gulls, formerly represented by a couple of hundred pairs, began to increase on the bog at Murthly to an alarming extent. Their nests were everywhere in the reed tufts, and about this time the teal began to decrease. James Conacher, the keeper of the Moss, at once put it down to the gulls, who, he said, killed the ducklings as soon as their mothers brought them down to the bog, and said, moreover, that we should have no quantity of duck until a war of gull extermination had taken place. On talking the matter over with the head keeper, one James Keay, a very superior and observant man, he said that he had noticed that all the young teal that were killed lay dead near two places, and in an area of 30 yards square. This seemed plainly to point to the work of individuals, and on subsequently watching the places Keay saw a gull that had a nest close by actually seize a young teal, lift it into the air for a moment, and drop it dead. This gull and its partner were shot, and no more young ducks were found dead in that vicinity during the season; but the next year the gulls of certain nests were found to have again started the murders, and they were marked down and shot, after which no more ducks were killed for some time, and the teal increased greatly. All the young teal killed by the gulls were put to death in the same way, the skulls were nipped and crushed at the back, and they were not touched again. In June, 1890, another pair began duck killing, and near the nest of these birds Keay found the remains of 16 teal, 3 tufted ducks, and 2 mallard nestlings.

Plumages.—The downy young and the sequence of molts and plumages are so similar to those of our American bird that it seems unnecessary to repeat them here.

Food.—Bewick (1847), says of its food:

Buffon remarks that the young are seen in clusters on the pools, feeding on cresses, wild chervil, etc., and no doubt, as they grow up, they feed like other ducks, on the various seeds, grasses, and water plants, as well as upon the smaller animated beings with which all stagnant waters are so abundantly stored.

Macgillivray (1852) says:

Its food consists of seeds of grasses, slender rhizomata, which it pulls up from the mud, insects, mollusca, and worms.

Behavior.—Of its flight and vocal powers, Mr. Millais (1902) writes as follows:

During the day the teal is one of the most silent and inactive of birds. It will sit for hours motionless, apparently lost in a brown study or with the head buried in the scapulars. Out on the estuary a pack rests on the tidal heave without a sign of movement until night comes and with it the desire for food. In the daytime, during the early autumn, even in our much disturbed islands, teal are sometimes extremely tame, and will permit the approach of man within a few yards before flying away, and there are always certain holes in the large bogs where teal may be found and closely approached with certainty unless they have been previously disturbed. On being flushed they shoot up straight into the air, sometimes very rapidly, and often swaying slightly and rendering themselves a by no means easy mark—in fact, I once heard a friend, who had ineffectually expended 100 cartridges in one day, declare that *rising* teal were far more difficult to kill than snipe. Be that as it may, I can remember certain windy days when *driven* teal were wild and "dodgy," and were quite as difficult to bag as the snipe with whom they flew. Teal can suddenly turn in the midst of a straight forward flight and either dive downward, or, what is far more difficult for the gunner to accept, shoot straight upward, and only present as a target a practically invulnerable stern. It is a pretty sight on a sunny day to watch a flock of teal about to settle; they wheel and swing almost as much as flocks of dunlins, the dark backs and the light breasts alternately shining; and it is not until they have thoroughly surveyed their prospective resting place and its approaches that they come to a halt. Whilst on the wing one male occasionally utters his low double whistle, but teal are silent birds at all times, and the female rarely calls unless frightened, such as when the brood is threatened, when she emits a subdued little "quack."

Fall.—On the fall migration, teal are inclined to wander and, as most of our records have occurred during the late fall, winter, and early spring, they are probably stragglers from Palaearctic regions, or from the Aleutian Islands, that have strayed from their normal migration routes. Green-winged teal are found regularly in the Aleutian Islands in winter, but whether these are migrants from the mainland of Alaska, *Nettion carolinense*, or the resident breeding birds, *Nettion crecca*, I can not say. Probably most of the European teal that breed in the Aleutian Islands migrate in the fall down the Asiatic coast to Siam and India, with the birds from Siberia. Dr. John C. Phillips (1911[a] and 1912) relates an interesting incident which illustrates the tendency of this species to return each spring to the locality of its birth. A young European teal which was hatched and reared on his grounds in Wenham, Massachusetts, in 1910, returned to the same pond in 1911 and again in 1912.

DISTRIBUTION.

Breeding range.—Northern parts of the Eastern Hemisphere. Iceland, the British Isles, throughout Europe and Asia, north to 70° N.

and south in decreasing numbers to the Mediterranean, the Azores, Turkestan, Mongolia, and the Amur Valley. East to the Kurile and Aleutian Islands, as far east as Unalaska.

Winter range.—South to the Canary Islands, Madeira, Abyssinia, Sokotra, Persia, India, Ceylon, China, Japan, Formosa, and the Philippine Islands. North as far as open water is to be found. Apparently resident in the Aleutian Islands or migrating westward into Asia.

Casual records.—Accidental in Spitsbergen, Greenland, Labrador (Coues record, July 23, 1860, and Hamilton Inlet, no date given), Nova Scotia and northeastern United States (Maine, Casco Bay, April 6, 1893; Massachusetts, Muskeget Island, March 16, 1890, and Sagamore, February 20, 1896; Connecticut, East Hartford, November 14, 1889; New York, Cayuga Lake, April 10, 1902 and Merrick, Long Island, December 17, 1900; Virginia, Potomac River, 1885). California records are indefinite.

Egg dates.—Great Britain: Nine records, May 3 to June 13. Iceland: Thirteen records, May 1 to July 8; seven records, May 24 to June 8. Aleutian Islands: One record, June 8.

NETTION CAROLINENSE (Gmelin).

GREEN-WINGED TEAL.

HABITS.

Spring.—Following close on the heels of the pintail and the mallard, the hardy little green-winged teal is one of the earliest migrants to start in the spring for its northern breeding grounds. It begins to leave its winter haunts in the lower Mississippi Valley in February, proceeding slowly northward, and the first arrivals appear in its summer home in northern Alaska early in May. Dr. F. Henry Yorke (1899) says:

> The first issue arrives a day or two after the pintails and follows up the rivers, lakes, and sloughs, usually preferring the edges of muddy banks. This issue stays only a short time and departs before the second arrives, usually about four or five days intervening; the second issue spreads over the country and is often joined by the third, staying for several weeks before they travel northward. At times the third issue is delayed, probably, by overflowed lands in the south, where food is found in abundance; in such cases the third issue rushes by, or stays only a day or two late in the season.

Courtship.—Many of the birds are paired before the breeding grounds are reached as there is ample time for courtship during the leisurely migration; the warmth of returning spring stirs the amorous instincts of the males and prompts them to strut before the females, displaying their handsome colors in fantastic attitudes. The performance is the same as that of the European teal, as described by Millais (1902) an excellent account of which is given under that species.

Mr. George M. Sutton has sent me the following interesting notes on the courtship of the green-winged teal:

Suddenly, and apparently without any premeditation, two of the males started toward the center of the pond, one directly behind the other; the two birds in ridiculously similar attitudes, both with bills pointing somewhat upward, and head drawn back and down in a stiff and uncomfortable looking manner. Strange as it may seem, so similar were the birds in their every movement, that the feet, in swimming, seemed to stroke in perfect harmony, and the picture presented was one of unusual beauty. They swam deliberately, and in a direct line until they reached the center of the pond. It was not evident until now that a female bird had anything to do with the matter; but a sudden change in the attitude of the approaching males informed the observer that a very restless and inattentive female was the cause of the whole performance. Both males, still in perfect harmony of movement, were describing a circle about the female, swimming about 2 feet from each other, and at a tantalizingly deliberate speed. I could not help feeling that the atmosphere was growing tense, because the males seemed to be fairly quivering in their effort to curb spirit. The female, which had thus far been utterly unobservant, now became rather quiet and attentive, and the males, still in perfect accord, began a remarkable series of bobbings, opening their bills rather widely, and uttering a soft, not unmusical *pheep, pheep*, one call to every movement of the head. The call was given on the forward thrust of the head, and as the two beautiful birds wheeled about, in a circle possibly only 6 feet in diameter, it was only natural to call to mind two little boats emitting whistles at regular intervals, and indulging in quaint maneuvers, so totally unducklike the creatures seemed. Matters were even more complicated when it became evident that the males were now churning the water lustily with their feet, though their speed remained about the same. Again suddenly, the harmony of movement ceased; the female rushed at one savagely and seemingly without warrant, whereupon he, the favored one, stood up in the water, lifted his wings somewhat, and with rapidly churning feet made a most unusual noise, sounding like water thrown rapidly at some object in a fine stream, which in some manner, inexplainable thus far, must have been connected with a stream of water which seemed to pass from the bill to the rapidly treading feet. I could not get close enough to see clearly, and had no glasses, so the details of this very odd antic I have never been able to explain, but there seemed to be a fine stream of water shot out of the bird's bill, accompanied by the queer sound mentioned above. The sound was of a startling quality, and directly after its delivery the bird's feet quit their violent churning motion, and the bird sank to rest for a time. The other male, which had ceased operations for a momentary preening of feathers, would then "occupy the floor" and repeat the antic. Occasionally both males would do the stunt exactly at the same time. During the period of their greatest activity this act of standing in the water and treading was repeated about every 20 seconds. The part played by the female in these odd dramatics, was as far at least as the observer could see, a minor one. She was followed continually, but she did not follow in return. Her attitude was one of inattentiveness, save when one of the males came too near, or strange to say, on one occasion, when the performers lagged in their enthusiasm. At this time (observed only once) she tore madly across the pond and back again, in a manner indescribably nimble, past the males, whence her courters in steamboat formation, with the accompaniment of *pheep, pheep*, followed her.

The attitude of the males toward each other was one of dignified tolerance while the female was at hand. They apparently vied vigorously in matters of elegance of movement, but there was no sign of combat. Once, when the female rushed at the male, he surprised me by giving her a vicious jab in return, which seemed to subdue her for the time.

Nesting.—The green-winged teal is widely distributed during the nesting season, but throughout much of its breeding range, especially the eastern part of it, is only sparingly represented. We found a few scattering pairs in the Magdalen Islands, where it was undoubtedly breeding. Maynard (1896) found a nest here and gives the following good account of it:

On the southern side of Amherst Island, one of the Magdalen group, are several salt-water ponds which were formerly lagoons, but which the shifting sand of the beaches have cut off from the water of the gulf. These miniature lakes are surrounded by a thick growth of trees, composed mainly of spruce and hemlock, which have been so dwarfed by the severe climate that they rarely attain the height of 10 feet. I was making my way along the border of one of these ponds, on the 16th of June, in company with my friend, Mr. Gilman Brown, when a female green-winged teal rose within a yard of our feet, and stepping forward, we discovered a nest containing eight greenish eggs which were placed in a depression of the sandy soil on a few twigs, and surrounded with a ring of gray down, thus presenting a very pretty appearance. The spot was concealed by the overhanging branches of a little spruce, and had the bird remained quiet, we should have passed without discovering her treasures. The female was quite shy, and after circling about a few times disappeared.

Although the breeding range of the green-winged teal extends much farther north than that of the blue-winged teal, its center of abundance in the nesting season is in the vast prairie regions of the Northern States and western Canada. In both North Dakota and Saskatchewan we found a few pairs breeding with the other ducks in these great wild-fowl nurseries. Here the nests are usually concealed in the long grass near the borders of the lakes and sloughs or on the islands. The nest is generally well made in a hollow on dry ground and often at a considerable distance from the water. The hollow is first deeply lined with soft grasses and weeds, sometimes with a few fine twigs and leaves, on which is placed a thin layer of down from the bird's breast; more down is added as incubation advances, which, mixed with the loose dry grass, forms a convenient blanket to conceal and protect the eggs when the mother teal leaves them. Often the nests are placed in clumps of willows or under bushes on high land, a quarter of a mile or more from any water and occasionally a nest is found under a log.

Mr. R. C. McGregor (1906) records a nest found in the Krenitzin Islands, Alaska, which "was on the ground beneath the overhanging trunk of a twisted willow; it was thinly furnished with down about the top and the eggs rested on the ground. Inside diameter of nest about 5.5 inches; depth 3.5 inches." Henshaw (1875) reports a nest found in southern Colorado as follows:

A nest belonging to this species was found under a sagebush, perhaps 30 feet from the water's edge. A deep hollow had been scooped in the sand, and lined warmly with fine grasses and down, evidently taken from the bird's own breast, which was plucked nearly bare. The hen bird was sitting; in fact, so artfully was the nest placed that it was only when I had almost trodden upon it, and the old bird had shuffled out at my feet and made good her retreat behind some thick bushes, that I discovered it. Returning a couple of hours later, I found she had again taken possession of her treasures, nor did she leave till I had approached within 3 feet of her.

The down in the nest is exactly like that of the European teal, which is fully described under that species.

Eggs.—The green-winged teal lays from 6 to 18 eggs, the average number being from 10 to 12. The eggs are absolutely indistinguishable from those of the blue-winged teal; in shape they are ovate or elliptical ovate; in color they are dull white, cream color, or very pale olive buff. The measurements of 93 eggs in various collections average 45.8 by 34.2 millimeters; the eggs showing the four extremes measure 49 by 36, 48 by 37, 42 by 31.5 and 43 by 31 millimeters.

Young.—Incubation, which lasts 21 to 23 days, is performed by the female alone, as the male deserts her as soon as the eggs are laid. The entire care of the young also devolves on the mother teal, who performs her duty with exemplary devotion. The young all hatch within a few hours of each other and after a short rest the down becomes dry and they gain sufficient strength to make the perilous overland journey, which is often a long one, to the nearest water. Mr. Ernest Thompson Seton (1901) has drawn a very vivid picture of this momentous event in which he has portrayed some of the many dangers to which the little ducklings are exposed. Like other young ducks, they need no food for the first few hours but their mother soon teaches them to feed on the insects and other soft animal, as well as vegetable, food which they can pick up around the edges of the ponds and among the aquatic vegetation. They also learn to heed her warning cry when danger threatens, when to hide and how to escape their many enemies, but often they would not escape except for her bravery and zeal in their defense, even at the risk of her own life.

Plumages.—The downy young greenwing differs from the bluewing in having a smaller and shorter bill, with a hooked nail, and in its generally darker coloration. The upper parts are "mummy brown" or "Prout's brown," darkest on the crown and rump; the under parts shade from "buckthorn brown" or "clay color," on the sides of the head and throat, to "cinnamon buff" or "light buff," on the breast and belly; the side of the head is distinctly marked by a broad loral and postocular stripe of dark brown and a similar auricular stripe below it, from the eye to the occiput; a broad superciliary stripe of buff extends from the bill to the occiput, but it is interrupted by an extension of the dark crown nearly or quite down to the eye; the color of the back is relieved by buffy spots on the thighs, scapulars, and wings. I have never had any experience with hatching and raising this species, but presume that its plumage grows about as it does in the other surface-feeding ducks. The plumage appears first on the flanks and scapulars.

In the juvenal or first plumage the sexes are practically alike, except that in the young female the colors in the wing are somewhat duller, the pattern being the same in both sexes, and there is less spotting on the belly. But changes soon take place and the advance toward maturity is rapid. The young male becomes lighter below and the red plumage of the head appears in October; by December the young bird has become practically indistinguishable from the adult.

The adult male sometimes begins to molt into the eclipse plumage in June, usually in July, and in August this plumage is complete, all the contour feathers and scapulars having been molted; the flight feathers are renewed in August and the second complete molt of the contour feathers begins late in September and is completed in October or November. In the eclipse plumage old and young males and females look very much alike, but old males generally have fewer, more clearly defined, and well rounded spots below and old females are usually uniformly and thickly mottled on the under parts.

Food.—The green-winged teal enjoys a varied diet which it obtains in various ways in different parts of its habitat. In its summer home it loves to dabble in the shallow water about the edges of the sloughs, ponds, and creeks, with its body half immersed, its feet kicking in the air and its bill probing in the mud for aquatic insects or their larvae, worms, small mollusks and crustaceans, or even tadpoles. In such places it also feeds on the soft parts of various water plants and their seeds. In harvest time it wanders to the grain fields and picks up the fallen grains of corn, wheat, oats, barley, and buckwheat, where it also feeds on various other seeds, grasses, and vegetable matter. At this season and in the winter, when it lives in the southern rice fields feasting on the fallen harvest, it grows very fat and its flesh becomes very desirable for the table, equaling the finest of the ducks. It ordinarily feeds during the daytime, but in sections where it is much disturbed it is forced to become a night feeder. As it is active on land and can walk or run long distances, it often resorts to the dry uplands and woods to feed on berries, wild grapes, chestnuts, acorns, and other nuts, all of which help to improve the flavor of its flesh and make it a much sought game bird.

Neltje Blanchan (1898) says:

Nothing about its rankness of flavor when it has gorged on putrid salmon lying in the creeks in the northwest, or the maggots they contain, ever creeps into the books; and yet this dainty little exquisite of the southern rice fields has a voracious appetite worthy of the mallard around the salmon canneries of British Columbia, where the stench from a flock of teals passing overhead betrays a taste for high living no other gourmand can approve. When clean fed, however, there is no better table duck than a teal.

Mr. Douglas C. Mabbott (1920) says of its food:

Of the contents of 653 green-winged teal stomachs examined, more than nine-tenths (90.67 per cent) consisted of vegetable matter. By far the largest item of food contributed by any one family of plants came from the sedges, and this amounted to nearly two-fifths (38.82 per cent) of the total food. Next to the sedges, pondweeds are the favorite food supply, contributing 11.52 per cent, while grasses follow closely with 11; then smartweeds, 5.25; algae, 4.63; duckweeds, 1.9; water milfoils, 1.11; arrow grass, 0.91; and burreed, 0.85 per cent. The remaining 14.68 per cent is made up of a great number of smaller items.

Insects formed 4.57 per cent of the total food of the green-winged teal, the remainder of the animal food consisting of mollusks, 3.59 per cent; crustaceans, 0.92; and miscellaneous, 0.25; the total amounting to 9.33 per cent.

Behavior.—Nearly every writer on American ornithology has commented on the swiftness of flight of the green-winged teal, in which it certainly excels. In proportion to its size, and perhaps actually, it is the swiftest of the ducks, though its diminutive size might lead to an overestimate of its speed. It has been credited with a speed of 160 miles an hour, but this is undoubtedly an exaggeration. Mr. J. H. Bowles (1909) has well described its flight, as follows:

Moving at a rate of certainly not less than 100 miles an hour, the evolutions of a large flock of these birds are truly startling. They fly in such close order that one would think their wings must interfere, even on a straight course; yet of a sudden the whole flock will turn at a right angle, or wheel and twist as if it were one bird. The looker-on can only wonder what the signal may be which is given and obeyed to such perfection, for the least hesitation or mistake on the part of a single bird would result in death or a broken wing to a score.

On land this pretty little teal is quite at home; it walks gracefully and easily and it can run quite swiftly. It often travels long distances on foot in search of food or when moving from one pond to another; in making such overland journeys it sometimes moves in a compact flock, giving the pothunter a chance for a raking shot. In the water it swims easily and swiftly. Unlike most of the surface-feeding ducks it is an expert diver and can swim for a long distance under water to reach some needed shelter, where it can hide with only its head or its bill exposed. A wounded bird often escapes in this way and seems to have vanished. Mr. T. Gilbert Pearson (1919)—

recalls on one occasion seeing a wounded green-winged teal fall, which, on striking the water, instantly dived. After watching a few minutes for it to reappear, he waded out to the point where it had disappeared and found the bird about 2 feet beneath the surface, clinging with its bill to a water plant.

The note of the male greenwing is a short mellow whistle or twittering call and that of the female a faint reproduction of the quack of the mallard. Mr. Aretas A. Saunders writes me that the note of the male—

is a high-pitched, short, staccato whistle, and is accompanied by a lower-pitched trilled note, uttered less frequently. The effect of this note, when repeated frequently, and heard from a distance is much like the peeping of spring hylas. It also somewhat suggests the piping noise made by a spring flock of evening grosbeaks.

The green-winged teal is associated on its breeding grounds with various other species of ducks and it flocks with other ducks, particularly the blue-winged teal, on its migrations. Still it can hardly be called a sociable species and seems to prefer to travel in large flocks of its own species. In the winter many kinds of ducks frequent the favorite breeding grounds together and usually live peaceably with each other.

Fall.—The fall migration begins with the first cold weather, often quite early in the season, but the birds linger on the way wherever they can find attractive feeding grounds in the wild rice patches or on cultivated lands; probably the early migrants are wanderers from near-by breeding grounds. The first snowstorms bring along the main late flight and the northern winter has set in before the last of the migrants are driven south by the ice and snow, together with the northern mallards, the last of the surface-feeding ducks.

Game.—From the sportsman's viewpoint the green-winged teal is an important member of the long list of American wild fowl. Its abundance assures him plenty of sport; its swift flight, with its sudden turnings and rapid twistings, tests his marksmanship to the limit; and its plump little body, fattened on the best of grains, nuts, and succulent herbs, provides a dainty morsel for the table. I quote two well-drawn pen pictures of this bird as a hunter sees it. The first, written by the illustrious Audubon (1840) tells of the advent of the first migrants, as follows:

Nothing can be more pleasing to an American sportsman than the arrival of this beautiful little duck in our Southern or Western States. There, in the month of September, just as the sun sinks beneath the horizon, you may find him standing on some mote or embankment of a rice field in Carolina, or a neck of land between two large ponds in Kentucky, his gun loaded with No. 4, and his dog lying at his feet. He sees advancing from afar, at a brisk rate, a small dark cloud, which he has some minutes ago marked and pronounced to be a flock of green-winged teals. Now he squats on his haunches, his dog lies close, and ere another minute has elapsed, right over his head, but too high to be shot at, pass the winged travelers. Some of them remember the place well, for there they have reposed and fed before. Now they wheel, dash irregularly through the air, sweep in a close body over the watery fields, and in their course pass near the fatal spot where the gunner anxiously awaits. Hark! two shots in rapid succession! The troop is in disorder, and the dog dashes through the water. Here and there lies a teal, with its legs quivering; there, one is whirling round in the agonies of death; some, which are only winged, quickly and in silence make their way toward a hiding place, while one, with a single pellet in his head, rises perpendicularly with uncertain beats, and falls with a splash on the water. The gunner has charged his tubes, his faithful follower has brought up all the game, and the frightened teals have dressed their ranks, and flying, now high, now low, seem curious to see the place where their companions have been left. Again they fly over the dangerous spot, and again receive the double shower of shot. Were it not that darkness has now set in, the carnage might continue until the sportsman should no longer consider the thinned flock worthy of his notice. In this manner, at the first arrival of the green-winged teal in the western country, I have seen upwards of six dozen shot by a single gunner in the course of one day.

And then, second, the pen of that keen sportsman, Dr. F. Henry Yorke (1891), describes the departure of the late flight, in the following words:

The last issue of bluewings had collected, circled high in the air, and, following their instinctive impulse, had traveled southward. The second issue of mallards had come and gone, after staying with us a short time. The pintails, widgeons, green-winged teal (first issue), redheads, canvasbacks, and bluebills had also departed, and Grass Lake was almost "duckless." Even the mudhens had almost disappeared, and only a few scattered individuals, or small flocks of belated widgeons or pintails could be seen. Once in a while a few mallards turned up, but they were old, wary birds, "not to be caught with chaff." The only chance we could get was when a "stranger" flock of mallards came in, drifting down from the last issue, just preceding the frosts.

A week like this about the end of October is not an unusual occurrence. The sun shines warm after the cold nights, and the hazy atmosphere of our "Indian summer" induces idleness to a very reprehensive degree. But there was nothing to do, and we waited for a blast from old Boreas to awaken the ducks and put new life into ourselves.

Suddenly the herald of winter was heard. A fierce storm of rain or snow swept down from the North, where the icy grip of winter already held the lakes, and all nature was awake again. The laggard ducks came streaming in, mallards, pintails, and widgeon. Bluebills rushed down the flyways, and the game little green-winged teal, whipping and pitching in all directions, made his second appearance. This time the ducks meant business. While the weather was more uncertain, they came and returned, loth to leave their happy nesting grounds in the far north; but now Jack Frost was after them, and they were bent on a long and inevitable journey although some of them dropping here and there, they would stay until they were absolutely frozen out before they betook themselves to the mild clime and the open waterways of the sunny South.

Winter.—This hardy little duck winters as far north as British Columbia and Montana in the west, in the central States of the prairie region, in southern New England and even in Nova Scotia, when it can find open water in the spring holes and the streams near the coast. But its main winter home is in the great wild-fowl resorts of the Southern States and Mexico, where it can find safe retreats and abundant feeding grounds. It prefers the rice fields of the interior and the inland sloughs and ponds, but it also visits the coastal estuaries and the mouths of streams occasionally.

DISTRIBUTION.

Breeding range.—Northern North America practically across the continent, but very sparingly in the east. South to the Gulf of St. Lawrence (Magdalen Islands), southern Quebec (Monacougan), casually in western New York (Niagara River and Montezuma marshes), rarely in southern Ontario (Toronto, Point au Pins, Oshawa, and

Gravenhurst) and northeastern Michigan (Neebish Island), formerly
Wisconsin (Lake Koshkonong), formerly northeastern Illinois, south-
ern Minnesota (Rice and Jackson Counties), formerly central western
Iowa (Sac County), northern Nebraska (Cherry County), southern
Colorado (San Luis River), northern New Mexico (San Miguel
County), northern Utah (Bear River and Great Salt Lake), north-
western Nevada (Washoe Lake), and south central California (Tulare
Lake and formerly Ventura County). North to the limit of trees in
northern Alaska (Kotzebue Sound and Yukon River), northern
Mackenzie (Fort Anderson), Great Slave Lake (Fort Rae), northern
Manitoba (Churchill), and James Bay. Ungava and Labrador
records doubtful. Replaced in Aleutian and perhaps Pribilof Islands
by the European teal.

Winter range.—Southern North America, mainly in southwestern
United States and Mexico. East to the Bahama Islands, Cuba,
Jamaica, and the Lesser Antilles (Carriacoú, Grenada, Tobago, etc.).
South to British Honduras (Belize River) and southern Mexico
(Michoacan). North, more or less regularly, to southern British
Columbia (Chilliwack and Okanagan), central Montana (Great Falls),
northern Nebraska (Cherry County), northern Missouri (Kansas
City), southern Illinois (Mount Carmel), Kentucky and the coast of
Maryland (Chesapeake Bay); rarely or irregularly farther north to
southern Alaska (Sitka), the Great Lakes, and southern New York
(Long Island); and casually to Massachusetts (Boston) and Nova
Scotia (Halifax).

Spring migration.—Average dates of arrival: Pennsylvania, southern, March 16; Connecticut, southern, April 6; Quebec, Montreal,
April 27; Prince Edward Island, April 26; Missouri, central, February
26; Illinois, central March 7; Iowa, Keokuk, March 3; Minnesota,
Heron Lake, March 24 (earliest March 6); South Dakota, central,
March 20; North Dakota, northern, April 6; Manitoba, Aweme, April
16; Saskatchewan, southern, April 19; Alaska, Yukon River, May
3 (earliest). Late dates of departure: Florida, Wakulla County, April
11; North Carolina, Raleigh, April 13; Louisiana, Hester, April 6;
Texas, northern, April 16; Iowa, Keokuk, April 30 (average April 7).

Fall migration.—Early dates of arrival: Pennsylvania, Erie, September 1; Long Island, Mastic, September 4; Virginia, Alexandria,
September 22; Florida, Wakulla County, September 26. Average
dates of arrival: Quebec, Hatley, October 11; Maine, Bangor, October 8; Pennsylvania, Erie, September 15; Virginia, Alexandria, September 29; Iowa, Keokuk, September 21; Kansas, central, September
12; Texas, central, September 22; California, central, September 17.
Late dates of departure: Alaska, Taku River, September 26; Ontario,
southern, November 7; Massachusetts, Essex County, December 2;
Iowa, Keokuk, November 27.

Casual records.—Accidental in Bermuda (October 10, 1874, November, 1874, and fall of 1875), southern Greenland (Julianshaab, Godthaab, etc.), Great Britain (Hampshire about 1840, Yorkshire, November, 1851, and Devonshire, November 23, 1879), Hawaiian Islands, and Japan.

Egg dates.—Alaska and Arctic America: Four records, June 4, 10, 18 and July 1. Saskatchewan and Alberta: Thirteen records, May 21 to June 21; seven records, May 25 to June 17. Colorado and Utah: Twenty-five records, May 6 to August 17; thirteen records, May 17 to June 22.

QUERQUEDULA DISCORS (Linnaeus).

BLUE-WINGED TEAL.

HABITS.

Spring.—Not until spring is well advanced and really hot weather has come in its winter haunts does this tender warm-weather bird decide to leave the sunny glades of Florida and the bayous of Louisiana, where it has spent the winter or early spring, dabbling in the shallow, muddy pools, and marshes. The early migrants are probably hardier individuals that have wintered farther north, but the later migrants linger in the Gulf States through April and even into May. Dr. F. Henry Yorke (1899) designates three distinct spring flights, as follows:

The first issue of this, our tenderest, duck arrives in latitude 37° from March 25 to April 1, staying about six or eight days. The second follows a few days after the first has departed northward, up to and past the boundary line. A short period elapses when they likewise travel north to the southern part of Minnesota and its parallel. The third soon follows, and stays an indefinite period, working up through Illinois, Indiana, Wisconsin, and eastward about the last week in April if the we athe permits, the Ohio, Missouri, and Mississippi, with their tributaries, furnishing the fly ways.

Dr. P. L. Hatch (1892) thus describes the arrival of this species in Minnesota:

No other species of the ducks is so cautious upon its arrival as the blue-winged teal, a trait by which the old hunter determines its identity at once. In parties of 8 to 10 or a dozen they will circle around, descending again and again only to rise again and go farther up or lower down the stream to repeat the same demonstrations of indecision, many times over, and just as unexpectedly they suddenly drop out of sight between the treeless banks. They are, as a general thing, several days later in their spring arrivals, and as much earlier than the greenwings in autumn. This is not true in every migration, for I have once or twice known them to come a little before the other, and several times simultaneously; but in my observations, extending over many years in succession, it has proved a noticeable characteristic in its migrations. They are seldom seen on the large clear lakes; but on small ponds, mud flats, and sluggish streams where various pondweeds and aquatic roots afford in abundance its favorite vegetable food.

Courtship.—The courtship of the blue-winged teal is largely performed on the wing much after the manner of the black duck, a nuptial chase as it were, of which Mr. Ernest E. Thompson (1890) says:

I have frequently remarked that during the breeding season this species may be seen coursing over and around the ponds in threes, and these when shot usually prove a male and two females. After dark they may be identified during these maneuvers by their swift flight and the peculiar chirping, almost a twittering, that they indulge in as they fly.

Nesting.—The breeding range of the blue-winged teal has been materially reduced in area during the past 50 years by the increasing settlement of the Middle West, the encroachments of agriculture on its breeding grounds, and by the constant persecution by gunners of an unsuspicious and desirable game bird. Although it formerly bred abundantly throughout all the Middle and Northern States east of the Rocky Mountains, it is now mainly restricted to the prairie regions of the northern United States and Canada, with only a few scattering pairs left in the eastern and southern portions of its breeding range. We found a few pairs breeding in the East Point marshes in the Magdalen Islands, and only a few are left in eastern Canada and south of the Great Lakes. In North Dakota it was still abundant in 1901; this, with the pintail and shoveller, were the three commonest ducks; almost every little pond hole, creek, or grassy slough contained one or more pairs of blue-winged teal, and we could see the pretty little ducks swimming in pairs, close at hand among the vegetation or springing into the air as we drove past.

Here their nests are generally well concealed in the long prairie grass growing around the borders of the sloughs and small pond holes, almost always on dry ground but not far from the water; they are sometimes located in moist meadows bordering such places, where the grass is long and thick enough to conceal them. I found one nest in an open place where the dead grass had been beaten down quite flat; it was beautifully concealed from view under the grass. They also nest sparingly with the baldpates and lesser scaup ducks on the islands. The nest of the blue-winged teal is well built; a hollow is made in the ground and filled with a thick soft lining of fine grass mixed with down, on which the eggs are laid, and the grass is arched over it for concealment; as incubation advances more down is added until a thick blanket is provided, which the female uses to cover the eggs when she leaves them. The nests are so well concealed that comparatively few are found, considering the abundance of the species.

In Saskatchewan in 1905 and 1906 the blue-winged teal was one of the most abundant of the ducks; we found 16 nests in all on dates ranging from June 13 to July 9; the nests were on the islands and in the meadows near the lakes, similar in location and construction to those we found in North Dakota. On that wonderful duck island in Crane Lake 10 out of the 61 ducks' nests found were of this species; only the gadwall, of which we found 23 nests, exceeded it in abundance.

Rev. Manley B. Townsend has sent me his notes on a nest which he found in a slough near Crystal Lake, in Nebraska, on June 10, 1910. He writes:

One June day we made a systematic search of the swamp for nests, and were rewarded in richest measure, finding numerous nests. As we picked our cautious way through the swamp we came to a small dry area, some 30 feet back from the open water. Out from under our feet burst a large bird with a startled "quack" and went hurtling off over the pond. It was a female blue-winged teal. There, beneath a tuft of grasses, in a hollow on the ground, was the nest, built of grasses and lined with dark-brown mottled down pulled from the mother's own breast. In the midst of the downy bedclothes rested 10 beautiful, cream-colored eggs—an exquisite casket of jewels destined to develop into living gems far lovelier than any rubies or diamonds ever dug from the earth. The beauty of such a spectacle can not be adequately described and must be seen to be appreciated. On leaving the nest, the bird is accustomed to nicely cover her treasures with the warm comforters to prevent too rapid evaporation of the heat. We had unexpectedly "jumped" her and she had left in too great a hurry to perform that customary function. Two weeks later we found the nest empty, but the whole family were out there on the pond, bobbing about as buoyant as corks, learning how to make a living and survive in a wonderful but dangerous world.

Several observers have reported nests in close proximity to railroad tracks, which seems to be a favorite location.

Mr. Robert B. Rockwell (1911) has made some extensive studies of the nesting habits of ducks in the Barr Lake region of Colorado; he writes:

By far the most abundant nesting duck throughout the Barr district was the pretty little blue-winged teal. No matter what type of ground our searches carried us over, we were sure to be startled by the occasional flutter of wings, as a dainty little gray-clad mother left her nest like a flash upon our too close approach. We found nests of these birds in the dense cat-tail growth along sloughs; on the soggy, spongy seepage ground under the big dykes; at the edge of beaten paths near the lake shore; by roadsides back from the water; among the dry weeds and sand of the prairie, far from the water's edge; amid the dense rank grass on a tiny island; in alfalfa fields, on grassy flats, and in cavities in and upon muskrat houses.

The nests exhibited a wide diversity in construction. The predominating type was a neat basketlike structure composed of fine soft dead grass, sometimes set well into a dense clump of rank grass on the surface of the ground, and sometimes sunken into a cavity until the top of the nest was flush with the surface of the ground. These nests were usually liberally lined with down; much thicker on the sides and rim of the nest than on the bottom. In fact several were examined which had no down whatever underneath the eggs. The quantity of down varied greatly in different nests, but apparently increased in quantity as incubation advanced.

A less common type of nest was made entirely of bits of dead cat-tail blades deepset into a cavity in the ground. This type of nest was usually found in marshy places, where this material was more available, and in these there was much less of the downy lining. The concealment of these nests was likewise less effective, and taken as a whole this type of nest was altogether inferior. We found a few built in wet places where the foundation of the nest was actually wet, but we did not find a single nest where the eggs were the least bit damp; and the large majority were in perfectly dry locations in close proximity to water.

The concealment of the better built nests, especially those in the center of a tussock of rank grass, was well-nigh perfect; in fact in most cases we were unable to see either the brooding bird or the eggs from a distance of 5 to 6 feet even when we knew the exact location of the nest. Upon leaving the nest during incubation the parent covered the eggs with the downy rim of the nest and the concealment thus afforded was remarkable.

Several radical departures from the characteristic habits were encountered. One bird had built her nest on a little flat amid some short blue grass which afforded her no concealment whatever. As she brooded her eggs she was plainly visible at a distance of 20 yards or more. She allowed me to approach to within 4 or 5 feet and set up my camera for an exposure; and then instead of springing lightly into the air as usual, she ambled awkwardly off the nest, waddled slowly between the legs of my tripod, uttering lazy little quacks of protest, and finally after walking a distance of 30 yards or more took flight.

While plowing our way through a dense cat-tail swamp in water above our knees we frightened a teal from a nest in a muskrat house. A careful search finally revealed the eggs fully a foot back from the entrance of a deep cavity in the side of the house. To our surprise the nest contained four eggs of the teal and five eggs of some big duck, all of which were incubated.

Another queer nest was found, which was a shallow depression on the side of a dilapidated muskrat house, which had been originally built between a fence post and its diagonal brace. The lower barbed wire of the fence prevented the top of the house from collapsing, while the side weathered away, leaving a cavity well protected by the overhanging top. In this cavity without a sign of lining or a bit of concealment lay the 10 conspicuous white eggs. They could be readily seen from a distance of 20 yards.

The down in the blue-winged teal's nest is larger and lighter colored than in that of the green-winged teal; it varies in color from "hair brown" to "drab," and it has large whitish centers. '

Eggs.—The blue-winged teal lays from 6 to 15 eggs, but the numbers most commonly found in full sets are 10, 11, and 12. All ducks are more or less careless about laying in each other's nests. This seems to occur less frequently with the teals than with the larger species, but the nest mentioned above by Mr. Rockwell (1911), containing "four eggs of the teal and five eggs of some big duck," shows that the little teal is sometimes imposed upon.

The eggs of the blue-winged teal vary in shape from ovate to elliptical ovate; the shell is very smooth, but only slightly glossy. In color they are dull white, light-cream color, creamy white, or pale olive white. They are not distinguishable from those of other teals; but if the female is flushed from the nest, she can be distinguished from the green-winged teal by the blue wing-coverts, but not so easily from the cinnamon teal.

The measurements of 93 eggs in various collections average 46.6 by 33.4 millimeters; the eggs showing the four extremes measure **49.5** by 35, 47.2 by **36.2**, **43.5** by 32, and 45.6 by **31.3** millimeters.

Young.—As the male deserts the female soon after the eggs are laid, incubation is performed solely by her. Incubation does not begin until after the last egg is laid, one egg having been laid each day until the set is complete. The period of incubation is from 21 to 23 days. The young hatch almost simultaneously, or at least within a few hours; they remain in the nest until they have dried off and are strong enough to walk, when they are led to the nearest water and taught by their devoted mother to feed. Their food at this age consists mainly of soft insects, worms, and other small, tender, animal food, but they soon learn to forage for themselves and pick up a variety of vegetable foods as well. The young are guarded with tender care by one of the most devoted of mothers; when surprised with her brood of young she resorts to all the arts and strategies known to anxious bird mothers to draw the intruder away from her brood or to distract his attention, utterly regardless of her own safety, while the young have time to hide or escape to a place of safety. The young are experts at hiding, even in open situations, where they squat flat on the ground and vanish; but they usually run or swim in among tall grass or reeds, where it is almost useless to look for them. All through the remainder of the summer, until they are able to fly, she remains with them teaching them where to find the choicest foods and how to escape from their numerous enemies; they learn to know her warning calls, when to run and when to hide, and by the end of the summer they are ready to gather into flocks for the fall migration.

Plumages.—In the downy young the colors of the upper parts vary from "mummy brown" to "Dresden brown," darker on the crown and rump, lighter elsewhere, the down being much darker basally; the under parts are "maize yellow," shaded locally with "buff yellow," due to the darker tips of the down; the sides of the head are "yellow ocher" or pale "buckthorn brown" in young birds, but these colors soon fade and all the colors grow paler as the young bird increases in size. The color pattern of the head consists of a dark-brown central crown bordered on each side by a broad superciliary stripe of yellow ocher, below which is a narrow postocular stripe, a loral patch, and an auricular spot of dusky. On the back the brown is broken by four large spots of yellowish, one on each side of the rump and one on each scapular region. Young blue-wing teal closely resemble young shovellers, but the latter are paler colored, with all the brown areas more extensive, with less of the rich buff and yellow tints and with longer and more broadly tipped bills.

The young develop more rapidly than those of the larger ducks, as they are late breeders and early fall migrants. The first feathers to appear on the downy young are the mottled feathers of the sides, below and above the wings; these come when the young bird is hardly one-third grown, sometimes by the end of June. The growth of feathers spreads over the breast first, then over the back and head, the down disappearing late on the rump and last on the hind neck; by the end of July the young teal is nearly fully grown and the whole of the spotted juvenal plumage has been acquired except the wing quills which are still in their sheaths. During August the wings and and tail are acquired and before the end of that month the young birds can fly. Before the wings are grown the sexes are practically indistinguishable and both resemble the adult female except that they are lighter colored below and often nearly immaculate white on the belly.

During the fall and winter the young teal makes slow progess toward maturity; the blue lesser wing coverts and the green speculum are acquired as soon as the wings are grown, but they are duller than in adults; other changes come slowly until spring, when the first nuptial plumage is assumed, hardly distinguishable from the adult nuptial plumage, but the colors are all duller and the long blue-edged scapulars are not yet developed.

The first eclipse plumage is assumed in July and August; and at this first complete postnuptial molt the young bird becomes indistinguishable from the adult, when about 14 months old.

The eclipse plumage in the adult involves the change of all the contour feathers and the scapulars; it does not begin until July, is complete in August, when the flight feathers are molted, and lasts through September. In this plumage the male closely resembles the female, but can always be recognized by the wings, in which no marked seasonable change takes place. Adults are slow in shedding the eclipse plumage, individuals varying greatly in this respect. The full body plumage is seldom acquired before the middle of winter and sometimes not until March, so that the gradual changes taking place might be regarded as a prolonged prenuptial molt.

Hybrids among the teals are not common, but Mr. William G. Smith (1887) records a specimen, which he took in Colorado, "the whole body color of the cinnamon teal, with the head the color, and snow-white cheek marks distinctly, of the bluewing."

Mr. Frederic H. Kennard (1919) has described, under the subspecific name *albinucha*, a supposed southern race of the blue-winged teal, the sole distinguishing character being a continuation of the white crescents over the eyes in thin superciliary lines down to the nape, where they join to form a white nuchal patch. It does not seem to have been proven that *all* southern breeding teal are so marked, and I have seen several northern breeding teal partially so marked. Mr. Stanley C. Arthur (1920) records a case where a bird in captivity lost this marking after molting into a new spring plumage. This marking may prove to be merely a high stage of plumage, assumed by the most vigorous birds. Mr. Arthur's bird died soon after assuming the normal spring plumage, which may mean that waning vitality was the cause of its losing its white adornment.

Food.—The blue-winged teal is decidely a surface feeder; it feeds in shallow, muddy pond holes overgrown with aquatic vegetation, about the reedy shores of lakes and sloughs, and even in wet meadows, particularly along the banks of grassy ditches and creeks, where it is usually concealed from view; its food is usually obtained on the surface or within reach of its submerged head and neck, but occasionally its tail is tipped up and its body half immersed. Its food consists largely of tender aquatic plants.

In the fall it visits the grain fields occasionally and eats some wheat and barley. It eats wild rice wherever it can find it and, on its winter feeding grounds, it lives and feasts in the extensive rice fields. Its animal food includes tadpoles, worms, snails, and other small mollusks, water insects, and larvae. Dr. J. C. Phillips (1911) found that the stomachs of birds shot in Massachusetts contained "many young snails, various insects, and seeds of burreed, pondweeds, smartweed, and various sedges and grasses. Animal matter, 88 per cent; vegetable, 12 per cent; mineral, 8 per cent."

Mr. Douglas C. Mabbott (1920) sums up the food of the blue-winged teal as follows:

About seven-tenths (70.53 per cent) of the blue-winged teal's food consists of vegetable matter. Of this about three-fourths is included in four families of plants. Sedges (Cyperaceae), with 18.79 per cent; pondweeds (Naiadaceae), 12.6; grasses (Gramineae), 12.26; and the smartweeds (Polygonaceae), 8.22. The remainder of the plant food is made up of algae, 2.95 per cent; water lilies (Nymphaeaceae), 1.37; rice and corn, 0.98; water milfoils (Haloragidacae), 0.71; bur reeds (Sparganiaceae), 0.38; madder family (Rubiaceae), 0.35; and miscellaneous 11.92 per cent.

Animal matter constitutes 29.47 per cent of the total food of the blue-winged teal, which is more than three times the percentage of animal food eaten by the green-wing. Over half of this (16.82 per cent) is mollusks, the remainder being made up of insects, 10.41 per cent, crustaceans, 1.93, and miscellaneous, 0.31 per cent.

Behavior.—From the water the blue-winged teal springs into the air with surprising agility, and when under way is one of the swiftest of the ducks in flight; it has been credited with attaining a speed of 90, 100, or even 130 miles an hour, but probably these speeds are all overestimated, as there is very little accurate data on which to base an estimate. Doctor Yorke (1899) says: "They travel at the rate of about 130 miles an hour, exceeded only by the green-winged teal." This seems incredible.

Audubon (1840) says:

The flight of the blue-winged teal is extremely rapid and well sustained. Indeed, I have thought that, when traveling, it passes through the air with a speed equal to that of the passenger pigeon. When flying in flocks in clear sunny weather, the blue of their wings glistens like polished steel, so as to give them the most lively appearance, and while they are wheeling over the places in which they intend to alight, their wings being alternately thrown in the shade and exposed to the bright light, the glowing and varied luster thus produced, at whatever distance they may be, draws your eyes involuntarily toward them. When advancing against a stiff breeze, they alternately show their upper and lower surfaces, and you are struck by the vivid steel blue of their mantle, which resembles the dancing light of a piece of glass suddenly reflected on a distant object. I have never observed them traveling in company with other ducks, but I have seen them at times passing over the sea at a considerable distance from land. Before alighting, and almost under any circumstances, and in any locality, these teals pass and repass several times over the place, as if to assure themselves of the absence of danger, or, should there be cause of apprehension, to watch until it is over. They swim buoyantly, and generally in a close body, at times nearly touching each other.

Nuttall (1834) says that "when they alight," they "drop down suddenly among the reeds in the manner of the snipe or woodcock."

About the vocal powers of this teal there is very little to be said. Dawson (1903) has covered the ground very well in the following words:

In addition to the whistling of the wings, the teals have a soft lisping note, only remotely related to the typical anatidine *quack*, and is uttered either in apprehension or encouragement.

While feeding and at other times these teal are usually silent; the lisping or peeping of the male are more often heard when the birds are in flight than at other times and are probably used as signals, as to dangers or the presence of food. The female has a faint quacking note.

On their breeding grounds blue-winged teal are associated with various other species, notably shovellers, pintails, gadwalls, and mallards. On their migrations they usually fly in flocks by themselves, but often resort to the same feeding grounds as other surface-feeding ducks. Doctor Yorke (1899) says:

They mix a great deal with the coots, eagerly devouring the seeds of the teal moss, which the former by diving tear up by the roots, and the long sprays covered with seeds float upon the surface of the water.

In Florida and Louisiana they seem to associate with the larger shore birds, feeding with them in the shallow lagoons. They are always gentle and harmless towards other species. Their only enemies are the predatory birds and animals, among which the human hunter is most destructive.

Fall.—As soon as the young are able to fly, or even before that, they begin gathering into flocks preparing for the fall migration, which begins with the first early frosts in August and is mainly accomplished during September, for these delicate birds are very sensitive to the approach of autumn and are the earliest ducks to migrate. Doctor Yorke (1899) has described this movement very well; he writes:

About the early part of August the local ducks of each State begin to work northward; during September they flock together and form the first flight, passing over the same grounds. The collecting or flocking together of the local birds, which form the first fall issue, presents an interesting sight. For nearly two days the ducks will be noticed as getting very uneasy, whipping about without the regularity which had hitherto been customary upon their feeding, playing, and roosting grounds. On the day of their departure, after feeding, they will flock to some large common playground; where, instead of quietly resting, as usual, they assume a stage of activity. About 3 in the afternoon, instead of drifting back to their feeding grounds as usual in little flocks, singles, and pairs, they form flocks and sweep up and around the open water and alight again. The flocks soon increase in size and after two or three circles around the open water, each time rising higher and higher, they proceed south in well-defined and distinct flocks, each under a leader, and soon vanish in the distance, never returning that fall. Three or four days of no shooting occurs, except upon those which were too weak and incapacitated for a long flight, before the second issue arrives, which stays a few days. A cold snap brings down the third, the weather determining the length of their stay. The second and third depart at night or late in the evening, but evince no disposition to assemble as the first. They are the second of our warm-weather birds to leave, closely following the wood ducks.

Game.—The little blue-winged teal is a favorite with the sportsmen; it comes at the beginning of the season, when he is eager to try his skill at one of the swiftest of ducks; it decoys readily, especially to live decoys; it flies in large, compact flocks, which offer tempting shots as they twist and turn or swing and wheel in unison; it is unsuspicious of its hidden foe, is easily killed with small shot and makes a fine table bird. We used to look for it about the full of the moon in September and could always count on finding plenty of birds in the shallow ponds, marshes, and grassy creeks; but, unfortunately, it has been steadily decreasing since the early eighties and is now quite scarce in Massachusetts. In the good old days, when these birds were abundant, they were an easy mark for the youthful gunner, as they huddled together in a compact flock on the water, and a large number could be killed at a single discharge of the old muzzle-loader.

Dr. L. C. Sanford (1903) writes of shooting blue-winged teal as follows:

In late August we find them fully fledged, frequenting the marshes of the West where the wild rice grows. They are relentlessly hunted from time of first arrival. During the hours that are sacred to the duck marsh, the time after dawn and toward dusk, they are found. At first many are killed by pushing through the grass as they jump up in front of the skiff or on their line of flight between the ponds. At the approach of evening the first line appears over the tops of the rush grass, flying low and with a speed possessed only by a teal. Another minute and they have passed; the rush of their wings told how closely they came; but no one but an old hand could have stopped one. The next flock follow, the gunner rises in time, and they sheer off. crowding together in an attempt to turn; but a well-placed shot drops several birds. So they come on until dark, when the soft whistling overhead tells of ducks still looking for a spot to feed and spend the night in peace.

Mr. Dwight W. Huntington (1903) pays the following tribute to their speed:

After some days' shooting at the sharp-tailed grouse, I went one day to a famous duck pass in North Dakota, when the teal were flying from the Devils Lake to a smaller one to breakfast. As soon as I had made my blind, they began to come singly and in pairs, sometimes three or four together or a small flock, and although they came in quick succession and the shooting was fast enough to heat the gun, I believe it was an hour or more before I killed a bird. I was almost in despair, when I fired at a passing flock, holding the gun a yard or more before the leading birds, and at the report a single teal, some distance behind the others, fell dead upon the beach. I at once began shooting long distances ahead of the passing ducks, and before long I had a large bag of birds.

A few days afterwards an officer from the garrison near by, a good shot in the upland fields and woods, went with me to my duck pass to shoot at teal. We made our blinds some two gun shots apart and soon began to shoot. The birds came rapidly as before, and my friend gave them two barrels as they passed, but was entirely out of ammunition before he killed a bird. His orderly came to my blind for shells, and with them I sent a message to shoot three times as far ahead as he had been doing, and he was soon killing birds.

Winter.—They are still abundant in some parts of the South, where they make their winter home in the great rice fields and extensive marshes, feeding on the ripened grains that fall upon the water, feasting and growing fat. Here they are safe enough as long as they paddle about and remain hidden in the innermost recesses of the rice fields and inaccessible swampy pools; but the sportsmen soon learn their haunts and habits, build their blinds near their favorite feeding grounds or fly ways and shoot them as they fly about in search of food and shelter. Constant persecution has thus materially reduced their numbers, but since such extensive sanctuaries have been established in Louisiana, it is to be hoped that they will have a safe haven of rest, in the fall at least; this may also result in larger numbers sojourning there for the winter, rather than passing on farther south, as the majority of this species now does.

DISTRIBUTION.

Breeding range.—Mainly central North America, more rarely toward the east and west coasts. East rarely or casually to Gulf of St. Lawrence (Magdalen Islands) and New Brunswick (St. John County). South casually to southeastern Maine (Washington County), southern Rhode Island (Sakonnet), and southern West Virginia (Brooke County); more recently to northern Ohio (Ottawa and Sandusky Counties), southwestern Indiana (Gibson County), southern Illinois (Union County), central Missouri (Missouri River valley), central Kansas (Emporia and Wichita), northern New Mexico (Lake Burford), central Utah (Fairfield), and northern Nevada (Truckee valley and Washoe Lake). West only to the Sierra Nevada and Cascade Mountains. North to central British Columbia (Lac la Hache and Cariboo), southern Mackenzie (Great Slave Lake), northern Saskatchewan (58°N), central Manitoba (Lake Winnipegosis), and probably sparingly in the interior of Ontario and Quebec. Has bred in Louisiana (Marsh Island), Oklahoma (Fort Reno), and Texas (San Antonio). Birds which breed south of United States are probably subspecifically distinct.

Winter range.—Southern North America and northern South America. East to the Atlantic coast of the United States from Maryland southward, the Bahamas, the Greater and Lesser Antilles and the coasts of Venezuela, French Guiana, and Brazil. South to central Chile (Ovalle). West to the Pacific coasts of Chile, Peru, Ecuador, Colombia, Central America, Mexico, and the United States. North irregularly to southern California (Santa Barbara and Los Angeles Counties), southern Illinois and Indiana, and eastern Maryland; but not common in winter north of Mexico, Texas, Louisiana, and South Carolina.

Spring migration.—Early dates of arrival: North Carolina, Raleigh, March 23; Pennsylvania, Erie, March 27; Massachusetts, Templeton, April 1; Prince Edward Island, April 20; Iowa, central, March 18; Minnesota, Heron Lake, April 7; North Dakota, central, April 12; Manitoba, Aweme, April 27; Alberta, Edmonton, May 1. Late dates of departure: Panama, February 7; Texas, San Antonio, May 14; Louisiana, New Orleans, May 21; California, Santa Barbara, May 7; Florida, Gainesville, April 29; North Carolina, Raleigh, May 6; Maryland, Baltimore, May 7.

Fall migration.—Early dates of arrival: Nova Scotia, Sable Island, August 19; Pennsylvania, Philadelphia, August 24; Virginia, Alexandria, August 18; Florida, Wakulla County, September 2; Alabama, Alabama River, September 20; California, Santa Barbara, August 25; Panama, October 14. Average dates of arrival: Virginia, Alexandria, August 31; Kansas, central, September 12; Mississippi, southern, September 16. Late date of departure: Nova Scotia, Sable Island, November 1; Prince Edward Island, October 8; Maine, Lewiston, November 7; New Jersey, Cape May, December 5; North Carolina, Raleigh, December 7; Ontario, Ottawa, October 27; Manitoba, Aweme, October 30; Illinois, Chicago, October 22; Iowa, southern, November 4; Missouri, central, November 13.

Casual records.—Occasional in Bermuda (October 22, 1854, April 30, 1875, etc.). Accidental in the British Isles (Dumfriesshire, 1858, Cheshire, about 1860, Anglesey, 1919, and County Cork, September, 1910) and in Denmark.

Egg dates.—Colorado and Utah: Twenty records, May 10 to July 21; ten records, May 31 to June 24. Manitoba and Saskatchewan: Nineteen records, June 4 to July 26; ten records, June 17 to July 4. Minnesota and North Dakota: Thirty-three records, May 8 to July 23; seventeen records, May 31 to June 13.

QUERQUEDULA CYANOPTERA (Vieillot).

CINNAMON TEAL.

HABITS.

The "western champion," as Dawson (1909) has aptly called this species, holds a unique position among American ducks, for it is the only member of the family that is confined to the western part of the continent with its center of abundance west of the Rocky Mountains and the only member of the family which has a regular breeding range in South America separated from that in North America by a wide gap of about 2,000 miles. The history of its discovery is also interesting. Coues (1874) says:

It has not often occurred that an abundant bird of North America has been first made generally known from the extreme point of South America, and for a long time recognized only as an inhabitant of that continent. Yet this species furnishes such a case, having been early named by King *Anas rafflesi*, from a specimen taken in the Straights of Magellan. It is, moreover, a singular fact, that it was first discovered in the United States in a locality where it is of very unusual and probably only accidental occurrence.

Subsequent to its discovery in Louisiana in 1849, it was afterwards rediscovered, as a North American bird, and found to be one of the most abundant species west of the Rocky Mountains, by the various survey expeditions to the Pacific coast during the next 20 years.

Spring.—The spring migration of the cinnamon teal is not a long flight, for its winter and summer ranges overlap and it is absent in winter from only the northern portion of its breeding range. The northward movement begins in March and continues through April. Dr. J. C. Merrill (1888) noted that, at Klamath, Oregon, "early in May several flocks of this beautiful teal arrived, and before the end of the month it was common in the marsh, mostly paired and not at all shy."

Courtship.—Mr. W. Leon Dawson has sent me the following notes on the courtship of this species:

Upon a little pond entirely surrounded by reeds I watched six or eight cinnamon teals disporting themselves and indulging in courting antics. A male would follow about very closely after his intended, and bob his head by alternately extending and withdrawing his neck in a lively fashion. Now and then the female would make some slight acknowledgment in the same kind. In at least one instance I think I appeared in the decisive moment, for from pretended indifferences a duck responded to long bobs of inquiry with emphatic bobs of approval given face to face, and immediately thereafter joined her favored suitor in chasing away discredited rivals. The males were repeatedly charging upon each other with open beaks, but it is hard to think that they could or would do each other bodily harm.

The teals, by the way, of both kinds, associate closely, so that the females of the two species are sometimes confused by the observer, and the males exhibit some jealousy toward each other, as though really fearing confusion of brides. A favorite play on the part of these teals is leap-frog. A bird will vault into the air and pass over another's head and down again with a great splash, and the other as likely as not will repeat the same trick, especially two males of two pairs playing together.

Dr. Alexander Wetmore (1920) says:

These single males persisted in paying attention to females already mated, much to the disgust of the paired drakes, who drove them away, bowing at them and chattering angrily. On one occasion six were seen making demonstration toward one female who paid no attention to them, but followed her mate. He swam first at one and then another after each chase returning to his mate and bowing rapidly, while occasionally she bowed to him in return. After a few minutes another mated pair of teal flew by and four of the males flew off in pursuit of them, leaving the first males only two to combat.

Nesting.—Mr. Dawson has also sent me more or less data on some two dozen nests of this species found in Washington and about a dozen found in California, from which I have made a few selections to illustrate the variations in nesting habits. A nest containing nine eggs, found at Stratford, Washington, on June 8, 1906, was located while dragging a rope in a pasture. The nest was a deep depression in the ground in a "loose clump of rye grass, lined sparsely with bits of grass and copiously with down. The down arches up at the hinder end and makes a little rear wall above the ground. Depth 5 inches; width 5 inside." On the same day he flushed a bird from a nest of 11 eggs in a "thick clump of yellow dock and mint. The rope gave a vicious tug at the clump else she never would have flown." On his return later to photograph the nest, he peered down into the vegetation and surprised the teal at home; she struggled wildly to escape and left the usual deluge of fresh excrement on the eggs, as evidence of her fright. The nest was "a shallow depression scantily lined with broken grass and trash, and heavily with dark down"; it measured "6 inches across and 3 deep inside." All of the other Washington nests were apparently on dry ground, concealed in tall grass or rank herbage, often on high land and many of them were from 75 to 200 feet away from the nearest water. The California nests were in more varied situations. On May 13, 1911, in Nigger Slough, near Los Angeles, he flushed a bird at close quarters in heavy saw grass;" the nest was "built up above damp earth" and contained 11 fresh eggs. On May 24, 1912, at Los Banos, he found two nests "buried in the heart of cat-tails built up to a height of some 6 inches out of a foot of water"; the "nests were really woven baskets placed in the depths of the reeds, an unusual situation for cinnamon teals."

One of these nests originally held eight eggs of the teal; but two eggs of the mallard had been added and three of the teal's eggs had been thrown out into the water; this nest "was built up of dried cat-tail and sedges, 5 inches high in the center and 9 at the edges, with a free way to the water after the manner of coots. It was about 7 inches across," and was wet and bulky.

Mr. Harry H. Sheldon (1907) refers to three nests which he "found in a grain field" near Eagle Lake, in the Sierra Nevada Mountains. Mr. Fred A. Schneider (1893) describes a nest which he found in a marsh near College Park, California, as "very neatly constructed an inch or two above the water and firmly fastened to the round marsh grass, which grew about 30 inches high and almost concealed the nest from view." It "was made entirely of marsh grass and lined profusely with gray down, especially around the edges. By cutting off the grass which supported the nest it could easily have been removed without danger of its falling apart."

Mr. Robert B. Rockwell (1911) found a number of cinnamon teal breeding in the Barr Lake region of Colorado, but positively identified only four nests as belonging to this species, in which he was—

unable to detect any radical departures from the habits already attributed to the bluewings except that two of the four nests were in very wet locations, where the eggs were in constant danger of becoming damp. These two nests were practically devoid of the downy lining while the other two nests, which were built in perfectly dry locations were warmly lined with down. One of the nests was on a dry prairie fully 100 feet back from the shore of the lake amid a fairly thick growth of weeds and grass.

The down in the nest of the cinnamon teal is much like that of the blue-winged teal, but lighter than that of the green-winged teal. It is "hair brown" to "drab" in color, with large conspicuous white centers. Two types of breast feathers are found in the nests, dusky with buff edges and tips, or dusky with whitish central markings.

Eggs.—The cinnamon teal lays from 6 to 14 eggs, the usual set being from 10 to 12. In shape they are ovate, elliptical ovate, elliptical oval or almost oval. The shell is smooth and only slightly glossy. The color varies from "pale pinkish buff" or "cartridge buff" to almost pure white.

The measurements of 90 eggs in various collections average 47.5 by 34.5 millimeters; the eggs showing the four extremes measure 53 by 35, 48 by 37 and 44 by 30 millimeters.

Young.—Although the male does not wholly desert the female during the nesting season, the duties of incubation seem to be wholly performed by her. The period of incubation does not seem to have been accurately determined, but it is probably not very different from that of closely related species. Mr. John G. Tyler writes me:

I have observed that the male of this species departs from the usual rule among the ducks and very often assists the female in caring for her brood of young. It is rather unusual to find young cinnamon teal that are not accompanied by both parents and the solicitude of the male bird increases with the age of the ducklings; in fact, the male is often far more demonstrative than his mate. In one instance I observed three males and a single female accompanied by 10 downy young, the males showing unmistakable evidence of their great distress at the near presence of a man while the female swam about near her family in a most unconcerned manner.

On the authority of Mr. A. M. Shields, of Los Angeles, Mr. Fred A. Schneider (1893) has published the following interesting account of the behavior of the young:

After being hatched, the mother duck (joined by her mate) escorts the young brood to the nearest body of water and manifests the greatest solicitude for the welfare of the little fellows, giving a signal upon the slightest approach of danger, which is followed by the almost instant disappearance of the entire brood, as if by magic. If on the shore they disappear in the grass; if in the water, they dive, and that is generally the last seen of them, for the time being at least, as they swim under water for great distances until reaching the edge of the stream or pond, when they imperceptibly secrete themselves among the water moss or grass. I once watched a little fellow as he made his way under the clear water. He went straight for a little bunch of floating moss, and by gazing intently I could just distinguish the least possible little swelling of the moss; a small hump, as it were, about the size of a marble. He had come to the surface (as intended) under the patch of moss, and his head and bill were responsible for the little hump in the moss.

Possibly one thing more than anything else helps the little fellows to disappear in such marvelously quick time and before you can realize it. The old duck flutters and falls around you just out of your reach and most successfully imitates a fowl badly winged, hardly able to rise from the ground. Her actions are bound to more or less avert your attention for a moment at least, and it is just that moment that the little fellows disappear, as the mother duck undoubtedly intended. After a short time, when the little ones are all securely hidden, the mother, feeling no further anxiety, gracefully recovers from her crippled condition, flies off a few hundred yards, and there awaits your departure, when she returns to her family, who soon gather around her one by one till they are all assembled and everything goes on as though nothing had happened—until the next intruder appears, when "Presto! change!" and the same actions are repeated.

Plumages.—The downy young of the cinnamon teal is "mummy brown" above, darkest on the crown, and the tips of the down are "buffy citrine," producing a golden olivaceous appearance on the back; the forehead, the sides of the head, including a broad superciliary stripe, and the under parts vary from "mustard yellow" on the head to "amber yellow" on the breast and "naphthaline yellow" on the belly; there is a narrow stripe of dark brown on the side of the head; and the color of the back is relieved by a yellowish spot on each side of the rump, scapular region, and edge of the wing.

The first feathers appear on the scapulars and flanks; these are brownish black, edged with "cinnamon brown." When the young bird is about half grown the tail appears and the under parts become feathered; the chest and flanks appear to be lustrous "Sanford's brown" and the belly silvery whitish, both mottled with dusky, each feather being centrally dusky. The bird is fully grown before it is fully feathered, the down disappearing last on the hind neck and rump; and the wings are the last to appear. A young bird nearly two-thirds grown is only partially feathered on the head; the back is wholly covered with glossy down, varying from "bister" to "sepia" and darkest on the rump; and the wing quills have not yet burst their sheaths.

In the full juvenal plumage, which in California is complete in July, the young male closely resembles the female, except that the wings are more like those of the adult male; the wings are duller colored and less complete than those of the adult male; the tertials and the scapulars are dusky, edged with "cinnamon brown," the former with a greenish sheen. During the winter and first spring the young male makes steady progress toward maturity; the "mahogany red" plumage comes in on the head, neck, breast, and flanks; the adult barred plumage appears on the upper back; and some of the gaily colored scapulars, blue on the outer web and having a buffy median stripe, are acquired. The young bird then in its first spring closely resembles the adult male, except that the belly still remains more or less dull brown, the colors are everywhere less brilliant and the wings and scapulars are less perfect.

Both old and young males then molt into an eclipse plumage. Beginning in June the head and neck become mottled with new buffy feathers, centrally dusky, which gradually replace the red; the red of the chest and flanks is gradually replaced by handsome feathers, centrally dusky but broadly edged and barred with rich shades of buff and brown in a variety of patterns; the faded brown plumage of the breast and belly are then invaded and gradually replaced by a new growth of buff, whitish-tipped feathers, each with two large spots or central areas of dusky; when absolutely fresh the long white tips of these new feathers give the under parts a silvery white appearance, but the tips soon wear off, leaving these parts as in the female. While this eclipse plumage is at its height, in August, the wings are molted, the secondaries first, with the greater and lesser coverts, and then the primaries; there is much individual variation in the time at which the showy tertials and scapulars are molted; the large, blue-tipped tertials are sometimes renewed before the eclipse plumage is complete and sometimes not until after it is shed; the long, pointed, white-striped scapulars are usually the last to be acquired. The tail is molted in August with the wings and the back plumage is renewed by a double molt simultaneously with that of the under parts; the eclipse feathers of the back are dusky, narrowly edged with buff. In September a new growth of "mahogany red" or "burnt sienna" feathers begins to replace the eclipse plumage on the breast and the renewal of the fully adult plumage spreads over the rest of the body, neck, and head, until, sometime in October or November, the full plumage is complete.

Food.—Mr. Tyler writes me that:

This duck seems to prefer, at all times, the shallow ponds and overflowed areas rather than deep canals and sloughs. The feeding operations are carried on entirely above the water and for the most part along the margin of the ponds or even out on the banks. I have never known them to dive in search of food and in fact believe that the female seldom, if ever, dives for any purpose whatever. The males, however, occasionally, but not often, plunge below the surface of the water during the mating season; this feat usually being accomplished in the presence of a rival.

Mr. Douglas C. Mabbott (1920) says:

Like the greenwing and the bluewing, the cinnamon teal lives mainly upon vegetable food, this comprising about four-fifths (79.86 per cent) of the total contents of the stomachs examined. And, like the other teals, its two principal and most constant items of food are the seeds and other parts of sedges (Cyperaceae) and pondweeds (Naiadaceae). These two families of plants furnished 34.27 and 27.12 per cent, respectively, of the bird's entire diet. The grasses (Gramineae) amounted to 7.75 per cent; smartweeds (Polygonaceae), to 3.22; mallows (Malvaceae), 1.87; goosefoot family (Chenopodiaceae), 0.75; water milfoils (Haloragidaceae), 0.37; and miscellaneous, 4.51.

The 41 cinnamon teals examined had made of animal matter 20.14 per cent of their food. This consisted of insects, 10.19 per cent; mollusks, 8.69 per cent; and a few small miscellaneous items, 1.26 per cent.

Behavior.—Mr. Tyler writes me:

Cinnamon teal are seldom found associated in large flocks but are most often encountered in pairs before the breeding season and in small family groups during the fall. So far as my observations go, the male is quite silent at all times and the only note that I have ever heard the female give is a very matter-of-fact "quack" which serves as an alarm note and is heard just as the bird takes wing.

Doctor Wetmore (1920) says:

The only note that I have ever heard from the male cinnamon teal is a low rattling, chattering note that can be heard only for a short distance.

I have had only limited opportunities of observing this beautiful species in life, but, judging from what I have seen and from what I have read about it, I should say that it differs very little in behavior from the blue-winged teal, to which it is closely related. In the shallow tule-bordered lakes and marshes of the far West, where this handsome little duck makes its summer home, it finds abundant shelter in the thick growth of tules and other luxuriant vegetation, in which to escape from its many enemies, prowling beasts and birds of prey. It is a prolific and persistent breeder and seems to maintain its abundance in spite of the frequent raids upon its eggs and young by predatory animals. Mr. Dawson's notes contain many references to raided nests, of which the following is a fair sample:

As I was returning at 2 p. m. from examination of a gadwall's nest I came upon two broken eggshells of a cinnamon teal. A little search revealed the nest about 6 feet from the nearest egg, and a glance showed the tragedy which had been enacted last night. The grass tussock gaped open and the dark down was scattered. A befouled and broken egg bore sad testimony to the mortal fright of the mother, although none of the remaining six were broken. A runt egg lay a foot or so from the nest, and I think the mother bird must have dropped it there long before the fatal night. A bit of blood on the down showed that it was the bird rather than the eggs the miscreant was after, and I found her lying dead upon her back only 6 feet away. There was a sharp deep wound over the heart—no other mark of violence—and dissection showed that although the heart itself had not been pierced, the neighboring blood vessels had and the blood was practically withdrawn.

This species, like many others, has always been able to cope successfully with its natural enemies, but against its chief enemy, man, it is powerless. The encroachments of civilization and agriculture have driven it from many of its former haunts by draining, cultivating, or destroying its breeding grounds, its shelters, and its feeding places. Many nests are destroyed and some birds are killed by mowing the fields in which it breeds. In the San Joaquin Valley, in 1914, I was disappointed to find that during dry season the land company, which controls vast areas, had drawn off the water for irrigation purposes and left its wonderful sloughs dry and almost duckless. But the worst enemy of all ducks is the unrestrained market hunter, of which Mr. Vernon Bailey (1902) says:

The young are protected in the tule cover until old enough to fly, but they have many enemies. The prowling coyote dines with equal relish on a nest full of eggs or an unwary duck, and there are hawks by day and owls by night. The teals could hold their own against these old time enemies, however, but a new danger has come to them in the form of the unrestrained market hunter. He goes to the breeding ground just before the young can fly and while the old ducks are molting and equally helpless, and day after day loads his wagon with them for the train. This wholesale slaughter has gone on until some of the breeding grounds have been woefully thinned not only of teal, but of other ducks. Without speedy and strenuous efforts to procure and enforce protective laws, many species of ducks that breed principally within our limits will soon be exterminated.

Fall.—As the cinnamon teal winters as far north as southern California and central New Mexico, the fall migration is short and merely means withdrawal from the northern part of its breeding range, during September and the first half of October. During the short southward flight, it flocks in large numbers into all suitable sloughs and lakes, where it is eagerly sought by the sportsmen and is fully as popular as its eastern relative, the bluewing, which it closely resembles in all its habits. I am tempted to quote in full the attractive and vivid picture which Doctor Coues (1874) has drawn of this bird in its fall and winter haunts. He writes:

I have in mind a picture of the headwaters of the Rio Verde, in November, just before winter had fairly set in, although frosts had already touched the foliage and dressed every tree and bush in gorgeous colors. The atmosphere showed a faint yellow haze, and was heavy with odors—souvenirs of departing flowers. The sap of the trees coursed sluggishly, no longer lending elastic vigor to the limbs, that now cracked and broke when forced apart; the leaves loosened their hold, for want of the same mysterious tie, and fell in showers where the quail rustled over their withering forms. Woodpeckers rattled with exultation against the resounding bark, and seemed to know of the greater store for them now in the nerveless, drowsy trees, that resisted the chisel less stoutly than when they were full of juicy life. Ground squirrels worked hard, gathering the last seeds and nuts to increase their winter's store, and cold-blooded reptiles dragged their stiffening joints to bask in sunny spots, and stimulate the slow current of circulation, before they should withdraw and sink into torpor. Wild fowl came flocking from their northern breeding places, among them thousands of teal, hurtling overhead and splashing in the waters they were to enliven and adorn all winter.

The upper parts of both forks of the Verde are filled with beavers, that have dammed the stream at short intervals, and transformed them, in some places, into a succession of pools, where the teal swim in still water. Other wild fowl join them, such as mallards, pintails, and greenwings, disporting together. The approach to the open waters is difficult in most places, from the rank growths, first of shrubbery, and next of reeds, that fringe the open banks; in other places, where the stream narrows in precipitous gorges, from the almost inaccessible rocks. But these difficulties overcome, it is a pleasant sight to see the birds before us—perhaps within a few paces, if we have very carefully crawled through the rushes to the verge—fancying themselves perfectly secure. Some may be quietly paddling in and out of the sedge on the other side, daintily picking up the floating seeds that were shaken down when the wind rustled through, stretching up to gather those still hanging, or to pick off little creatures from the seared stalks. Perhaps a flock is floating idly in midstream, some asleep, with the head resting close on the back and the bill buried in the plumage. Some others swim vigorously along, with breasts deeply immersed, tasting the water as they go, straining it through their bills, to net minute insects, and gabbling to each other their sense of perfect enjoyment. But let them appear never so careless, they are quick to catch the sound of coming danger and take alarm; they are alert in an instant; the next incautious movement, or snapping of a twig, startles them; a chorus of quacks. a splashing of feet, a whistling of wings, and the whole company is off. He is a good sportsman who stops them then, for the stream twists about, the reeds confuse, and the birds are out of sight almost as soon as seen.

DISTRIBUTION.

Breeding range.—Western North America and southern South America. In North America east to western Montana (Missoula County), eastern Wyoming (Lake Como), southwestern Kansas (Meade County), and south central Texas (Bexar County). South to southwestern Texas (Marathon), northern Mexico (Chihuahua), and northern Lower California (San Rafael Valley). West to practically all the central valleys of California, central Oregon (Paulina Marsh), and northwestern Washington (Tacoma). North to southern British Columbia (Revelstoke, Okanogan and Chilliwack). In South America, from central Argentina (Buenos Aires) south to the Falkland Islands, and from the Straits of Magellan north in the Andes to central Peru (Santa Luzia).

Winter range.—Southwestern North America and central South America. In North America east to southern Texas (Brownsville). South to south central Mexico (Jalisco and Puebla) and perhaps farther; has occurred in Costa Rica. North to central California (Stockton), southern Arizona (Tucson), central New Mexico, and probably southwestern Texas. In South America south to central Patagonia (Senger River) and southern Chile (Chiloe Island). North to southern Brazil (Rio Grande de Sul), southern Paraguay, Bolivia (Lake Titicaca), Peru (Corillos), and rarely to Ecuador (Quito) and

Colombia (Bogota and Santa Marta). These latter records may
have been stragglers from North America.

Spring migration.—Early dates of arrival: Nevada, Ash Meadows,
March 18; Idaho, Grangeville, April 11; British Columbia, Chilli-
wack, April 22; Colorado, Beloit, March 23, Loveland, April 13, and
Lay, April 20; Missouri, Lake City, April 15; Nebraska, Omaha,
April 10; Wyoming, Lake Como, May 5. Late date of departure:
Lower California, Colnett, April 8.

Fall migration.—Withdrawal from the northern portions of the
breeding range begins in September and is completed by the middle
of October. A late northern record is, North Dakota, Mandan,
October 10.

Casual records.—Has wandered on migrations as far east as
Alberta (Edmonton, May 12, 1917), Manitoba (Oak Lake), Wisconsin
(Lake Koshkonong, October 18, 1879, and October 9, 1891), Ohio
(Licking County Reservoir, April 4, 1895), New York (Seneca Lake,
about April 15, 1886), South Carolina (a somewhat doubtful record),
Florida (Lake Iamonia and Key West), and Louisiana (Lake Pont-
chartrain).

Egg dates.—California: Thirty-seven records, April 18 to July 14;
nineteen records, May 14 to June 17. Colorado and Utah: Forty-
two records, May 3 to July 8; twenty-one records, May 15 to June
3. Oregon and Washington: Thirteen records, May 8 to June 13;
seven records, May 26 to June 2.

CASARCA FERRUGINEA (Pallas).

RUDDY SHELDRAKE.

HABITS.

The fact that this Old World species has been taken several times
as a straggler in Greenland constitutes its slim claim to be included
in the list of North American birds. Its center of abundance seems
to be in eastern Europe and Asia.

Nesting.—Yarrell (1871) says of its breeding habits:

The ruddy sheld duck makes its nest in a hole; sometimes in the middle of a
cornfield or in a marmot's burrow on the plains; at others, in clefts of precipitous
rocks, as in Algeria and in Palestine, where Canon Tristram, found nests amongst
those of griffon vultures, etc. In southern Russia hollow trees are said to be selected,
the male bird keeping watch on a branch while the female is sitting; felled hollow
logs and deserted nests of birds of prey are also utilized; and, according to Colonel
Prjevalsky, the female sometimes lays her eggs in the fireplaces of villages abandoned
by the Mongols, becoming almost black with soot while sitting.

Eggs.—The ruddy sheldrake is said to lay from 8 to 16 eggs, but
probably the smaller numbers are commoner. The color is described
as white, creamy white or tinged with yellowish. A set of eight eggs
in my collection is nearly pure white in color; they vary in shape

between oval and elliptical oval; and the shell is smooth, with very little luster. The measurements of 71 eggs, given in Witherby's Handbook (1920) average 67 by 47 millimeters; the eggs showing the four extremes measure **72** by 49, 68.8 by **49.5**, **61.5** by 45.6 and 65 by **45** millimeters.

Young.—The period of incubation is said to be from 29 to 30 days. Yarrell (1871) says:

The male does not share the task of incubation, but afterwards he is very assiduous in his attentions to the young. The female is said to carry the nestlings to the water.

Plumages.—The same writer says on this subject:

A nestling from the Volga, in the collection of Mr. E. Bidwell, is dull white on the forehead, cheeks, and entire under parts; the crown of the head to the eye, nape, and back, brown, with broad streaks of white on the inner side of each pinion and on each side of the center of the rump. The young of the year are like the female, but rather duller in color; the inner secondaries and scapulars are brown, marked with rufous; and the wing coverts are grayish white.

Food.—Morris (1903) says of the food of this species:

They feed early in the morning, and again toward nightfall, in corn and stubble fields, resorting thither from the marshes, which they otherwise inhabit. Their food, water plants, water insects and their larvæ, worms, and the roe and young fry of fish.

Behavior.—Referring to its habits he writes:

These birds assemble in flocks, except when paired in the summer. They seem not to associate with other species. They are difficult to be tamed, but have been kept for ornament, and have even been known to breed in confinement, on being provided with burrows in the earth for the purpose. The male and female seem much attached to each other. They are very shy and restless birds.

Yarrell (1871) says:

The call note, when uttered on the wing, is described by Pallas as resembling a clarionetlike *a-oung*, whence the name of *Aangir* given to the bird by the Mongols, who hold it sacred; and *Ahngoot*, by the natives of the vicinity of Lake Van, in Armenia. According to a Hindoo legend, as given by Jerdon, the birds represent two lovers talking to each other across a stream at night—"Chakwa, shall I come? No, Chakwi. Chakwi, shall I come? No, Chakwa." In confinement the note is a sort of *kape* or *ka*, several times repeated. In its manner of walking this species resembles a goose, and it feeds in a similiar manner, grazing in the fields of young corn and picking up seeds of grass, grain, etc. In summer the birds go in pairs, but at other times they are gregarious, and Jerdon says that on the Chilka Lake he has seen thousands in one flock in April.

DISTRIBUTION.

Breeding range.—Mainly in southeastern Europe and central Asia East to Manchuria and China. South to the plateau of Tibet, Persia, and rarely to Algeria and northern Morocco. West rarely to southern Spain; more regularly to the Adriatic Sea. North to Roumania, Bulgaria, Macedonia, southern Russia and Siberia, Lake Baikal, and Mongolia.

Winter range.—Resident over much of its breeding range. East in winter to Japan. South to Formosa, Ceylon, India, southern Arabia, Egypt, Abyssinia, Sahara, Algeria and Morocco.

Casual records.—Wanders to Scandinavia, Great Britain, Iceland, and Greenland.

Egg dates.—Southern Russia: Eight records, May 7 to June 1.

TADORNA TADORNA (Linnaeus.)

SHELD DUCK.

HABITS.

Here we have the latest addition to the American list of ducks, the common sheld duck of Europe, which has recently been taken on the coast of Massachusetts. Mr. Albert P. Morse (1921) has recorded the important event, as follows:

An example of the common sheld duck, a female, was killed October 5, 1921, by Capt. Howard H. Tobey, of Gloucester, in Ipswich Bay off Annisquam, not far from the mouth of the Essex River. Through the kind efforts of Mr. Carl E. Grant, game warden at Gloucester, the specimen was secured for the Peabody Museum of Salem, and identified by State Ornithologist Forbush, who has reported its occurrence to the Auk. It has been mounted by J. W. Goodridge, of South Hamilton, and now adds interest to the Essex County collection of the Peabody Museum. The bird was described as being extremely wild, and its plumage showed no signs of the wear and tear or soilure indicative of captivity, so that this specimen can properly be regarded as a wanderer from the Old World.

Nesting.—Yarrell (1871) refers to the nesting habits of this species as follows:

The sheld duck breeds, as already stated, in some kind of burrow, which often describes an imperfect circle, the nest being sometimes 10 or 12 feet from the entrance. It is composed of bents of grass and is gradually lined, during the progress of laying, with fine soft down, little inferior to that of the eider duck and collected in some places for its commercial value. The eggs are of a smooth, shining white, and measure about 2.75 by 1.9 inches. The nest may sometimes be discovered by the print of the owner's feet on the sand, but the wary bird will often fly straight into the entrance without alighting outside. The old bird is sometimes taken by a snare set at the mouth of the burrow, and the eggs being hatched under domestic hens, the birds thus obtained are kept as an ornament on ponds.

On the North Frisian Islands, according to Mr. Durnford, the natives make artificial burrows in the sand hillocks, and cut a hole in the turf over the passage, covering it with a sod, so as to disclose the nest when eggs are required. There are sometimes as many as a dozen or 15 nests in one hillock within the compass of 8 or 9 yards. The eggs are taken up to the 18th of June, after which the birds are allowed to incubate: but the nest is never robbed of all the eggs. Naumann, who had already given a similar account of the way in which these birds are farmed in the island of Sylt, states that if no eggs are taken the same bird never lays more than 16; but if the first 6 eggs are left, and all those subsequently laid are taken, she will continue laying up to 30. Some German authorities state that nests have been found in the "earths" of the fox and the badger.

Eggs.—Macgillivray (1852) describes the eggs as follows:

The eggs, from 8 to 12, are of an oval form, rather pointed at one end, smooth glossy, and thin shelled, of a white color, slightly tinged with reddish, their length from 2$\frac{11}{12}$ inches to 2$\frac{8}{12}$ inches and their breadth an inch and ten or eleven twelfths. The male continues in the neighborhood of the nest during incubation, and is said occasionally to take the place of the female.

Witherby's Handbook (1920) gives the number of eggs as normally 8 to 15, but as many as 16, 20, 28, and even 32 have been recorded. The measurements of 100 eggs, therein recorded, average 65.7 by 47.3 millimeters; the eggs showing the four extreme measure **71** by 48.8, 69 by **50**, **60** by 44, and 62.8 by **43.3** millimeters.

Young.—The period of incubation is given by various writers as from 24 to 30 days or about 4 weeks. It is said to be performed mainly by the female, but apparently partially by the male as well. Bewick (1847) writes:

During this time the male, who is very attentive to his charge, keeps watch in the daytime on some adjoining hillock, where he can see all around him, and which he quits only, when impelled by hunger, to procure subsistence. The female also leaves the nest for the same purpose in the mornings and evenings, at which times the male takes his turn and supplies her place. As soon as the young are hatched, or are able to waddle along, they are conducted, and sometimes carried in the bill, by the parents to the full tide, upon which they launch without fear, and are not seen afterwards out of tide mark until they are well able to fly; lulled by the roarings of the flood, they find themselves at home amidst an ample store of their natural food, which consists of sand hoppers, sea worms, etc., or small shellfish, and the innumerable shoals of the little fry which have not yet ventured out into the great deep but are left on the beach or tossed to the surface of the water by the restless surge.

If this family, in their progress from the nest to the sea happen to be interrupted by any person, the young ones, it is said, seek the first shelter, and squat close down, and the parent birds fly off, then commences that truly curious scene, dictated by an instinct analogous to reason, the same as in the wild duck and the partridge; the tender mother drops, at no great distance from her helpless brood, trails herself along the ground, flaps it with her wings, and appears to struggle as if she were wounded, in order to attract attention, and tempt a pursuit after her. Should these wily schemes, in which she is also aided by her mate, succeed, they both return when the danger is over, to their terrified motionless little offspring, to renew the tender offices of cherishing and protecting them.

Food.—Mr. John Cordeaux (1898) says of its feeding habits:

As far as my own observation goes, on the Lincolnshire coast, the sheld duck appears to live exclusively on various mollusca and crustaceans; the stomach is remarkable for its very thick and strong muscular coat, capable of digesting any tough morsel. In the stomach of one I found some sand and many small shells of the genus *Buccinum.* The late Mr. Thompson opened the stomachs of 10 shot in Belfast Bay and took from one of them 9,000 specimens of *Skenea depressa* and *Montacuta purpurea,* and about 11,000 others, making a total of 20,000 shells *in the crop and stomach of a single sheld duck.* Mr. St. John says: "Its food appears to consist almost wholly of small shellfish, and more especially of cockles, which it swallows whole. It extracts these latter from the sand by paddling or stamping with both its feet; this brings the cockle quickly to the surface. I have often seen the tame birds of this species do the same in the poultry yard when impatient for or waiting for their food."

Witherby's Handbook (1920) gives the following list of food:

Chiefly mollusca (*Buccinum, Paludina, May, Skenea, Tellina, etc.*) crustacea (shrimps, prawns, and small crabs), with a small quantity of vegetable matter (algae and fragments of gramineae) and occasionally insects (*Carabus* and larvae of diptera.)

Behavior.—Macgillivray (1852) writes of the habits of the sheld duck, as follows:

It seems to continue in pairs all the year round, although frequently in winter and spring large flocks may be seen in which the families are intermingled. I have never met with it inland, or in fresh water near the coast; but have seen it feeding in wet pastures near the sea, although more frequently on wet sands, and am unable, from my own observation, to say of what its food consists. Various authors allege that it feeds on shellfish and marine plants; but this, judging from the structure of its bill and its general appearance, I felt inclined to doubt until I met with Mr. Thompson's statement. It walks with ease, in the manner of the wild geese, but with quicker steps, and flies with speed, in the manner of the mallard and other ducks, with more rapid beats of the wings than the geese. In spring and the early part of summer it has a habit of erecting itself, thrusting forward its neck, and shaking its head, as if endeavouring to swallow or get rid of something too wide for its gullet; but this appears to be merely an act of attention to the female. Being shy and vigilant, and frequenting open places, it is not easily approached unless when breeding.

Mr. Cordeaux (1898) quotes Mr. G. H. Caton Haigh as saying:

It is an extremely common bird on the coast of Merionethshire both as a winter visitor and a breeding species. In the former season it appears in flocks about the latter end of November, the numbers are very variable, but in severe weather is sometimes present in immense quantities. At such times it frequents the open sands, particularly in the estuaries, in company with widgeon and mallard. It is (excepting geese) the most wary of all the fowl, and will frequently not allow a punt to approach within 300 yards. In February another large increase takes place, when the breeding birds return to their summer haunts, and from thence to September they are one of the most numerous birds on the shore. From the middle of October to the end of November the sheld duck is entirely absent from the coast. The first clutches of young generally appear about the end of May or early in June, and heavy weather at this time produces great mortality amongst them. The old remain with the young for a very short time, and young broods are often to be seen alone, or with 40 to 50 young and one pair of old birds.

During winter the sheld duck feeds at night, but in summer it feeds at low water both during the day and night. Large numbers of nonbreeding birds spend the summer on the coast. They are very noisy birds, and the harsh quack or laugh of the female, and whistle of the male, is heard both day and night in spring, and there is much fighting amongst the males at this season. It is a poor diver, and rarely goes under water, even when wounded.

He writes further in regard to it:

The sheld duck is heavier and stands higher than the mallard, and it is much more a goose in manner than a duck, having an erect carriage and light active step, instead of waddle; their flight too, more resembles that of geese and swans. The young are so active that it is almost impossible to catch them. In winter, not unfrequently, great numbers visit the Lincolnshire coast, particularly in those seasons when a grain ship is wrecked and broken up on some of the outlying sand banks, at which time ducks congregate in large numbers from all parts to the feast. I have, at this season, known flocks of two to three hundred sheld ducks to be seen off the coast.

With us the sheld duck is in all seasons of the year inseparably connected with one of its most favorite haunts, the dreary flat coast of Lincolnshire, where the sea, at the ebb of spring tides, recedes for miles, and is scarcely visible from the dune except by a far-away glimmer along the horizon, or, if there is any *breeze*, by that long checkered line of black and white, like the squares of a chessboard, rising and falling alternately, in almost rhythmical pulsations, as the breakers on the sand banks flash into light or recoil into deep shadows.

DISTRIBUTION.

Breeding range.—Temperate portions of Europe and Asia. East to eastern Siberia. South to Mongolia, southern Siberia, Turkestan, Caspian and Black Seas, France, and Spain. West to the British Isles. North to 70° N. in Norway and 51° N. in the Ural Mountains.

Winter range.—Southern Europe and Asia. East to Japan. South to Formosa, China, Burma, northern India, Egypt, and northern Africa to the Tropic of Cancer. West to the British Isles. North to the Mediterranean basin and the Black and Caspian Seas.

Casual records.—A straggler in the Faeroes and Iceland. One record for North America (Ipswich Bay, Massachusetts, October 5, 1921).

SPATULA CLYPEATA (Linnaeus).

SHOVELLER.

HABITS.

The little shoveller is one of the best known and the most widely distributed ducks in the world; by its peculiar spatulate bill and by the striking color pattern of the drake it is easily recognized; it is universally common over nearly all of the continents of North America, Europe, and Asia, wandering south in winter to northern South America and Africa and even to Australia. It is essentially a fresh-water duck at all seasons, never resorting to the seacoasts except when forced to by stress of weather; it is a bog-loving species, fond of inland sloughs, marshes, streams, and ponds, where it can dabble in the shallows like a veritable mud lark. It is always associated in my mind with the shallow pond holes and sluggish creeks which are so characteristic of the wet, grassy meadows of the prairie region, where pairs of these handsome birds are so frequently seen jumping into the air, surprised by a passing train or wagon.

Spring.—The shoveller is not a hardy bird and is therefore not an early migrant in the spring; it comes along with the gadwall and the baldpate after the ice has entirely left the sloughs. The migration in the south is well under way before the end of March, but they do not wholly disappear from Louisiana until early in May and the first arrivals do not reach northern Alaska until about the middle of May. On the spring migration the birds are in small flocks, frequenting the ponds and rivers, usually not associating much with other species. Soon after their arrival on their breeding grounds they spread out among the sloughs, creeks, and marshes, breaking up into pairs or small parties of three or four.

Courtship.—The courtship of the shoveller does not amount to very much as a spectacular performance; Millais (1902) describes it, as follows:

The spring courtship on the part of the male shoveler is both quiet and undemonstrative, nor does his ladylove betray any particular emotion. He swims slowly up to her, uttering a low guttural croak, like the words konk, konk, and at the same time elevating his head and neck and jerking his bill upwards. The female then bows in recognition, and both proceed to swim slowly round in circles, one behind the other, with the water running through their bills.

A somewhat unusual circumstance in the matrimonial arrangements of this duck is the prevalence of polyandry where circumstances seem to call for it, and the amiability with which it is accepted by the united drakes. As a rule, where the sexes are equal in a breeding haunt the male and female pair and keep together in the usual way; but where there is a preponderance of males it is quite common to see a female with two males constantly in attendance, and these two husbands will remain with her, apparently in complete amity, until she has commenced to sit. The custom is, of course, quite common in the case of mallards, but with them there is a certain amount of jealousy on the part of the males, either of whom will drive off and, if possible, keep away altogether, his marital partner. Somewhat remarkable, too, is the fact that after two adult shovelers have paired, the additional male is generally a bird of the previous year whose plumage is only partially complete. Possibly this may be due to the misfortune of the young Lothario, who, finding that most of the young females of the previous year have gone off by themselves and will not pair, must content himself with such favor as he may find with an older and already mated bird. Certainly, on Loch Spynie, in the month of May, I have seen quite as many trios as pairs of shovelers, and in nearly every case the third bird was in immature plumage.

It also indulges in spirited courtship flights, in which two males often pursue a single female in an aerial love chase, exhibiting their wonderful powers of flight with swift dashes and rapid turnings until one of the males finds himself outclassed.

Nesting.—In North Dakota in 1901 we found the shoveller evenly distributed everywhere, one of the commonest ducks, frequenting the same localities as the blue-winged teal and equally tame. We saw them frequently flying about in pairs, up to the middle of June, from which I inferred that their sets were not complete until about that time. In that region the nesting ground of the shoveller was the broad expanse of virgin prairie, often far away from the nearest water, sometimes on high dry ground and sometimes in moist meadow land or near a slough or pond. The first nest that we found was in the center of a hollow in the prairie between two knolls, where the ground was moist but not actually wet, and where the grass grew thick and luxuriantly. The nest was well hidden in the thick, green grass, so that we never should have found it if we had not flushed the bird within 10 feet of us. It was merely a depression in the ground, well lined with dry grasses, and sparingly lined with gray down around the eggs; more down would probably have been added as incubation advanced. The 10 eggs which it contained were perfectly fresh when collected on June 3.

The second nest was found on June 7 while driving across the prairie in Nelson County. We had stopped to explore an extensive tract of low "badger brush," looking for the nest of a pair of short-eared owls which were flying about, as if interested in the locality. We were apparently a long distance from any water, and while returning to our wagon over a high dry knoll, flushed the duck from her nest, which was only partially concealed in the short prairie grass. The slight hollow in the ground was lined with dead grasses and a plentiful supply of down. It contained 11 eggs which were too far advanced in incubation to save. Although the shoveller frequently breeds in open and exposed situations at a long distance from water, I think it prefers to nest in the rank grass around the boggy edges of a slough or pond.

In southwestern Saskatchewan in 1905 and 1906 we found shovellers everywhere abundant, breeding on the islands, on the meadows near the lakes, and on the prairies. On that wonderful duck island in Crane Lake, on June 17, 1905, we found 7 nests of the shoveller—2 with 8 eggs, 1 with 9, 2 with 10, and 2 with 11; the nests were located in the long grass and under rosebushes, scattered indiscriminately among the nests of mallards, gadwalls, baldpates, green-winged and blue-winged teals, pintails, and lesser scaup ducks; this island has been more fully described under the gadwall. The nests were very much like those of the other ducks, hollows scooped out in the ground, sparingly lined with dry grass and weeds and surrounded by a rim of down; as incubation advances the supply of down increases until there is enough to cover the eggs when the duck leaves the nest.

I believe that the above-described nests illustrate the normal nesting habits of the shoveller, but Mr. Edward Arnold (1894) records a nest which "was built in a heavy patch of scrub poplars," in Manitoba. Mr. W. Otto Emerson (1901) thus describes a nest which he found in California in an exceedingly exposed situation in a salt marsh:

After working over the marsh for several hours I started back and when half way across I again saw a pair of ducks headed inland, but thought nothing of it until a single duck started up 10 feet from me and 300 yards from the mainland. On going to the spot there lay a nest in open sight on the bare ground among the saltweed. It was not over 4 inches off the ground and contained 14 eggs. The nest was composed of dry stems of the saltweed, lined with down and a few feathers from the parent bird, and measured 14 inches across the top with a depth of 5 inches.

The down in the shoveller's nest is larger than that of the teals, but smaller and darker than that of the pintail; it varies in color from dark "drab" to light "hair brown," with large grayish-white centers. The breast feathers in the nest are quite distinctive; they have large rounded gray centers, with broad buff and white tips and margins.

Eggs.—The shoveller is said to lay from 6 to 14 eggs, but the set usually consists of from 10 to 12 eggs. Only one brood is normally raised. In color and texture the eggs are strikingly like those of the mallard and pintail; I have never been able to detect any constant difference between the three in these respects, the individual variations in all three overlapping; but the shoveller's eggs are, of course, smaller and usually more elongated. In shape they are nearly elliptical ovate or elliptical oval. The color varies from a very pale olive buff to a very pale greenish gray. The shell is thin and smooth, with very little luster.

The measurements of 177 eggs in various collections average 52.2 by 37 millimeters; the eggs showing the four extremes measure **58** by 38.5, 54.5 by **39, 48** by 37, and 50.5 by **34.5** millimeters.

Young.—Morris (1903) gives the period of incubation as "three weeks"; others give it as from 21 to 23 days. Incubation is performed entirely by the female, though the male does not wholly desert her during the first part, at least, of the process and is often quite solicitous if the nest is disturbed. But before the broods are hatched the males congregate in small flocks in the sloughs and ponds, leaving the care of the young to their mates. The young are led to the nearest water by the female, carefully guarded and taught to feed on insects and soft animal and vegetable food. The young are expert divers; we had considerable difficulty in catching what specimens of grown young we needed. By the time that the young are fully fledged, the molting season of the adults is over, and the old and young birds are joined together in flocks.

Plumages.—Even when first hatched the young shoveller's bill is decidedly longer and more spatulate than that of the young mallard, and it grows amazingly fast, so that when two weeks old there is no difficulty in identifying the species. The color of the downy young above varies from "olive brown," or "sepia," to "buffy brown," darker on the crown, which is "clove brown" or "olive brown"; the color of the back extends far down onto the sides of the chest and on the flanks. The under parts vary from "maize yellow" or "cream buff" to "cartridge buff" or "ivory yellow"; this color deepens to "chamois" on the cheeks. There is a stripe of "olive brown" through the eye, including the loral and postocular region, also an auricular spot of the same. There is a light buffy spot on each side of the back, behind the wings, and one on each side of the rump. The buffy or chamois colored stripes above the eyes are well marked and often confluent on the forehead. All of these colors fade out to paler and grayer shades as the bird grows older.

The flank feathers are the first to appear, then the mottled feathers of the breast and belly, together with the scapular and head plumage, then the tail and lastly the wings. Birds in my collection, as large as blue-winged teal, collected July 17 and 18 in Manitoba, are still downy on the back of the neck and rump, with the wing quills just bursting the sheaths; they evidently would not be able to fly until fully grown in August. In this first plumage the sexes are alike, but the male is slightly larger.

Millais (1902) gives the following full account of the progress toward maturity:

By the middle of September we see the molt beginning, and from this date till the following February there is no surface-feeding duck whose plumage change progresses so slowly. In its ordinary course there is little difference between September and January, but toward the end of the latter month a big flush of new feathers takes place, either on the whole of the breast down to the vent, or amongst the feathers of the lower neck, where a few pure white feathers appear. In very advanced birds the molt extends over the whole of the lower neck and breast. By the middle of March, numbers of the dark-green feathers begin to show themselves on the cheeks, and in April there is an accession of white feathers on the scapulars. In May and June the whole plumage continues to trend toward maturity, and many new feathers which have come in the plumage on the scapulars and sides of the neck are changing color all the time, from a half compromise with the old first plumage to that of the adult bird. Nevertheless, the whole bird can not be said to be anything like complete, and still undergoes feather recoloration and molt until the full and complete molt of the eclipse takes place at the beginning of July.

The young drake then molts the wings for the first time in August, and, passing through the usual autumnal color change and molt, arrives at a plumage dull and incomplete, yet resembling that of the adult male. Thus we see that in gaining adult dress, this bird takes the same time as the widgeon, namely, about 17 months. His plumage, however, so far as my experience goes, is never absolutely perfect until the third season. In that year his full breeding dress seems to attain perfection earlier than at any previous season. Amongst those that I have kept in confinement from immaturity the bill seemed blacker, and all the colors of the plumage more brilliant, when they reached this age. Male shovelers of 21 months old generally have a number of arrowhead-brown bars on the sides of the white breast shield and upper scapulars. The presence of these broad-arrow marks on the white chest must, however, not be taken as indisputable evidence of immaturity, for many perfectly adult males retain year after year one or two of these markings, whilst others have a wholly white shield. It will nevertheless be found that these markings, together with a sandy-edged breast, are constant signs of difference between the young and the old males; for in the first spring the immatures of all the surface feeders, except the mallard, whose appearance is largely due to condition and feeding, always lack the color, size, and finish of the perfectly adult drake.

Similar changes take place in the young female, a complete new dress being acquired, except on the wings, by January, in which young birds can be distinguished from old birds by their dark shoulders and wings; the fully adult dress is not acquired until the following October or at the age of 17 months. Millais (1902) also says that the immature females do not breed during their first spring.

The midsummer eclipse plumage of the male is quite complete, and closely resembles the female plumage, except for the wings, which, of course, are molted only once in August, and for the breast and belly, which remains largely brown. The molt into the eclipse begins about the 1st of July and the change is very rapid. The molt out of the eclipse in the fall is more protracted; it sometimes does not begin until the middle of October and is not complete until December or later. Adult males can always be recognized by the wings.

Food.—In feeding the shoveller uses its highly specialized bill to advantage. All the surface-feeding ducks have the edges of the upper and lower mandibles more or less well supplied with rows of comblike teeth or lamellae through which the water and mud is sifted to obtain food; in some species these are somewhat rudimentary, but in the shoveller they reach their highest development because the shoveller is more essentially a surface feeder than any other duck, dabbling along the surface to sift out what small particles of food it can find, shovelling in the soft muddy shallows and straining out its food much after the manner of a right whale. The tongue, the roof of the mouth, and the soft edges of the broad bill are all well supplied with sensitive nerves of touch and taste, which helps the bird to retain what it wants to eat and to reject worthless material. The shoveller seldom tips up to feed by semi-immersion, but paddles quickly along, skimming the surface, with its head half submerged so that whatever is found is taken into the mouth, tasted by the sensitive tongue, and sifted out through the pectinated bristles of the bill if not wanted.

Millais (1902) relates the following incident to illustrate the activity of the shoveller in feeding:

To the observer who sees the shoveler casually by day he appears to be somewhat of a lethargic nature; but, when he cares to do so, he can move faster on the water than any of the fresh-water ducks. I have watched with pleasure the wonderful sight, calculation, and quickness of a male shoveler that I once kept in confinement on a small marshy pond at Fort George. About the last week in April a certain water insect, whose name I do not know, would "rise" from the mud below to the surface of the pool only to be captured by the shoveler, who, rushing at full speed along the water, snapped up the beetle the moment it came to the surface. How it could see the insect in the act of rising I could never make out, for it was invisible to me standing on the bank above, and I could only just catch a glimpse of it as the shoveler reached his prey and dexterously caught the beetle as it darted away again. After each capture the duck retired to the side of the pool again and there awaited the next rise—commonly about 25 feet away. While thus occupied he seemed to be in a high state of tension; the feathers are closely drawn up and he kept his neck working backwards and forwards, in preparation, as it were, for the next spring, exactly like a cat "getting up steam" for the final rush on a victim. Sometimes he seemed to get into a frantic state of excitement, darting here and there as if he saw beetles rising in every direction. I noticed also that while devouring his prey the pupils of his eyes were unusually contracted, and the golden circlets seemed to shine more brilliantly than usual.

The food of the shoveller consists of grasses, the buds and young shoots of rushes, and other water plants, small fishes, small frogs, tadpoles, shrimps, leeches, aquatic worms, crustaceans, small mollusks, particularly snails, water insects, and other insects, as well as their larvae and pupae.

Doctor Yorke (1899) adds the following to the list: "Teal moss (*Limnobium*), various water lilies, flags, duck-weeds, and pondweeds."

The shoveller is an exceedingly active flyer; it rises quickly from the water, mounting straight up into the air, and darting off with a swift though somewhat erratic flight. Its flight is somewhat like that of the teals and, like them, it frequently makes sudden downward plunges. It is not shy and shows a tendency to return to the spot where it was flushed. On migrations it flies in small flocks by itself, though in the fall it is often associated with the gadwall, baldpate, or lesser scaup duck. During the mating season it is usually seen flying in pairs, with the male leading, or in trios, with a female leading two males. The shoveller is easily recognized in flight; the striking colors of the drake can not be mistaken; and the females and young can easily be identified by the long slender necks and conspicuously large bills; I have seldom been in doubt when flushing a female shoveller from her nest.

The shoveller has a small throat and a weak voice. It is usually silent, but the female sometimes indulges in a few feeble quacks and the male makes a low guttural sound like the syllables *woh, woh, woh,* or *took, took, took;* this sound has been likened by some writers to the sound made by turning a watchman's rattle very slowly.

Mr. Robert B. Rockwell writes me:

From a good many years of observation as a duck hunter I am of the opinion that the shoveller is one of the most sociable species of wild duck. Single shovellers are very frequently seen in flocks of other species, especially teal, and the ease with which individuals and even good-sized flocks of these birds are decoyed is in itself good evidence that they are of a sociable disposition.

Throughout our Barr work the drake shovellers during the nesting season were seen in considerable numbers but were seldom seen swimming about alone, nearly always being in company with other species of ducks; nor did they seem to prefer the company of males of their own species particularly.

Fall.—The shoveller is one of the earliest migrants in the fall; the first autumnal frosts, late in August or early in September, are enough to start it drifting along with the blue-winged teal; the migration is well under way by the middle of September, and a month later it is practically over.

Game.—Mr.T. Gilbert Pearson (1916) says:

Shovellers feed mostly at night, especially in places where they are much pursued by gunners. I have often seen dozens of flocks come from the marshes at sunrise and fly out to the open water, far from any place where a gunner might hide. There, if the weather is fair and not too windy, they will often remain until the shades of night and the pangs of hunger again call them back to the tempting marshes. They do not gather in enormous flocks like some other ducks. I have never seen over 40 in one company, and very often they pass by in twos and threes. In hunting them the fowler usually conceals himself in a bunch of tall grass or rushes, on or near the margin of an open pond, and, after anchoring near-by 20 or 30 wooden duck dummies called decoys, sits down to wait the coming of the birds. Sometimes the ducks fly by at a distance of several hundred yards. It is then that the hunter begins to lure them by means of his artificial duck call. *Quack-quack, quack-quack,* comes his invitation from the rushes. The passing birds, unless too intent on their journey to heed the cry, see what they suppose to be a company of mallards and other ducks evidently profiting by a good feeding place, and, turning, come flying in to settle among the decoys. It is just at this moment, with headway checked and dangling feet, that they present an easy mark for the concealed gunner.

Audubon (1840) says: "No sportsman who is a judge will ever pass a shoveller to shoot a canvasback." I can not quite agree with this view for the shoveller never seems to get very fat and, to my mind, its flesh is inferior to that of several others. It lives largely on animal food which does not add to its flavor. Perhaps under favorable circumstances it may become fatter and more palatable.

Winter.—Its main winter range is in the Southern States and Mexico, where it frequents shallow inland waters and rarely is it driven to the coast by severe weather.

DISTRIBUTION.

Breeding range.—Temperate regions in the Northern Hemisphere. In North America east more or less regularly to the west coast of Hudson Bay and the eastern boundary of Manitoba. Casually east to west central New York (Cayuga County). South to northwestern Indiana (Lake County) and northern Illinois formerly; more recently to western Iowa (Sac County), central western Nebraska (Garden County), Kansas (probably locally), northwestern New Mexico (Lake Burford), central Arizona (Mogollon Mountains), and southern California (Los Angeles County); rarely and locally in Texas (Bexar County and East Bernard), and perhaps in northern Mexico. West to the central valleys of California, central Oregon (Tule Lake and Malheur Lake), northwestern Washington (Lake Washington), and central British Columbia (Fraser Valley and Cariboo District). North regularly to central Alberta (Edmonton) and the valley of the Saskatchewan River; irregularly farther north to the Bering Sea coast of Alaska (Kuskokwim River to Kotzebue Sound), the Anderson River region, and Great Slave Lake. In the Eastern Hemisphere it breeds from southern Europe and central Asia northward nearly or quite to the Arctic Circle; and from Great Britain east to Kamchatka and the Commander Islands.

Winter range.—Milder portions of both hemispheres. In America east to the Atlantic coast of southern United States, the Greater (Cuba, Jamaica, Porto Rico) and the Lesser Antilles (St. Thomas, Barbados, Trinidad, etc.). South to northern South America (Colombia). West to the Pacific coast of Central America, Mexico, and United States. North to coast of southern British Columbia (Vancouver and Puget Sound region); in the interior north to central California (Fresno), Arizona, New Mexico, eastern Texas (Galveston), the lower Mississippi Valley, rarely to southern Illinois (Cairo and Mount Carmel), and the coast of Virginia (Cobb Island); has occurred in winter at Lanesboro, Minnesota, and Atlantic City, New Jersey, but is not common north of South Carolina. Winters in Hawaiian Islands. In the Eastern Hemisphere winters south to the Canary Islands, Senegambia, Somaliland, Arabia, India, Ceylon, Borneo, southern China, Formosa, the Philippine Islands, and Australia.

Spring migration.—Early dates of arrival: Alberta, Edmonton, May 1; Mackenzie, Fort Chipewyan, May 7, and Fort Resolution, May 18. Average dates of arrival: Illinois, central, March 23; Iowa, central, March 23; Minnesota, Heron Lake, March 26; North Dakota, central, April 13; Manitoba, southern, April 21. Late dates of departure: Lower California, Colnett, April 8; Rhode Island, Point Judith, April 29.

Fall migration.—Early dates of arrival: Ontario, Beamsville, September 19; Rhode Island, Point Judith, September 24; Pennsylvania, Erie, September 6; Lower California, southern, October 18; Panama, October 16. Late dates of departure: Ontario, Rockland, November 2; New York, Branchport, November 12; Rhode Island, Point Judith, November 7

Casual records.—Accidental in Bermuda (December, 1844) and Labrador (Cartwright, September, 1901). Rare on migrations as far east as Maine.

Egg dates.—Minnesota and North Dakota: Forty records, May 9 to July 3; twenty records, May 31 to June 17. Manitoba and Saskatchewan: Eighteen records, June 1 to July 5; nine records, June 5 to 11. California and Utah: Eighteen records, March 28 to July 11; nine records, May 3 to 21.

DAFILA ACUTA TZITZIHOA (Vieillot).

AMERICAN PINTAIL.

HABITS.

Spring.—Northward, ever northward, clearly indicated on the distant sky, points the long slim figure of the pintail, in the vanguard of the spring migration, wending its way toward remote and still frozen shores. Vying with the mallard to be the first of the surface-feeding ducks to push northward on the heels of retreating winter, this hardy pioneer extends its migration to the Arctic coast of the continent and occupies the widest breeding range of any North American duck, throughout most of which it is universally abundant and well known.

Prof. George E. Beyer (1906) says that, in Louisiana, "winter visitant individuals, as with similar individuals of the mallard, move northward very early, probably never later than the middle of January," whereas the spring transients in that State "are the latest of all the ducks except the teals and the shoveller." This accounts for the two distinct flights of pintails with which gunners are familiar. Dr. F. Henry Yorke (1899) recognizes three distinct flights; he says:

The spring migration above the frost line commences with the first breaking up of winter; the ducks follow the open pools of water to be found in sloughs, lakes, and rivers, and with the yellow-leg mallard are the first of the nondivers to start for their northern nesting grounds. They arrive in three distinct issues, the first leaving, in bulk, at least, before the second arrives; these stay about a week before they proceed northward. An absence of pintails, for three or four days, generally follows before the third issue puts in an appearance, which stay a week or 10 days, according to the weather, then travel northward, breeding chiefly south of the Canadian line.

Mr. Edmonde S. Currier (1902) says of its arrival in Iowa:

If the great break-up of the ice comes late in the season, as the first week in March, which often happens after a severe winter, we find the eager sprigtails (*Dafila acuta*), and the first flight of mallards coming up, and then there is a bird life worth seeing. Although the number of ducks that pass here is rapidly falling off, still thousands are left.

The first flight of pintails is, with us, the greatest, and they always appear while the ice is running. Several days before the ice gives way an occasional flock will come up and circle around over the frozen river as if taking observations, and then disappear to the south. If a rain comes before the ice goes out, and forms pools in the bottom-land corn fields, they will settle in these until the rivers open, or a cold wave strikes us.

The pintail reaches its breeding grounds in northern Alaska early in May and sometimes before the end of April, while winter conditions are still prevailing. Dr. E. W. Nelson (1887) says:

One spring a small party was found about a small spring hole in the ice on the seashore the first of May, while a foot of snow still covered the ground and the temperature ranged only a few degrees above zero. As snow and ice disappear they become more and more numerous, until they are found about the border of almost every pool on the broad flats from the mouth of the Kuskoquim River north to the coast of Kotzebue Sound.

Courtship.—The courtship display of the shy pintail is not often seen, for even on their remote northern breeding grounds the males are ever alert and are not easily approached. The performance resembles that of the teals, where several drakes may be seen crowding their attention on a single duck, each standing erect on the water proudly displaying his snowy breast, with his long neck doubled in graceful curves until his bill rested upon his swelling chest and with his long tail pointed upwards; thus he displays his charms and in soft mewing notes he woos his apparently indifferent lady love until she expresses her approval with an occasional low quack.

A more striking form of courtship, and one more often seen, is the marvelous nuptial flight, which Doctor Nelson (1887) has so well described as follows:

Once, on May 17, while sitting overlooking a series of small ponds, a pair of pintails arose and started off, the male in full chase after the female. Back and forth they passed at a marvelously swift rate of speed, with frequent quick turns and evolutions. At one moment they were almost out of view high overhead and the next saw them skimming along the ground in an involved course very difficult to follow with the eye. Ere long a second male joined in the chase, then a third, and so on until six males vied with each other in the pursuit. The original pursuer appeared to be the only one capable of keeping close to the coy female, and owing to her dextrous turns and curves he was able to draw near only at intervals. Whenever he did succeed he always passed under the female, and kept so close to her that their wings clattered together with a noise like a watchman's rattle, and audible a long distance. This chase lasted half an hour, and after five of the pursuers had dropped off one by one the pair remaining (and I think the male was the same that originated the pursuit) settled in one of the ponds.

Nesting.—Mr. F. Seymour Hersey says in his notes on this species in northern Alaska:

There is probably no place within the breeding range of this widely distributed duck where it is more abundant than on the stretch of tundra bordering the Bering Sea coast of western Alaska. Almost every little tundra pond will contain a few birds—perhaps a pair or a female and two or three males—and parties of two to five or six are constantly flying from one pond to another.

The pintail very often makes its nest farther from water than any other of the northern breeding ducks, although the greater number nest near the shores of ponds. Before the set is complete, the eggs are covered with down, intermingled with leaves, sticks, dead grass, and mosses, and the female spends the day at a considerable distance from the nest. Incubation begins only when the set is complete. Early in June, 1914, while walking over the tundra some miles back from St. Michael I noticed a few pieces of down clinging to the base of some dwarf willow bushes. It aroused my suspicions and searching among the accumulated dead leaves and moss at the roots of the bush I soon disclosed an incomplete set of pintail's eggs. They were thoroughly concealed and had it not been for the few telltale bits of down would have remained undiscovered. The female later completed this set, and on June 10 the nest held nine eggs. This nest was at least a half mile from the nearest water.

At the mouth of the Yukon on June 17, 1914, two nests were found in the center of some clumps of willows in a marsh. The bushes were growing in a few inches of water through which a heavy growth of coarse grass protruded. About the base of the willows the dead grass of previous years was matted and in this dead grass the nests were made. This was the wettest situation that I ever knew this species to select in the north.

As might be expected of an early migrant, the pintail is one of the earliest breeders; in North Dakota it begins to lay by the 1st of May or earlier and we found that many of the broods were hatched by the first week in June. The nest is placed almost anywhere on dry ground, sometimes near the edge of a slough or pond, sometimes on an island in a lake, but more often on the prairie and sometimes a half a mile or more from the nearest water; it is generally poorly concealed and is often in plain sight. Once, while crossing a tract of burned prairie, I saw a dark object fully half a mile away, which on closer inspection proved to be a pintail sitting on a nest full of half roasted eggs; this was a beautiful illustration of parental devotion and showed that the bird was not dependent on concealment. A deep hollow is scooped out in the ground, which is sparingly lined with bits of straw and stubble, and a scanty lining of down is increased in quantity as incubation advances.

My North Dakota notes describe four nests of this species. The first nest, found on May 31, 1901, was concealed in rather tall prairie grass on the highest part of a small island in one of the larger lakes. On June 15 we found another nest in an open situation among rather sparse but tall prairie grass, which was in plain sight, the eggs being beautifully concealed by a thick covering of down. Another nest was shown to us by some farmers who were plowing up an extensive tract of prairie and had flushed the bird as they passed within a few feet of the nest; they left a narrow strip containing the nest unplowed, but something destroyed the eggs a few days afterwards; this nest was fully half a mile from the nearest water. The fourth nest was on the edge of a cultivated wheat field, near the crest of a steep embankment sloping down into a large slough; the nest was a deep hollow in the bottom of a furrow, 7 inches wide by 4 deep lined with bits of straw and weed stubble, with a moderate supply of down surrounding the eggs; it was very poorly concealed by the scanty growth of weeds around it; the eight eggs, which it contained on June 10, proved to be heavily incubated.

In Saskatchewan, in 1905 and 1906, we recorded 11 nests of pintails, 8 of which were found on one small island on one day, where this species was breeding with large numbers of gadwalls, blue-winged and green-winged teals, shovellers, mallards, baldpates, and lesser scaup ducks. One pintail's nest was prettily located under a wild rosebush among the sand hills near Crane Lake, 1 mile from the nearest creek and 2 miles from the lake.

Mr. Robert B. Rockwell (1911) found two nests of this species, in the Barr Lake region of Colorado, in decidedly exposed situations, which he describes as follows:

The first nest, found May 11, 1907, was probably the most unusually located nest of the pintail on record. It was just a trifle less than 18 feet from the rails of the main line of the Burlington route, over which a dozen or more heavy trains thundered every day, and well within the railroad right of way, where section hands and pedestrians passed back and forth continually. The mother bird had found a cavity in the ground, about 8 inches in diameter and 8 inches deep, and had lined it with grass; and the two fresh eggs which it contained on this date were deposited without any downy lining whatever. The female flushed as we passed along the track about 20 feet distant, thus attracting our attention. A week later (on the 18th) the nest was fairly well lined with down and contained nine eggs, one egg having apparently been deposited each day. On May 24 the nest contained 11 eggs and the parent was much tamer than on the two preceding visits, allowing us to approach to within 15 feet of her, and alighting within 20 yards of us upon being flushed.

Another peculiar nest was found May 30, 1908, containing 11 eggs which hatched during the first week in June. This nest was a depression in a perfectly bare sandy flat without a particle of concealment of any kind. The cavity was located in the most exposed position within hundreds of yards, and was fairly well lined with weed stems, grass, etc. and well rimmed with down. The brooding female was very conspicuous against the background of bare sand, and could be readily seen from a distance of 50 feet or more. This bird was rather wild and flushed while we were yet some distance from the nest.

Mr. Eugene S. Rolfe (1898) records, what I have never seen, a pintail's nest in a wet situation, which is very unusual; he says:

The nesting of the pintail differs little generally from other ducks that select high dry spots among the prairie grass, badger brush, or old stubble; but a young farmer this year piloted me to a clump of thick green bulrushes covering a space as large as a dining table in the midst of a springy bog, and in the center of this, built up 6 inches out of water (18 inches deep) on a foundation of coarse dried rushes, exactly after the manner of the redhead, canvasback, or ruddy, and lined with down, was a veritable nest of the pintail. The female was at home, and permitted approach within 6 feet; and I stood some moments watching her curiously and regretting the absence of my camera before I realized that this was the pintail in a very unusual situation.

The down in the pintail's nest most closely resembles that of the shoveller, but it is larger and darker. It varies in color from "hair brown" to "fuscous" or "clove brown" with whitish centers. The breast feathers mixed with the down are either of the characteristic banded pattern or are grayish brown with a broad white tip.

Eggs.—Only one brood is raised in a season and the number of eggs in the set averages less than with other surface feeding ducks. The set varies from 6 to 12 eggs, but it is usually less than 10. It is unusual to find the eggs of other ducks in a pintail's nest, but as the eggs closely resemble those of some other species, it may be a commoner occurrence than it is supposed to be. Mr. Edward Arnold (1894) records the finding of a golden eye's egg in a pintail's nest in

Manitoba. The eggs closely resemble, in color and general appearance, those of the mallard and the shoveller, but they average smaller than the former and slightly larger than the latter, the measurements overlapping in both cases. In shape they are usually elliptical ovate and the color varies from very pale olive green to very pale olive buff, which fades out to a mere tint.

Although the eggs of the pintail can not be separated with certainty from those of the above two species the nests of all three can usually be identified if a clear view of the female is obtained as she flies from the nest; the female pintail can be distinguished from female mallard by the absence of the purple speculum with its conspicuous white borders and by its long slender form; she can be distinguished from the shoveller by her larger size and her small bill; the female shoveller has a long neck, but a conspicuously large bill; the wing pattern is different, but the difference is difficult to detect in the rapidly moving wings of a flying duck.

The measurements of 102 eggs, in various collections, average 54.9 by 38.2 millimeters; the eggs showing the four extremes measure **60** by 38.5, 58.5 by **40.5, 50.5** by 37.2 and 53 by **35** millimeters.

Young.—The period of incubation is about 22 or 23 days and the incubation is performed wholly by the female; she is a very close sitter and is often nearly trodden upon before she will leave the nest; I have heard of one being knocked over with a stick or a plowman's whip as she fluttered off, and it is not a difficult matter to photograph one on her nest. The male does not, I believe, wholly desert the female during the process of incubation and he assists somewhat in the care of the young, though he is not as bold in their defense. The young remain in the nest for a day or so after they are hatched or until the down is thoroughly dried. The whole brood usually hatches within a few hours, for, although only one egg is laid each day, incubation does not begin until the set is complete. As soon as the young are strong enough to walk they are led by their mother to the nearest water, which is often a long distance away, and taught to feed on soft insect and aquatic animal food. I have seen some remarkable demonstrations of parental solicitude by female pintails; they are certainly the most courageous of any of the ducks in the defense of their young. Once in North Dakota as we waded out into a marsh a female pintail flew towards us, dropped into the water near us, and began splashing about in a state of great excitement. The young ducks were probably well hidden among the reeds, though we could not see or hear them. During all the time, for an hour or more, that we were wading around the little slough that pintail watched us and followed us closely, flying about our heads and back and forth over the slough, frequently splashing down into the water near us in the most reckless manner, swimming about in small circles

or splashing along the surface of the water, as if wounded, and often
near enough for us to have hit her with a stick, quacking excitedly
all the time. I never saw a finer exhibition of parental devotion than
was shown by her total disregard of her own safety, which did not
cease until we left the locality entirely. I have had several similar
experiences elsewhere. If alarmed, when swimming in the sloughs,
the young seldom attempt to dive though they can do so, if necessary;
they more often swim into the reeds and hide while the mother bird
attracts the attention of the intruder. Doctor Coues (1874) says
that during July in Montana—

the young were just beginning to fly, in most instances, while the old birds were for
the most part deprived of flight by molting of the quills. Many of the former were
killed with sticks, or captured by hand, and afforded welcome variation of our hard
fare. On invasion of the grassy or reedy pools where the ducks were, they generally
crawled shyly out upon the prairie around, and there squatted to hide; so that we
procured more from the dry grass surrounding than in the pools themselves. I have
sometimes stumbled thus upon several together, crouching as close as possible, and
caught them all in my hands.

Dr. Harold C. Bryant (1914) relates the following incident:

On May 21 a pintail with 10 downy young was discovered on the bank of a pond.
When first disturbed she was brooding her young on dry ground about 10 feet from
the water. The moment she flew the downy young assumed rigidly the same poses
they had variously held beneath the mother. Some were standing nearly erect
whereas others were crouching, but all were huddled close together. They remained
perfectly motionless while, leaving Kendall to watch, I went for the camera. I had
gone over a hundred yards before they moved. By the time I returned they had
wandered off about 10 yards. They marched in single file and every now and then
huddled close together posing motionless for a few moments.

Plumages.—The downy young is grayer and browner than other
young surface-feeding ducks and thus easily recognized. The crown
is dark, rich "clove brown"; a broad superciliary stripe of grayish
white extends from the lores to the occiput; below this the side of
the head is mainly grayish white, fading to pure white on the throat
and chin, with a narrow postocular stripe of "clove brown" and a
paler and broader stripe of the same below it. The back is "clove
brown," darkest on the rump, with grayish or buffy tips on the down
of the upper back; the rump and scapular spots are white, the latter
sometimes elongated into stripes. The lower parts are grayish white,
palest in the center. The chest, and sometimes the sides of the head,
are suffused with pinkish buff, but never with yellow. The colors
become duller and paler as the bird grows older. When the young
bird is about 3 weeks old the first feathers appear on the flanks and
scapulars and the tail becomes noticeable; about a week later
feathers begin to show on the rump, breast, head, and neck, and the
bird is fully grown before its contour plumage is complete; the flight
feathers are the last to be acquired. The length of time required to

complete the first plumage varies greatly in different individuals, but the sequence in which it appears is uniform.

Mr. J. G. Millais (1902) says of the sequence of plumages to maturity:

When in first plumage the young male and female are exceedingly like one another, especially at the commencement of this period; they also resemble the mother to a certain extent, but from her they can be easily distinguished by the small spots which cover the breast and belly, and the narrow brown edge of the feathers on the back and scapulars. The young male pintail, however, like the young mallard drake, almost as soon as he has assumed his first dress commences to color change in the back and scapulars. A gray tinge suffuses the brown plumage and slight reticulations appear on the feathers themselves, rendering it easy to notice the difference between him and the young female. He is also somewhat larger. By the middle of September the usual molt and the more advanced feather changes commence, and sometimes, in birds in a high state of condition, advance so rapidly, that young drakes of the year may attain the full plumage of the adult drake by the beginning of December. Most of them, however, retain a considerable proportion of the brown plumage until February, when the spring flush finishes off the dress. Even then young pintail drakes are not nearly so brilliant as 2 or 3 year old birds, and often show their youthfulness by their shorter tail, dull coloring on the head, and reticulated black bars traversing the white stripes on either side of the neck.

There is considerable individual variation in the length of time required by young birds to throw off the last signs of immaturity, but old and young birds become practically indistinguishable before the first eclipse plumage is assumed and entirely so after it is discarded. Some male pintails begin to show the first spotted feathers of the eclipse plumage early in June and during July the molt progresses rapidly and uniformly over the whole body, head, and neck until the full eclipse is complete in August, and the males are indistinguishable from females except by the wings and the difference in size. The wings are molted only once, of course, in August; and, after the flight feathers are fully grown, early in September, the second molt into the adult winter begins; this molt is usually not completed until November or December, the time varying with different individuals. I have never detected any signs of a spring molt in male pintails, but Mr. Millais calls attention to the fact that females which have pure white breasts in the winter become more or less spotted during the nesting season.

Food.—The pintail is a surface feeder, dipping below the surface only with the fore part of its body, with its tail in the air, maintaining its balance by paddling with its feet, while its long neck is reaching for its food. Here it feeds on the bulbous roots and tender shoots of a great variety of water plants, as well as their seeds; it also finds some animal food such as minnows, crawfish, tadpoles, leeches, worms, snails, insects, and larvae. Dr. F. Henry Yorke (1899) states that it feeds on wheat, barley, buckwheat, and Indian corn. Audubon (1840) says of its animal food:

It feeds on tadpoles in spring and leeches in autumn, while, during winter, a dead mouse, should it come in its way, is swallowed with as much avidity as by a mallard. To these articles of food it adds insects of all kinds, and, in fact, it is by no means an inexpert flycatcher.

Dr. P. L. Hatch (1892) says that, in Minnesota, the pintails may be found in spring "along the recently opened streams, and in the woodlands where they spend much of their time in search of acorns, insects, snails, and larvae of different kinds, which are under the wet leaves and on the old decaying logs with which the forests abound." Mr. Edward A. Preble (1908) found it feeding on small mollusks (*Lymnaea palustris*) in northern Canada, and Mr. F. C. Baker (1889) dissected 15 stomachs in Florida, all of which contained "shells of *Truncatella subcylindrica* (Say)." Mr. Douglas C. Mabbott (1920) sums up the food of the pintail as follows:

Vegetable matter constitutes about seven-eighths (87.15 per cent) of the total food of the pintail. This is made up of the following items: Pondweeds, 28.04 per cent; sedges, 21.78; grasses, 9.64; smartweeds and docks, 4.74; arrow grass, 4.52; musk grass and other algae, 3.44; arrowhead and water plantain, 2.84; goosefoot family, 2.58; water lily family, 2.57; duckweeds, 0.8; water milfoils, 0.21; and miscellaneous vegetable food, 5.99 per cent.

The animal portion, 12.85 per cent, of the food of the pintail was made up of mollusks, 5.81 per cent; crustaceans, 3.79 per cent; insects, 2.85 per cent; and miscellaneous, 0.4 per cent.

Behavior.—The pintail is built on graceful, clipper lines and is well fitted to cleave the air at a high rate of speed; it has been credited by gunners with ability to make 90 miles an hour; this may be rather a high estimate of its speed, but it is certainly very fleet of wing and surpassed by few if any of the ducks. Mr. Walter H. Rich (1907) says:

The pintails flight will at once remind the bay gunner of that of the "old squaw," so well known along the Atlantic coast. The same chain lightning speed and darting and wheeling evolutions are common to both species.

Dr. E. W. Nelson (1887) who had good opportunities for studying this species in Alaska, gives the following graphic account of one of its remarkable flight performances:

During the mating season they have a habit of descending from a great altitude at an angle of about 45,° with their wings stiffly outspread and slightly decurved downward. They are frequently so high that I have heard the noise produced by their passage through the air from 15 to 20 seconds before the bird came in sight. They descend with meteorlike swiftness until within a few yards of the ground, when a slight change in the position of the wings sends the birds gliding away close to the ground from 100 to 300 yards without a single wing stroke. The sound produced by this swift passage through the air can only be compared to the rushing of a gale through tree tops. At first it is like a murmur, then rising to a hiss, and then almost assuming the proportions of a roar as the bird sweeps by.

The pintail can generally be distinguished in flight by its long, slim neck and slender build, which is conspicuous in both sexes; the tail is also more pointed than in other species, even without the long tail feathers of the full plumaged male. The pintail springs upward from the water, much like a teal, and gets under way at once; a flock of pintails flushed suddenly will often bunch together so closely as to give the gunner a chance for a destructive shot.

The pintail is a graceful swimmer, riding lightly on the surface, with its tail pointing upward, its general attitude suggestive of a swan and with its long neck stretched up, alert to every danger, the first to give the alarm and always the first of the shy waterfowl to spring into flight. The hunter must be very cautious if he would stalk this wary bird. Though not a diver from choice, the pintail can dive when necessity requires it. It often escapes by diving while in the flightless stage of eclipse plumage.

Mr. Hersey's notes on this species in Alaska record the following interesting observation:

While the pintail is not a diving duck it can dive readily if wounded and in other emergencies. On one occasion a female followed by two males flew past and I shot the female. She dropped into a nearby pond but when I reached the shore had crawled into the grass and hidden. Circling the pond, which was but 30 or 40 feet in width by about the same number of yards in length, I soon reached my bird. Without hesitation she dove and crossed to the other side under water. The water was fairly clear and not more than 30 inches deep and the bird's movements could be plainly watched. The body was held at an angle, with the neck extended but not straight and the head slightly raised. The wings were partly opened but were not used and the feet struck out alternately as in running rather than with a swimming motion. The bird reminded me of a frightened chicken crossing the road in front of an automobile but the speed was much slower through the water than in the case of the chicken. The bird did not run on the bottom of the pond but was perhaps 6 or 7 inches from the bottom. On reaching the opposite shore she came up directly into the concealment of the grass. This proceeding was repeated in exactly the same manner several times before I secured the bird.

The following incident, described by Mr. Frank T. Noble (1906) will illustrate a strange habit which this and nearly all ducks have of disappearing beneath the surface when wounded; he had shot two pintails, one being—

killed outright, the other, a big drake, being hard hit and with one wing broken. Before the latter could be shot over, he made a dive with considerable difficulty and disappeared from view. We waited perhaps half a minute for him to appear again, but not doing so we paddled to the spot, where we found the water thereabouts to be scarcely 3 feet deep, and the bottom to be thickly covered with various kinds of lily pads and grasses. A few moments of careful search and the duck was discovered on the bottom, grasping with its bill the tough stem of a cowslip. The body of the bird floated upward posteriorly, somewhat higher than the position of the head, and the long tail feathers were a foot or more nearer the surface than the former. The bird's feet were outstretched, but he was motionless until molested, then he kicked and fluttered vigorously, all the time retaining his hold upon the bottom, and it required considerable force to break him away from his queer anchorage.

Mr. J. G. Millais (1902) says that:

The nuptial call of the drake is identical with that of the teal. The female only occasionally utters a low quack, but she sometimes makes a call something like the growling croak of the female widgeon. The notes of both sexes are always quite distinct.

The ordinary note of the male pintail is a low mellow whistle, and I doubt if it ever utters the quacking note which should be attributed to the female; the rolling note, similar to that of the lesser scaup duck, may be common to both sexes; Dr. E. W. Nelson (1887) says that this note "may be imitated by rolling the end of the tongue with the mouth ready to utter the sound of *k*."

The pintail associates freely on its breeding grounds with various species of ducks, particularly with the mallard, gadwall, blue-winged teal, baldpate, shoveller, and lesser scaup duck. It usually flocks by itself, however, on migrations. Its most formidable enemy is man; for with the sportsman the pintail is a favorite. Its eggs are also sought for food, in some localities quite regularly, for the nests are easily found and the eggs are very palatable. Mr. Robert B. Rockwell (1911) has published a photograph of a bull snake robbing a pintail's nest in Colorado. I have seen nests in Saskatchewan which showed signs of having been robbed by coyotes.

Fall.—Although the pintail is one of our earliest migrants in the spring, it seems much less hardy in the fall and is one of the first of the ducks to seek the sunny South as soon as the first frosty nights proclaim the approach of autumn. Doctor Yorke (1899) says of the fall migration:

In the fall migration they differ from other cold-weather birds of the nondivers in returning south before the cold weather sets in; in fact, the first frost finds those which bred in the United States rapidly wending their way toward the frost line. The first issue to come down in the fall usually leaves the northern part of Minnesota and North Dakota about the end of August. They associate a good deal with the baldpates and gadwalls, using the same feeding, roosting, and playgrounds in the fall, not associating with them in the spring owing to their having gone north several weeks before them, and feeding to a large extent upon grain and corn fields. The second fall issue generally overtakes the first before they reach the frost line. They collect in some quiet piece of water, migrate at night and never return that fall. They do not assume their full plumage north of the frost line.

Game.—As a game bird the pintail ranks about third among the surface-feeding ducks, next in importance to the mallard and black duck; its wariness and its swiftness on the wing test the cunning and skill of the sportsman; its wide distribution, its abundance and its excellent table qualities give it a prominent place as a food bird. Late winter and early spring shooting was popular in the Middle West before the laws prohibited it, where the birds arrived early, as soon as the ice began to break up in the marshes and sloughs; here the birds were shot on their morning and evening flights to and from their feeding grounds from blinds or boats concealed in their fly ways, no decoys being necessary.   Pintails will come readily to live mallard decoys during the daytime on their feeding grounds and they will respond to duck calls if skillfully handled, offering very fine sport where they are not shot at too much.

Dr. Leonard C. Sanford (1903) says:

In portions of the West where they frequent the ponds and smaller lakes they are much more easily killed than on larger bodies of water.   The pintail arrives on the coast of North Carolina late in October, and are found in numbers through the brackish sounds.   Decoys attract them occasionally, but never in as large numbers as the other ducks, for they are always wary and quick to suspect danger.   These birds can be distinguished afar.   The white under parts of the male and their long necks mark them at once.   The flight is high in lines abreast, but almost before the flock is seen they are by and out of sight.   When about to decoy no bird is more graceful; they often drop from a height far out of range and circle about the stool, watching carefully for the slightest motion; finally they swing within range and plunge among the wooden ducks.   After realizing the mistake, they spring up all together, and are out of shot almost before you realize the chance is gone.

Winter.—Like many other fresh-water ducks of the interior the pintail winters largely on the warm seacoasts of the Southern States, though it is also abundant among the inland ponds and marshes below the frost line.   It is particularly abundant in Florida, as the following account by Mr. C. J. Maynard (1896) will show:

On one occasion, while I was making my way down Indian River, numbers of these ducks were passing over my head southward.   They flew in straggling flocks, consisting of from twenty to some hundreds of specimens, and one company followed another so closely that there was an almost unbroken line.   They continued to move in this manner all the morning; thus many thousands of individuals must have passed us.   Shortly after noon they began to alight along the beaches in such numbers that they fairly covered the ground, and were so unsuspicious that my assistant, who had left the boat some time previous, walked within a few yards of them, and killed three or four with a single discharge of a light gun which was merely loaded with a small charge of dust shot.   This occurred in early March and the birds were evidently gathering, preparatory to migrating northward, for in a few days they had all disappeared.

While wintering on the seacoast, especially where it is much molested, the pintail often spends the day well out on the ocean, flying in at night to feed in the shallow tidal estuaries on the beds of *Zostera* or on the mud and sand flats where it finds plenty of small mollusks.

DISTRIBUTION.

Breeding range.—The species is circumpolar.   The North American form breeds east to the west coast of Hudson Bay, and James Bay (both coasts), and rarely east of Lake Michigan.   It has been known to breed in New Brunswick (Tobique River, 1879) and in southern Ontario (Rondeau, Lake Erie) and southeastern Michigan (St. Clair Flats).   South to northern Illinois (formerly, but now scarce even in Wisconsin), central Iowa (Hamilton and Sac Counties), central western Nebraska (Garden and Morrill Counties), northern Colorado (Larimer County and Barr Lake region), northern Utah (Bear River marshes), and southern California (Riverside County). West to the central valleys of California (Los Angeles, Kern, Merced, Sutter, and Butte Counties), central Oregon (Klamath and Malheur Lakes), western Washington (Pierce County), central British Columbia (Cariboo), and the Bering Sea coast of Alaska. North to the Arctic coast of Alaska (Point Barrow), northern Mackenzie (Fort Anderson), and the Arctic coast west of Hudson Bay. Replaced in northern Europe and Asia by a closely allied subspecies.

Winter range.—East to the Atlantic coast of the United States, the Bahamas, Cuba, and Porto Rico, and rarely to the Lesser Antilles (Guadeloupe, Martinique, and Antigua). South to Jamaica and Panama. West to the Pacific coast of Central America, Mexico, and the United States. North along the Pacific slope to southern British Columbia (Chilliwack and Okanagan Lake); in the interior north to northeastern Colorado (Barr Lake), Oklahoma, central Missouri (Missouri River), southern Illinois (Mount Carmel), southern Ohio (Ohio River), Maryland (Chesapeake Bay), and eastern Virginia (Cobb Island). Said to winter regularly in southern Wisconsin and casually as far north as southeastern Nebraska (Lincoln) and southeastern Maine (Calais). Winters in Hawaiian Islands.

Spring migration.—Early dates of arrival: Pennsylvania, Erie, February 23; New York, northwestern, February 25; Newfoundland, Grand Lake, April 20; Illinois, Chicago, March 12; North Dakota, Larimore, March 20; Manitoba, Raeburn, April 5; Mackenzie, Fort Simpson, April 28; Alaska, Kowak River, May 14; and Demarcation Point, May 24. Average dates of arrival: Illinois, southern, February 26; Missouri, central, February 26; Iowa, Keokuk, February 18; Illinois, Chicago, March 20; Minnesota, southern, March 9; Minnesota, northern, April 8; North Dakota, Larimore, April 3; Saskatchewan, Qu' Appelle, April 10; Manitoba, Raeburn, April 8; Mackenzie, Great Slave Lake, May 1; Alaska, St. Michael, about May 1.

Fall migration.—Early dates of arrival: Quebec, Montreal, September 3; Long Island, Mastic, August 21; Massachusetts, eastern, September 11; Pennsylvania, Erie, September 6; Virginia, Alexandria, September 13; Florida, Wakulla County, September 11; Texas, Corpus Christi, August 18; California, Santa Barbara, August 25; Lower California, southern, August 29. Late dates of departure: Alaska, Point Barrow, September 7; Kowak River, September 14; and St. Michael, October 10; Mackenzie, Fort Franklin, September 27; Long Island, East Rockaway, December 24.

Casual records.—Has occurred in Porto Rico (Cartagena Lagoon, April 8, 1921), Bermuda (winter 1847–48 and October 26, 1875),

Greenland (Godthaab and "northern"), and Labrador (Hopedale, Davis Inlet, etc.).   Recorded from Laysan Island.

Egg dates.—Alaska and Arctic America: Fifty-five records, May 23, to July 16; twenty-eight records, June 10 to 24.   California, Colorado, and Utah: Twenty-two records, April 30 to June 29; eleven records, May 15 to 30.   Manitoba and Saskatchewan: Twenty records, May 16 to July 3: ten records, June 4 to 14.   North Dakota: Twenty-three records, May 11 to June 27; twelve records, May 23 to June 10.

POECILONETTA BAHAMENSIS (Linnaeus).

BAHAMA PINTAIL.

HABITS.

This beautiful duck has been recorded only once in North America. Mr. W. Sprague Brooks (1913) reported the capture of a specimen by Mr. Gardner Perry at Cape Canaveral, Florida, in March, 1912.   Mr. Perry generously presented the specimen to the Museum of Comparative Zoology at Cambridge.

The bird was in company with a small flock of green-winged teal, and the wind at the time was southeast.   It seems a strange fact that this bird has not been recorded from Florida before, a region that has so long received the attentions of sportsmen and naturalists.

The Bahama pintail, or Bahama duck, as it is also called, has long been known as a wide ranging species, from the Bahama Islands to southern South America.   Mr. W. H. Hudson (1920) refers to it in his Argentine Ornithology under the name of "white-faced pintail"; he seems to object to the use of the name "Bahama pintail."   He says that it is one of the commonest ducks in Brazil; he also says:

The brown pintail is our most abundant species in Argentina, and I have noticed in flocks of great size, sometimes of many thousands, of that duck, that a single white-faced duck in the flock could be detected at a long distance by means of that same snowy whiteness of the face.

Mr. Outram Bangs (1918) has recently shown that there are two recognizable subspecies, of which he says:

Specimens from the Guianas and the lower Amazon are quite like West Indian examples, and are true *Poecilonetta bahamensis* (Linn.).   Those from southern South America—southern Brazil, Paraguay, Argentina, etc.—though little different in color, are much larger, and represent a recognizable subspecies for which there are several names.   I have seen no intergrades, but doubtless these occur in middle Brazil or Bolivia.

Very little seems to have been published on the habits of the Bahama pintail and I have been able to learn only a few meager facts about it.   Dr. Glover M. Allen (1905) says of its haunts in the Bahama Islands:

On the south side of Great Abaco, stretching for many miles east and west, is a tidewater region locally known as "the Marls."   Long reaches of shallow water alternate with clayey flats a few inches above the tide level. These flats are thinly covered with a growth of small mangroves, grasses, and a few other halophytes, while here and there are little pools surrounded by taller mangrove bushes. In this sort of country we found a good number of these handsome ducks. Most of those seen were in pairs, but one flock of 15 birds was started from a small pond among the mangroves, July 6.

Nesting.—Mr. Charles B. Cory (1880) writes:

This pretty little species was quite abundant at Inagua, frequenting the large salt ponds of the interior. On May 27, while shooting on a small island in the lake back of Mathewstown, I observed a number of these birds, and shot several, all of which were in full breeding dress. While passing through a small marsh I discovered the nest of this species, the old bird flying away as I approached. It was simply a mat of grass placed on the ground, and contained nine eggs of a pale brown color, Another nest, taken a few days later, contained eight eggs, slightly darker than the first set.

Young.—Dr. Alexander Wetmore (1916) was fortunate enough to find a brood of young in Porto Rico, of which he writes:

May 26 the birds were common at the Laguna de Guanica, where they would flush singly or in pairs from a growth of water plants covering a large area of the lagoon and, after circling over the open water, return to the same cover. Once in the short grass of a marsh a female flew out, but almost immediately fell back as though with a broken wing, repeating the performance several times. At the same time the low *peep, peep* of young birds was heard and two about 5 days old were caught. A third promptly dived and apparently never came up, while the others rapidly scattered to safety through the grass. The down of these little birds was not at all soft, but had a peculiar stiff bristly feeling.

Plumages.—Mr. W. E. Clyde Todd (1911) says that a young bird, collected on Watling Island on March 23,—

is assuming the juvenal dress, and already resembles the adult below. In the downy stage the general color is dull brown, with a white stripe on the flanks and an illy defined pale superciliary stripe. The throat and cheeks are white also, as in the adult.

Food.—Doctor Wetmore (1916) says of the food of this species:

Stomachs of eight adults and two downy young which were collected were examined by W. L. McAtee, of the Biological Survey. The adults had eaten nothing but vegetable matter. Seeds of ditch grass (*Ruppia maritima*) were found in every stomach and formed 16.25 per cent of the total bulk the largest amounts being 180 and 125, seeds, respectively. Foliage and antheridia of algae (*Chara, sp.*) made up 83.75 per cent and formed the great bulk in all the stomachs. The two downy young had eaten animal matter (amounting to 3.5 per cent), composed of remains of a water boatman (*Corixa, sp.*), bits of a water creeper (*Pelocoris, sp.*), and young snails. Grass seeds, foxtail grass (*Chaetochloa, sp.*), barnyard grass (*Echinochloa crus-galli*), and a species of guinea grass (*Panicum, sp.*) formed 94 per cent of the food of these ducklings, and a few other seeds 2.5 per cent.

He seems to think that they need protection, for he says:

These birds were much disturbed by egg hunters who were continually searching the marshes, and many were shot by gunners during the breeding season. They should be free from molestation from March 1 to December 1 at least. In a few years their range will be even more restricted than at present, because of the draining and clearing of swamps and marshes, and unless protected they will disappear entirely.

DISTRIBUTION.

Breeding range.—The Bahama Islands (Abaco, Andros, the Caicos, the Inaguas, Long Island, etc.), Porto Rico, some of the Lesser Antilles, the Guianas, and northern Brazil. The southern limit of the northern race is not determined. Replaced farther south by a larger race. Some form of this species breeds on the Galapagos Islands.

Winter range.—Same as above, apparently not migratory.

Casual record.—Accidental in Florida (Cape Canaveral, March, 1912).

AIX SPONSA (Linnaeus).

WOOD DUCK.

HABITS.

Spring.—While wandering through the dim cathedral aisles of a big cypress swamp in Florida, where the great trunks of the stately trees towered straight upward for a hundred feet or more until the branches interlaced above so thickly that the sunlight could not penetrate, we seemed to be lost in the gloom of a strange tropical forest and far removed from the familiar sights and sounds of the outside world. Only the frequent cries of the omnipresent Florida red-shouldered hawk and an occasional glimpse of a familiar flycatcher or vireo, migrating northward reminded us of home. But at last the light seemed to break through the gloom, as we approached a little sunlit pond, and there we saw some familiar friends, the center of interest in a pretty picture, framed in the surroundings of their winter home, warmed by the genial April sun and perhaps preparing to leave for their northern summer home. The sunlight filtering through the tops of the tall cypresses which surrounded the pool shone full upon the snowy forms of 50 or more white ibises, feeding on the muddy shores, dozing on the fallen logs, or perched upon the dead stumps or surrounding trees; the air seemed full of them as they rose and flew away. But with this dazzling cloud of whiteness there arose from the still waters of the pool a little flock of wood ducks, brilliant in their full nuptial plumage, their gaudy colors flashing in the sunshine, as they went whirring off through the tree tops. What a beautiful creature is this Beau Brummel among birds and what an exquisite touch of color he adds to the scene among the water hyacinths of Florida or among the pond lilies of New England!

The wood duck is a strictly North American species and principally a bird of the United States, for its summer range extends but a short distance north of our borders, except in the warmer, central portions of Canada, and even in winter it does not migrate far south of us. It is one of the most widely distributed species, breeding throughout most of its range and wintering more or less regularly over much of its habitat in the United States. For these reasons its migrations are not easily traced except in the Northern States and Provinces. It is a moderately early migrant, coming after the ice has left the woodland ponds and timbered sloughs. Dr. F. Henry Yorke (1899) says:

They arrive in three distinct issues, after sunset and through the night, suddenly appearing in the morning upon their accustomed haunts. The first stays but a brief period, and departs for the north to breed; the second puts in an appearance a few days later, but soon leaves to nest in the northern parts of the United States; the third arrives directly after the second leaves and scatters over the Middle States to nest. This issue forms the local ducks of each State it breeds in.

Dr. P. L. Hatch (1892) writes:

Arriving simultaneously with the other earlier species, none other braves the last rigors of the departing winter in the closing days of a Minnesota March with greater spirit. And when they come, like the rains of the Tropics, they pour in until every pool in the woodlands has been deluged with them.

Courtship.—Audubon's (1840) account of the courtship is very attractive; he writes:

When March has again returned, and the dogwood expands its pure blossoms to the sun, the cranes soar away on their broad wings, bidding our country adieu for a season, flocks of waterfowl are pursuing their early migrations, the frogs issue from their muddy beds to pipe a few notes of languid joy, the swallow has just arrived, and the bluebird has returned to his box. The wood duck almost alone remains on the pool, as if to afford us an opportunity of studying the habits of its tribe. Here they are, a whole flock of beautiful birds, the males chasing their rivals, the females coquetting with their chosen beaux. Observe that fine drake, how gracefully he raises his head and curves his neck! As he bows before the object of his love, he raises for a moment his silken crest. His throat is swelled, and from it there issues a guttural sound, which to his beloved is as sweet as the song of the wood thrush to its gentle mate. The female, as if not unwilling to manifest the desire to please which she really feels, swims close by his side, now and then caresses him by touching his feathers with her bill, and shows displeasure toward any other of her sex that may come near. Soon the happy pair separate from the rest, repeat every now and then their caresses, and at length, having sealed the conjugal compact, fly off to the woods to search for a large woodpecker's hole. Occasionally the males fight with each other, but their combats are not of long duration, nor is the field ever stained with blood, the loss of a few feathers or a sharp tug on the head being generally enough to decide the contest. Although the wood ducks always form their nests in the hollow of a tree, their caresses are performed exclusively on the water, to which they resort for the purpose, even when their loves have been first proved far above the ground on a branch of some tall sycamore. While the female is depositing her eggs, the male is seen to fly swiftly past the hole in which she is hidden, erecting his crest, and sending forth his love notes, to which she never fails to respond

Nesting.—The wood duck has earned the common name of "summer duck" on account of its breeding and spending the summer so far south; it has also been called the "tree duck" from its habit of nesting in trees. Its favorite nesting site is in a fairly large natural cavity in the trunk or large branch of a tree; it has no special preference for any particular kind of tree and not much choice as to its location; it probably would prefer to find a suitable hollow tree near some body of water, but it is often forced to select a tree at a long distance away from it and sometimes very near the habitations of man. The size and depth of the cavity selected vary greatly, and its height from the ground may be anywhere from 3 or 4 feet to 40 or 50. If it can not find a natural cavity that suits its taste, the wood duck occasionally occupies the deserted nesting hole of one of the larger woodpeckers, such as the ivory-billed or pileated woodpecker, or even the flicker; sometimes the former home of a fox squirrel or other large squirrel is selected, in which case the old nesting material, dry leaves and soft rubbish, is left in the cavity and mixed with the down of the duck. Such material is often found in the nest of the wood duck, but I doubt if it is ever brought in by the bird.

A few quotations from the writings of others will give an idea of the variety of nesting sites chosen. Audubon (1840) gives the best general idea of the nesting habits of the wood duck as follows:

The wood duck breeds in the Middle States about the beginning of April, in Massa chusetts a month later, and in Nova Scotia or on northern lakes, seldom before the first days of June. In Louisiana and Kentucky, where I have had better opportunities of studying their habits in this respect, they generally pair about the 1st of March, sometimes a fortnight earlier. I never knew one of these birds to form a nest on the ground, or on the branches of a tree. They appear at all times to prefer the hollow broken portion of some large branch, the hole of our large woodpecker (*Picus principalis*), or the deserted retreat of the fox squirrel, and I have frequently been surprised to see them go in and out of a hole of any one of these, when their bodies while on wing seemed to be nearly half as large again as the aperture within which they had deposited their eggs. Once only I found a nest (with 10 eggs) in the fissure of a rock on the Kentucky River a few miles below Frankfort. Generally, however, the holes to which they betake themselves are either over deep swamps, above canebrakes, or on broken branches of high sycamores, seldom more than 40 or 50 feet from the water. They are much attached to their breeding places, and for three successive years I found a pair near Henderson, in Kentucky, with eggs in the beginning of April, in the abandoned nest of an ivory-billed woodpecker. The eggs, which are from 6 to 15, according to the age of the bird, are placed on dry plants, feathers, and a scanty portion of down, which I believe is mostly plucked from the breast of the female.

Wilson (1832) describes a nest which he found, as follows:

On the 18th of May I visited a tree containing the nest of a summer duck, on the banks of Tuckahoe River, New Jersey. In was an old grotesque white oak, whose top had been torn off by a storm. It stood on the declivity of the bank, about 20 yards from the water. In this hollow and broken top, and about 6 feet down, on the soft decayed wood, lay 13 eggs, snugly covered with down, doubtless taken from the breast of the bird. This tree had been occupied, probably by the same pair, for four successive years.

Mr. William B. Crispin gave me his notes on a nest which he found near Salem, New Jersey, on April 25, 1908; it was in a natural cavity in a sour gum tree 40 feet from the ground; the 16 eggs were 3 feet below the opening in a nest of down mixed with dry leaves, which were probably taken there by squirrels the previous season; two gray squirrels were living just a few inches below the nest.

Mr. Henry R. Buck (1893) describes a nest found in a hole in a large apple tree near Hartford, Connecticut, as follows:

This tree was hardly 5 rods from an occupied house, and perhaps three times as far from a well-traveled road leading to the city. There was nothing to hide it from the road, and only a few trees in the immediate neighborhood. The trunk was hollow and had a wide split in one side from a height of 6 feet nearly to the ground.

I have a set of 14 eggs in my collection, taken in Norton, Massachusetts, on May 8, 1892, from a hollow apple tree; the cavity was 3 or 4 feet deep and the eggs lay in their bed of down 3 feet below the opening and only 2 feet above the ground. I found a nest in Taunton, Massachusetts, on May 19, 1917, containing 9 eggs, about 40 feet from the ground in a dead pine tree in a grove of tall trees near a house; the tree was so rotten that the cavity, which had once been a flicker's nest, had broken open and much of the down had fallen out and was scattered around the grove; a few feet below the duck's nest was a gray squirrel's nest in a cavity, with several half grown young in it. I was shown another nest, near Taunton, in a natural cavity in a large elm, about 30 feet from the ground; the tree stood close to a much-traveled road and in the front yard of a farm house. Mr. R. S. Wheeler found a nest in a barn near the Sacramento River, California; the birds entered through a hole in the boards and built a nest in the hay. Mr. Herbert K. Job found a nest similarly located in a barn located near Kent, Connecticut. Mr. Arthur T. Wayne (1910) found a nest in South Carolina on April 25, 1906. "The eight eggs were nearly hatched, and were laid in a sleeping hole of the pileated woodpecker, in a living sweet gum tree, 40 feet above the ground and more than a mile from the nearest reservoir." Mr. T. G. Pearson (1891) found a nest in Florida, on April 13, containing 13 fresh eggs. "The nest was in a hollow stump 30 feet from the ground. The entrance had been made by a yellow-shafted flicker, and it really seemed impossible for a duck to pass in and out of a hole of such small size. The nest was lined with a thick layer of downy feathers from the breast of the old bird." Mr. Walter B. Sampson (1901) records a nest, which he found in California on April 29, 1900, "in a deserted home of a red-shafted flicker and placed about 25 feet up in a white oak tree"; this nest contained the remarkable number of 21 eggs.

The down in the wood duck's nest is grayish white or "pallid mouse gray," with nearly pure white centers. More or less rubbish

from the cavity is mixed with it and the breast feathers found in it
are pure white.

Eggs.—The wood duck raises but one brood in a season in any
part of its wide range. The set usually consists of from 10 to 15
eggs, but sometimes only 6 or 8 eggs are laid and occasionally much
larger sets have been found, ranging from 18 to 29 eggs. Mr. George
D. Peck (1911) mentions a remarkable set that he found in Iowa,
containing 31 eggs of the wood duck and 5 eggs of the hooded merganser. There are other cases on record where these two species
have contended for the use of the same hole or have occupied it
jointly, as mentioned under the latter species. The eggs are nearly
oval in shape, with a slight tendency toward ovate. The shell is
smooth, hard, and somewhat glossy. The color is dull white or
creamy white, perhaps pale buffy white in some cases or a color
resembling old ivory white.

The measurements of 99 eggs in various collections average 51.1
by 38.8 millimeters; the eggs showing the four extremes measure
55.5 by 41, 53.5 by 42, 48 by 38.5 and 50.5 by 37.3 millimeters.

Young.—The period of incubation is from 28 to 30 days. This
duty is performed wholly by the female, but the male is more or less
in attendance on her during this period and returns to help her care
for the young. The young are provided with sharp claws which
they use in climbing from the nest up to the entrance of the cavity,
a distance of often 3 to 4 feet and sometimes as much as 6 or 8 feet.
Much has been written about how the female conveys the young
from the nest to the water in her bill, between her feet or even on
her back, and several writers claim to have seen the first method
employed. I am inclined to think that this method of conveyance
is used only when circumstances make it necessary; if the nest cavity is not too high, or if it overhangs the water, or if there is soft
open ground below it, I believe that the young are usually coaxed
or urged to jump or flutter down and are then led by the old bird
to the nearest water; certainly such is often the case.

Mr. J. H. Langille (1884) describes it very well as follows:

When the young are about 24 hours old, if the limb containing the nest be over
the water, they may find their way severally to the edge, and dropping into their
favorite element, begin life's perilous career. If the nest be a little distant from
the water, as is generally the case, the mother may seize them by the wing or neck,
and convey them to it, or, landing them thus on the ground, may lead them thither
in a flock. More commonly, however, the mother having thoroughly reconnoitered
the place for some time, and now uttering her soft cooing call at the doorway, the
little ones scramble up from the nest with the aid of their sharp toenails, and huddle
around the mother a few minutes. The mother, now descending to the ground, calls
again to the young, and they drop one by one on to the soft moss or dried leaves,
their tiny bodies so enveloped in long down, falling scarcely harder than a leaf or a
feather. Again they huddle around the mother bird; and, as the distance of the nest

from the water is sometimes as much as 60 or 70 rods, and generally more or less on an elevation, they need the maternal guidance to their favorite element.

Audubon (1840) says:

If the nest is placed immediately over the water, the young, the moment they are hatched, scramble to the mouth of the hole, launch into the air with their little wings and feet spread out, and drop into their favorite element; but whenever their birthplace is at some distance from it, the mother carries them to it one by one in her bill, holding them so as not to injure their yet tender frame. On several occasions, however, when the hole was 30, 40, or more yards from a bayou or other piece of water, I observed that the mother suffered the young to fall on the grasses and dried leaves beneath the trees, and afterwards led them directly to the nearest edge of the next pool or creek. At this early age, the young answer to their parents' call with mellow *pee, pee, pee*, often and rapidly repeated. The call of the mother at such times is low, soft, and prolonged, resembling the syllables *pe-ee, pe-ee*.

The young are carefully led along the shallow and grassy shores, and taught to obtain their food, which at this early period consists of small aquatic insects, flies, mosquitoes, and seeds. As they grow up, you now and then see the whole flock run as if it were along the surface of the sluggish stream in chase of a dragon fly, or to pick up a grasshopper or locust that has accidentally dropped upon it. They are excellent divers, and when frightened, instantly disappear, disperse below the surface, and make for the nearest shore, on attaining which they run for the woods, squat in any convenient place, and thus elude pursuit.

Mr. E. G. Kingsford (1917) has seen the wood duck carry its young to the water and thus relates his personal experience:

Early in July, 1898, while tented on the bank of the Michigamme River, township 43, north range 32 west, section 1, Iron County, Michigan, I had the good fortune to see it done. The nest was in a hollow pine that stood directly back of the tent and about 200 feet from the water, and the hole where the old duck went in, was 50 or 60 feet from the ground. After seeing the old duck fly by the tent, to and from her feeding grounds up the river many times during the time of incubation, one morning before sunrise she flew by from the tree to the river with a little duck in her beak which she left in an eddy a short distance upstream. She then made 10 or 12 trips to the nest and each time took a little duck in her beak by the neck to the water, where they all huddled in a little bunch. It was all done in a few minutes, and she evidently took them to the water very soon after they hatched, as they were only little balls of down. In going to and from work, we passed the little bunch many times. On our approach the old duck would fly away and leave the little ones huddled in a bunch near the shore where the water was quiet.

Mr. E. F. Pope in a letter to Mr. Edward H. Forbush says:

Once while fishing on the Nueces River in southeastern Texas, I observed a female wood duck bringing part of her brood of 10 ducklings down from a white-oak stub 28 feet above the water. There were three or four of the young already in the water when I appeared on the scene. She emerged from the cavity in the stub with a young duck on her back and simply dropped straight down into the water, using her wings to check the speed of her descent. When she arrived within a foot or two of the surface she suddenly assumed a vertical position which caused the duckling to slide from her back into the water. She rose quickly, circled a time or two, reentered the stub, and at once repeated the performance until the whole brood of 10 were on the water.

Mr. W. S. Cochrane, State game warden of Arkansas, also in a letter to Mr. Forbush, describes a similiar performance; after watching for three hours, he saw the female carry down the young on her back, as follows:

She visited the nest several times and after circling around the woods returned and rested on the edge of the nest which was in a hollow stub of the oak. After resting there about 10 minutes she flew down toward the water with her wings slightly elevated, and when about 10 feet from the water she began flying in an upward position, allowing one of the young which she was carrying on her back to slide off over her tail into the water. She went through this performance 14 times.

Mr. A. B. Eastman (1915) gives us the following account of the behavior of the young:

One day a friend and I were out on a little camping and canoe trip and on rounding a sudden bend in the creek above the pond, we came upon a mother duck and about seven little ones. A sudden note from the mother caused a prompt disappearance of the ducklings into the depths below. The courageous mother, however, instead of beating a hasty retreat, as one would most naturally expect, came flying toward the canoe and flopped down just in front of us, beating the water with her wings and trying by every means to make us believe that a crippled duck was just within our grasp. Seeing no signs of the little ones we started to follow the mother as if intending to catch her. She skillfully decoyed us up by the creek until around another bend when we were, in her estimation, a safe distance from her little brood. She then suddenly and miraculously recovered and quickly disappeared among the heavy growth of hardwood timber which clothes the banks of the creek. We promptly returned to the scene of the first encounter. The little ones had evidently recovered from their fright as we saw three of them swimming around. On seeing us, two of them dove, while the other made slowly for the bank, half submerged like a grebe. As soon as it landed we made a dash for the spot and the little fellow led us a merry chase through fallen timber, across ditches and through thicket and tangle. We finally corralled him, however, and made him pose as a photograph, much against his will. After taking a good look at the youngster, we set him down near the creek bank, and by the way he took to the water, we could imagine him congratulating himself on his fortunate escape from his terrible captors.

Mr. Manly Hardy, in his manuscript notes sent to Major Bendire, relates the following incident:

I once came suddenly upon a female with six half-grown young. As I approached the young ran into the tall grass while the mother flew away. I captured one while they were trying to escape to a bend in the stream above. An hour or more after, while approaching the stream above by a road through the woods from which I could see and not be seen, I saw the old one who had evidently been below looking for the missing one, flying high in the air until she was nearly opposite me, when dropping into the water she uttered a sharp call note upon which three young came out from the bushes on the right-hand bank and swam toward her—this evidently not pleasing her, she uttered a different note when they turned and swam back. She then, without moving, gave the first call note again, when two swam out from the left bank and came to her. Taking these with her she swam up abreast of where the others had disappeared, called them out and swam upstream with the united family. It was plain that she could count enough to know if one was missing, also that she had different notes by which she called her young or sent them away from her.

Plumages.—The downy young wood duck is much darker above and paler below than the young mallard; the lower mandible and the smaller tip of the upper mandible are of a rich yellowish shade, which will serve to distinguish it from other ducks. The crown is a very deep rich "seal brown" or "bone brown," or halfway between these colors and black; a stripe of the same color extends from the eye to the dark color of the occiput and there is a lighter auricular spot; the back shades from "bister" anteriorly to the same color as the crown posteriorly; the hind neck is of a darker shade of "bister"; the sides of the head and neck, including a superciliary stripe and the lores are "cream color" shaded locally with "Naples yellow"; the throat and under parts are "ivory yellow" to "Marguerite yellow," the colors of the upper and under parts mingling on the sides; there is a pale yellowish spot on each wing and on each side of the rump.

The plumage appears first on the scapulars and flanks, then on the tail, breast, and belly, then on the back and head, the last of the down showing on the hind neck and rump when the bird is nearly fully grown; the wing feathers are the last to grow. In this juvenal plumage the back varies from "argus brown" to "raw amber" with a metallic luster of purple, bronze, or green; the wings are similar to those of the adult female; the under parts are whitish, mottled with dull brown and tinged with bright brown or buff on the chest and flanks. The sexes look very much alike, but the wing of the male is more brilliant than the female's and the head pattern is different in the two sexes, each being a suggestion of the adult pattern; the crown is "clove brown" in both sexes but in the male it has a greenish luster; the white around the eye is more conspicuous in the female; the sides of the head are dull gray and the throat is white in both sexes, but in the male the white extends up into the cheek and side of the neck, as in the adult. The sexes soon begin to differentiate and the progress toward maturity is rapid. In the young male the mottled belly is replaced by white during September and October; the rich chestnut brown comes in on the chest and the vermiculated flank feathers are acquired. During October the adult color pattern of the head is assumed, and many of the brilliant, bronze, green, blue, and purple feathers appear in the back, scapulars, and tail, so that by November the young male has assumed a plumage which is practically adult, though the full brilliancy and perfection of plumage is not acquired until the following year.

The adult male begins to molt into the eclipse plumage in June or July, and the wings are molted in July or August, while the eclipse plumage is at its height. This plumage much resembles that of the young male, except that the belly of the adult is nearly pure white, instead of mottled as in the young, and the back retains nearly as much of the metallic colors as in the full plumage. The wings are molted only once and are always distinctive; also the brilliant colors of the eyes, feet, and bill are retained during the eclipse stage, though they lose a little of their brilliancy. The molt out of the eclipse occurs in August and September; I have seen an adult male in full plumage again as early as September 12 and another that had not finished the molt on October 16.

Food.—The wood duck obtains most of its food on or above the surface of the water, though it can tip up to feed on shallow bottoms if necessary, and it feeds largely on land. A large part of its food consists of insects which it finds on the surface of the water or on the leaves and stems of aquatic plants, such as beetles, mayflies, locusts, and various creeping insects. Here it also obtains small fish, minnows, frogs, tadpoles, snails, and small salamanders. Nuttall (1834) says:

I have seen a fine male whose stomach was wholly filled with a mass of the small coleoptera, called *Donatias*, which are seen so nimbly flying over or resting on the leaves of the pond lily. These birds are therefore very alert in quest of their prey, or they never could capture these wary insects.

Probably the greater part of the food of the wood duck, during the fall and winter particularly, is vegetable, of which a great variety is consumed. The bulbs of *Sagittaria* and other water plants, as well as the seeds and leaves of many varieties, are taken with the animal food in summer. Later in the season the wild rice marshes are visited and many wild fruits such as grapes and berries are found on dry land. The grain fields are apparently never visited, but the southern rice fields are favorite feeding grounds in fall and winter. The wood duck is particularly fond of acorns, chestnuts, and beechnuts, which it picks up on the ground in the woods, turning over the fallen leaves to find them. Messrs. Beyer, Allison, and Kopman (1909) state that, in Louisiana, "an undoubted factor in determining the abundance of the wood duck is the presence of the water chinquapin (*Nelumbium luteum*). As a food of the wood duck the seeds of this plant are extremely important."

Mr. Douglas C. Mabbot (1920) says of the food of this duck:

More than nine-tenths (90.19 per cent) of the food of the wood duck consists of vegetable matter. This high proportion of vegetable food is very similar to that taken by the mallard. With the wood duck it is quite evenly distributed among a large number of small items, chief among which are the following: Duckweeds, 10.35 per cent; cypress cones and galls, 9.25; sedge seeds and tubers, 9.14; grasses and grass seeds, 8.17; pondweeds and their seeds, 6.53; acorns and beechnuts, 6.28; seeds of water lilies and leaves of water shield, 5.95; seeds of water elm and its allies 4.75; of smartweeds and docks, 4.74; of coontail 2.86; of arrow arum and skunk cabbage, 2.42; of bur marigold and other composites, 2.38; of buttonbush and allied plants, 2.25; of bur reed, 1.96; wild celery and frogbit, 1.31; nuts of bitter pecan, 0.91; grape seeds, 0.82; and seeds of swamp privet and ash, 0.72 per cent. The remaining 9.4 per cent was made up of a large number of minor items.

The wood duck's animal food, which amounted to 9.81 per cent of the total consisted chiefly of the following items: Dragon flies and damsel flies and their nymphs, 2.54 per cent; bugs, 1.56; beetles, 1.02; grasshoppers and crickets, 0.23; flies and ants, bees, and wasps, 0.07; miscellaneous insects, 0.97; spiders and mites, 0.63; crustaceans, 0.08; and miscellaneous animal matter, 2.71 per cent. Thus, nearly two-thirds of the animal food consisted of insects.

Behavior.—No duck is so expert as the wood duck in threading its way through the interlacing branches of the forest, at which its skill has been compared with that of the passenger pigeon. I have stood on the shore of a woodland pond in the darkening twilight of a summer evening and watched these ducks come in to roost; on swift and silent wings they would glide like meteors through the tree tops, twisting, turning, and dodging, until it was almost too dark for me to see them. Ordinarily its flight is swift and direct, usually high in the air. The short neck and white breast are good field marks for the female and the color pattern of the male is conspicuous at a long distance; it is said to resemble the baldpate in flight. When migrating it flies in small flocks, probably family parties.

The wood duck is a swift and agile swimmer and can dive if necessary. Audubon (1840) says of its movements:

On the ground the wood duck runs nimbly and with more grace than most other birds of its tribe. On reaching the shore of a pond or stream, it immediately shakes its tail sidewise, looks around, and proceeds in search of food. It moves on the larger branches of trees with the same apparent ease; and, while looking at 30 or 40 of these birds perched on a single sycamore on the bank of a secluded bayou, I have conceived the sight as pleasing as any that I have ever enjoyed. They always reminded me of the Muscovy duck, of which they look as if a highly finished and flattering miniature. They frequently prefer walking on an inclined log or the fallen trunk of a tree, one end of which lies in the water, while the other rests on the steep bank, to betaking themselves to flight at the sight of an approaching enemy. In this manner I have seen a whole flock walk from the water into the woods, as a steamer was approaching them in the eddies of the Ohio or Mississippi. They swim and dive well, when wounded and closely pursued, often stopping at the edge of the water with nothing above it but the bill, but at other times running to a considerable distance into the woods, or hiding in a canebrake beside a log. In such places I have often found them, having been led to their place of concealment by my dog. When frightened, they rise by a single spring from the water, and are as apt to make directly for the woods as to follow the stream. When they discover an enemy while under the covert of shrubs or other plants on a pond, instead of taking to wing, they swim off in silence among the thickest weeds, so as generally to elude your search by landing and running over a narrow piece of ground to another pond. In autumn, a whole covey may often be seen standing or sitting on a floating log, pluming and cleaning themselves for hours. On such occasions the knowing sportsman commits great havoc among them, killing half a dozen or more at a shot.

Mr. P. A. Taverner has contributed the following on the notes of a captive bird:

Its only notes seem to be little whistles. One of its most peculiar notes is uttered when it is disturbed and consists of a series of little *chick, chick, chick's* low and hardly discernible at a distance of 30 feet. Accompanying these little monosyllables is a low thump that seems to be uttered immediately before the *chick* but seeming to be made by different organs than are used vocally. It has the peculiar intensity of the sound made by the springing in and out of the bottom of a tin or other can. It may be made during the utterance of the *chick*, for though quite loud positively it is so illusive that it is hard to tell exactly just when it is made. It does seem however to be made quite independently of the other sounds, though it is never heard alone.

Another note he gives when he is quiet and usually when quite alone. I have heard it several times in the dead of night. It is comparatively loud and consists of a series of from half a dozen to a dozen whistles like *H-o-o-w-c-e-e-t*. They follow each other rather rapidly and are without accent, the *H-o-o* gliding smoothly into the *w-e-e-e-t* without change in inflection. The whole having much the timbre of the sound made by drawing the finger nail sharply over and across the grain of heavily shot silk. Another note is made when he seems to be talking to himself and is something like *Chick a wangh*, the *angh* being rather drawn out, and the first syllable short. It is not loud either, in fact none of the notes it makes seem fitted for any more than the most private conversation. The only other note that I have heard it utter is a little short *cheep, cheep*.

Mr. Elon H. Eaton (1910) describes the note as follows:

The call of the drake is a mellow *peet, peet*, but when frightened it utters a harsher note which is usually written *hoo eek, hoo eek*. The note of the duck, when startled, is a sharp *cr-r-e-ek, cr-r-e-ek, cr-r-e-ek*, somewhat like the drake's alarm note.

The intimacy of the wood duck with the hooded merganser on its breeding grounds has been already referred to above, as well as under the latter species. It also associates with the hooded merganser somewhat at other seasons, as similar haunts are congenial to both species. On migrations it usually flocks by itself and is not much given to frequenting the open resorts of other ducks. It is more essentially a bird of the wooded bottoms, narrow sluggish streams, heavily timbered reservoirs, and forest swamps.

Young wood ducks have many natural enemies to contend with, such as large pickerel, pike, and snapping turtles, which attack them from below and drag them under water to drown them. I quote again from Audubon's (1840) matchless biography of this species:

Their sense of hearing is exceedingly acute, and by means of it they often save themselves from their wily enemies the mink, the polecat, and the raccoon. The vile snake that creeps into their nest and destroys their eggs, is their most pernicious enemy on land. The young, when on the water, have to guard against the snapping turtle, the garfish, and the eel and, in the Southern Districts, against the lashing tail and the tremendous jaws of the alligator.

The wood duck has always been able to hold its own against its natural enemies, but it has yielded to the causes of destruction brought about by the hand of man and by the encroachments of civilization. The wholesale cutting down of forests and draining of swampy woodlands has destroyed its nesting sites and made its favorite haunts untenable. Its beautiful plumage has always made it an attractive mark for gunners, collectors, and taxidermists, and its feathers have been in demand for making artificial trout flies. Almost anyone who has found a wood duck's nest has been tempted to take the eggs home to hatch them, as these ducks are easily domesticated and make attractive pets. It is so tame and unsuspicious that it is easily shot in large numbers and it has been extensively caught in traps. From the great abundance, noted by all the earlier writers, its numbers have been reduced to a small fraction of what they were; in many places, where it was once abundant, it is now unknown or very rare; and it has everywhere been verging towards extinction. Fortunately our attention was called to these facts by Dr. A. K. Fisher (1901) and Mr. William Dutcher (1907) before it was too late, and now that suitable laws have been enacted for its protection in many States, it has been saved from extinction and is even on the increase in some places.

Fall.—The fall migration starts early. Doctor Yorke (1899) says of it:

The first fall issue consists of local ducks, which migrate during the early part of the month of September. The second comes down from the Northern States about the end of September, while the last comes down in the early part of October. The second and third do not stay nearly so long as the first issue, which is the largest and collects in quantities on favorite grounds. The second and third collect in a different manner; they drop into willows, buck brush and on rivers and timber-clad ponds, in singles, pairs, or little flocks, about nightfall, and depart before morning; these places are used by them nightly during their migrations, until all have gone south, and appear to be regular stopping places. The ducks of the third issue are full fledged upon their arrival.

Game.—As a game bird the wood duck has always been popular, as it is a clean feeder, often very fat, and a delicious table bird. It will come readily to live decoys or even to well-made wooden decoys, if properly handled; it is such a swift flier and so clever in avoiding places that it has found to be dangerous that considerable skill and strategy is necessary to hunt it successfully. One of the best methods of hunting it is to lie in wait for it, properly concealed, on one of its fly ways between its feeding grounds and its roosting places, but to succeed in this the hunter must make a thorough study of its movements and learn all he can about the nature of the country in which it lives. Wood ducks usually roost for the night in small open pools in the woods, where they are sheltered and secure. About an hour before sunrise, or as soon as it begins to be light, they leave these pools and fly to their feeding grounds in the wild rice marshes, in sluggish streams and ponds filled with aquatic vegetation or along the wooded banks where they can pick up seeds, nuts, and acorns. If necessary, they will rise and fly over the tops of the forest trees, but they prefer to fly along the open lanes, streams or passageways which are usually found connecting the ponds in the regions they frequent. Here they fly low in regular flight lines and if the gunner places his blind in some narrow passageway between the trees in

such a fly way, he is practically sure of good sport for about three
hours in the morning and again for an hour or two on the return
flight at sunset.

Mr. Dwight W. Huntington (1903) describes another method of
hunting them:

At English Lake I shot them from a light boat, jumping them in the wild
rice. The punter pushed the boat (which contained a revolving office chair for the
gunner) rapidly. The birds often arose at short range and presented easy marks.
They were very abundant on the Kankakee at certain bends in the river, where they
fed on acorns which dropped from the oaks into the water. A friend one day killed
over 70 of these birds over decoys, and I often made fairly good scores shooting from
a blind, but my fondness for moving about and exploring the marshes and ponds for
other ducks and a change of scene always prevented my making very large bags.

Winter.—Long before the autumn frosts have begun to close the
northern ponds the tender "summer duck" has moved southward
toward its winter home in the rice fields of the Southern States, the
wooded sloughs and timbered ponds of Louisiana, and the cypress
swamps of Florida, mingling with the summer birds of these congenial climes. A few hardier individuals winter farther north, where
they can find sheltered ponds and streams with an abundant food
supply. They do not, like many other ducks, frequent the seacoast
in winter; if found near the coast at all, they are in the fresh-water
ponds and streams, protected from the winter winds.

DISTRIBUTION.

Breeding range.—United States and southern Canada, entirely
across the continent. Breeds locally in almost every State and
southern Province, where suitable conditions exist. South to Cuba,
the Gulf of Mexico, south central Texas (San Antonio), probably
Colorado, Utah, and Nevada, and in southern California (Ventura
County). North to southern British Columbia (lower Fraser Valley
and Okanogan Lake), northwestern Montana (Flathead Lake), rarely
Great Slave Lake (Fort Providence), southern Manitoba (Lake Winnipeg), southeastern Ontario (Parry Sound and Muskoka districts) and
central eastern Labrador (Hamilton Inlet).

Winter range.—Mainly in southern United States. South to Ja-
maica (rarely) and central Mexico (valley of Mexico and Mazatlan).
North to southern British Columbia (Chilliwack), central Missouri
(Missouri River) southern Illinois (Mount Carmel), and southern
Virginia (Petersburg). Winters casually north to Michigan (Kalama-
zoo County), and Massachusetts.

Spring migration.—Early dates of arrival: New York, central,
March 16; Ohio, northern, March 10; Ontario, Ottawa, March 26;
Iowa, central, March 7; Minnesota, Heron Lake, March 24; Manitoba,
southern, April 2. Average dates of arrival: New York, central,

March 25; Massachusetts, eastern, March 24; Quebec, Montreal, April 24; Iowa, central, March 20; Ohio, northern, April 1; Michigan, Petersburg, March 15; Ontario, southern, April 17, and Ottawa, April 22: Minnesota, Heron Lake, April 4; Manitoba, southern, April 15.

Fall migration.—Average dates of departure: Ontario, Ottawa, October 27; Quebec, Montreal, November 1; Maine, southern, October 27; Massachusetts, Essex County, December 16; Iowa, southern, November 9. Late dates of departure: Ontario, Ottawa, November 7; Maine, southern, November 2; Massachusetts, Charles River, December 28; Iowa, southern, November 21; Ohio, Loraine Reservoir, December 3.

Casual records.—Accidental in Bermuda (December 16, 1846).

Egg dates.—New England and New York: Eighteen records, May 4 to June 17; nine records, May 13 to 22. New Jersey: Four records, April 17 to May 16. Florida: Four records, April 8 to May 14. Illinois and Iowa: Seven records, April 17 to June 4. Minnesota, Michigan and Wisconsin: Seven records, May 10 to 30.

NETTA RUFINA (Pallas).

RUFOUS-CRESTED DUCK.

HABITS.

This beautiful European duck has but a slight claim to a place on our list. About all we know of it, as an American bird, is contained in the following statement by Mr. Robert Ridgway (1881):

About nine years since (February 2, 1872), Mr. George A. Boardman, of Calais, Maine, sent to the Smithsonian Institution a mounted specimen of a duck obtained in Fulton Market, New York City, and supposed to have been shot on Long Island Sound, which he was unable to determine satisfactorily, but which he supposed to be a hybrid between the redhead (*Aethyia americana*) and some other species. The specimen was in immature plumage, with the feathers of the first livery much worn, while those of the new molt, which were generally interspersed, indicated a very different garb when the molt should have been completed. At the time the specimen was received at the Smithsonian, I (also supposing it to be a hybrid) made comparisons with nearly, if not quite, all the American species of ducks, but was unable to get the slightest clue to its parentage. It was then put back in the case and not again thought of until a few days ago, when in removing the specimens with a view to their rearrangement I happened to take the one in question in one hand and an adult female of the European rufous-crested duck (*Fuligula rufina*) in the other; and having the two thus in a very favorable position for comparison, I at once perceived a striking similarity in general appearance and in the form of the bill, which induced me to extend the examination to an adult male, the result being that no question remained of the bird in question being an immature male of *F. rufina*, a species hitherto not detected in North America.

Yarrell (1871) observes that "it may be doubted if the presence of this solitary individual in the United States was due to natural causes"; it seems likely that the specimen may have come over in some shipment of foreign game or may have escaped from some private preserve.

As the bird is entirely unknown in this country I shall have to cull its life history from the writings of European authors. Mr. John G. Millais (1913) writes:

Essentially a southern species, the red-crested pochard comes north to Germany or to England in October and November before the frost and snow and leaves for the south on the first signs of Arctic conditions, whilst a few come north in March and April and wander about in small parties before seeking their breeding places in May. In Europe they do not appear to be very gregarious, as they are in India, where they arrive in flocks of thousands in late October and November. Hume mentions finding them in "flocks of many thousands and acres of water paved with them," whilst Reid says: "One morning in December I came across countless numbers in a jheel in the Fyzabad district, closely packed and covering the whole surface of the water, with their red heads moving independently, while the breeze kept their crests in motion; a distant spectator might have mistaken them for a vast expanse of beautiful aquatic flowers."

The red-crested pochard is essentially a duck of the fresh water and is never found upon the open sea. The ponds and lakes they like to frequent are reedy, sedge-lined sheets of water with a considerable area of deep water in the centre.

Courtship.—The red-crested pochards arrive at their breeding places at the end of March or early in April, but the females do not begin making their nests till the end of the latter month. The courtship of the male is somewhat showy but not very varied. It throws up the body from the water, depressing the bill to the fore part of the neck, and at the same time displaying, i. e., erecting and spreading the whole of the beautiful feathers of the crest, the body raised to about an angle of 75°. During this sudden act of show the bird utters a low squeaking whistle, and as the body falls to the water again the crest resumes its normal position, and, releasing the air in the chest, it utters a low grunt or groan. Sometimes when in full show the male will frequently swim round the female with depressed bill and expanded crest, but beyond this I have seen no other efforts at display. The female resorts to attitudes similar to the other diving species, such as swimming round the male with lowered body and extended head and neck held out along the water. She also frequently utters her harsh guttural cry at this season.

Nesting.—Dr. Baldmus, who took 10 nests in 1866–1870 in central Germany, states that—"The nest is always placed in the rushes or flags, usually on a small island in a pond or on the flags, and, like all ducks' nests, it has a foundation of rotten stems of rushes and dead leaves, on which a warm bed of down is placed, this down being plucked from the breast of the female. When the female leaves the nest quietly, she covers her eggs, as do all the ducks, even our common tame species. During the time the female is sitting the males are to be seen on the water with those of *ferina*, *leucopthalmus*, and *clypeata*, but generally somewhat apart from them."

Mr. W. Eagle Clarke (1895) describes a nest of this species, which he found on an island in a shallow *etang* in the southwestern Camarque, as—

placed in the center of a thick tangled mass of purslane (*Atriplex portulacoides*) so dense that it was reached by a covered way, 2 feet in length, worked in the shrub where it rested on the soil; the nest was on the ground and consisted of a broad rim of down, with a few short dry tamarisk twigs, and contained 10 fresh eggs. A few yards farther on another duck of this species was disturbed—this time from under an immense shrub of seablite, quite 4 feet in height and as many in diameter. The nest in all respects resembled the last, and contained 17 eggs of two distinct types, and probably the production of different females. The eggs of one set were white and were all singularly malformed. The normal eggs are of clear pea green and a trifle smaller than those of the pochard. The down in the nest closely resembles that of the eider duck in tint. Both nests were about 6 yards from the water, and the birds wriggled off at our feet.

Eggs.—Mr. Millais (1913) says that the usual number of eggs is from 7 to 10, but that 14, in one case 17, eggs have been found in a nest. The—

eggs are decidedly lighter and more thin shelled than those of the common pochard. When fresh, the eggs are a clear green stone-color with a gloss, but lose their bright tints and gloss after being blown, then becoming a dull grayish olive or greenish gray. The texture is smooth, fine, and clear, but somewhat fragile for a duck's egg. In form they are a broad oval with both ends the same size.

Rev. F. C. R. Jourdain furnishes the measurements of 74 eggs, which average 58.1 by 41.8 millimeters; the eggs showing the four extremes measure 61 by 42, 56.5 by 43.5, and 53.5 by 39.6 millimeters.

Plumages.—Mr. Millais (1913) describes the downy young as:

Upper parts, dull olive gray; under parts, buff or yellowish gray; a buff spot on either shoulder; a yellowish-gray stripe passes over each eye, and in front of and behind the eye runs a dark stripe which divides behind the eye; irides, dark brown; bill reddish brown, with the nail white; feet, ash-gray with a green tinge; webs and toes narrowly edged with yellowish white.

Of the young male he says:

Somewhat similar to the female, only darker and with an indication of a crest. The center of the feathers of the under parts are brown instead of gray, and the back and front of the breast is a much darker brown. The young male and female are easily recognized by the usual immature feathers on the lower breast, vent, and tail coverts, and frayed tail. The principal molt commences in November with a few black feathers on the lower parts and scapulars, and proceeds rapidly in February and March.

By April the young male has gained the whole of the adult plumage, although it is not nearly so rich or bright as that of a 2-year-old male. By the beginning of May the only sign of immaturity is a dark brown line as broad as a pencil on the upper surface of the bill, but this mark disappears as soon as the young male molts into full eclipse dress at the end of May, when the bird may be said to be adult at 10 months. Young males bred by Mr. St. Quintin in June, 1910, were just losing the last sign of immaturity on May 1, 1911, when I visited Scampston. The irides are brown or reddish brown, becoming red in March; the feet and legs, at first olivaceous orange, also become orange red in spring.

According to his colored plate, the male has a well-marked eclipse plumage, of which he writes:

If we do not accept the summer plumage of the long-tailed duck as an eclipse dress for reasons to be explained later, the red-crested pochard and the gadwall are the earliest species to change from spring into the eclipse dress. A male in my possession began to dress the first brownish-gray feathers on the flank as early as May 10 and another in the Scampston collection was in full molt on May 20. The eclipse dress of the male is very like the plumage of the female, but they may be at once distinguished by the wings, and the brighter color of the eye, eyelids, bill, and feet, and by the darker color of the under parts. The crest also is much longer and more pronounced.

Food.—Regarding the food of the rufous-crested duck, he quotes Naumann, as follows:

These ducks like to feed on tender roots, buds, shoots, the tips of leaves, flowers, and seeds of various kinds of plants growing rampant at the bottom of the water— thus on any kinds of *Potamogeton* of *Myriophyllum*, and *Ceratophyllum*. As the last named grow in a considerable depth of water, shooting up high toward the surface, and often form thick green plantations under the water, such places in the pool are the favorite haunt of these ducks. They are continually diving down in search of such undergrowth, and occasionally, in addition to the vegetable food, they catch the living creatures to be found on them. In places where this undergrowth approaches nearer to the surface, they attempt to get at it by merely tipping up the hinder part of the body and reaching for it by stretching the neck down vertically. They also fish for much which they enjoy when merely swimming on the surface, but they employ the last two methods less often than that of diving under completely for their food. For this reason they like deep water, and come extremely seldom to the bank for the sake of getting food. Along with the above-mentioned substances their stomachs always contain a quantity of sand and small pebbles the size of peas.

Behavior.—The same writer says:

The red-crested pochard frequently comes ashore on lakes where they are seldom disturbed; but if much harassed, they keep to the deep water the whole day. On land they stand and walk in a manner quite different from other diving ducks, and seem to be able to walk and run with less roll and greater ease than other species. At such times the neck is very much drawn up, with the bill depressed, and when moving fast it takes the form of the letter S, whilst the body is held more or less horizontal. If they approach anything suspicious or are suddenly frightened, the body is suddenly held up. They seldom leave the banks of a lake except during the nesting season. In swimming, or when about to dive for food, the body is held low in the water, the tail trailing on the surface, the neck is stiff and almost upright, and the bill held depressed. The bird at such times has an air of intent alertness, as if minutely searching the depths below. Although skillful divers, they do not stay below the surface as long as other species, 30 seconds being a long dive. Unlike other diving ducks they show a distinct preference for shallows at certain seasons, and especially in places where they are undisturbed. Here they may be seen paddling for hours round the edge of a lake, frequently tipping up the hinder parts after the fashion of mallard or pintail, and reaching for delicacies with their long necks. Their flight is similar to other diving species, and it is accompanied by a faint whistling sound, and is strong and well sustained. They have some difficulty in rising if there is no wind.

Mr. Millais (1913) says further:

The usual call generally uttered by the female is a harsh *kurr*, and it is not often emitted except in moments of excitement. Hume says the male utters a "sharp sibilant note—a sort of whistle," but does not state at what season he has heard it. This note is commonly uttered by the male in courtship, but I have never heard it at other times.

Game.—Those who have had experience in studying these ducks are of opinion that they are shy and difficult to approach when in large flocks, but as easily killed as other diving ducks when found on small pieces of water. Mr. Stuart Baker, who has had many opportunities of studying the species, says:

"From a sporting point of view, the red-crested pochard is all that can be desired. About as smart as they make them, he seems to have special aptitude for judging the length of range of different guns; and a flock may be caught once but seldom twice, whatever the distance the gun may reach. They swim so fast that they can by this means generally escape, and they are often very loath to rise when they can thus get out of shot."

As a table bird the red-crested pochard resembles the common pochard and the widgeon in its flesh. That is to say, when it feeds on fish and shellfish it is rank and uneatable, and when it eats vegetable matter it is excellent. "On account of their great timidity," says Naumann, "you can only creep up to them to shoot them unseen and against the wind, if they are swimming near enough to the bank; but as they almost always choose the open center of larger pieces of water as their places of sojourn, there is in this case no other method than to approach them openly in a boat, which can certainly only succeed with solitary specimens if they have not yet suffered any pursuit at that place, and then not always, whilst larger flocks generally take to flight when within a hundred paces of you. Occasionally when flying round afterwards they get near enough to the boat for a lucky shot. In the morning and evening twilight they are sometimes by chance brought down when flying over. It is easy to creep up to solitary specimens which have wandered off to a small pond if some care is exercised, but they can seldom endure to be shot at by a gun approaching them openly. Birds which have been lamed by a shot are generally lost to the pursuer on larger pieces of water, even with the help of a good water dog, as they do not easily tire of diving under, and, if they can reach some sedge, are very clever at hiding themselves in it, and in so doing only keep the head as far as the eye above the surface of the water. They can easily be caught in large decoy nets placed at their favorite places in the water."

DISTRIBUTION.

Breeding range.—Portions of Europe, Asia, and Africa. From southern Germany and Spain eastward through southern Russia to northern Persia and eastern Turkestan, and southward into northern Africa.

Winter range.—The basins of the Mediterranean, Black, and Caspian Seas, the Persian Gulf, India, Burma, and Ceylon.

Casual records.—Wanders to the British Isles, France, Belgium, Holland, Denmark, Germany, and China. One record for North America (Fulton Market, New York, 1872).

Egg dates.—Southern Russia: Eight records, April 26 to June 12.

NYROCA AMERICANA (Eyton).

REDHEAD.

HABITS.

The redhead or American pochard ranks as one of our most important game birds, for it is well known and widely distributed; from its main breeding grounds in central Canada and the northern Central States it spreads out its migrations to both coasts and appears, at some season of the year, in nearly every State.

Spring.—Dr. F. Henry Yorke (1899) says of the spring migration:

The first spring flight of this well-known duck passes the frost line whilst the ice still remains upon our lakes, water only existing in open holes or channels; the birds follow closely after the canvasback and like that bird appear in good-sized flocks. They stay but a short time, working rapidly toward the north and going to the far end of the British possessions. The second issue arrives about a week after the first has departed; if abundance of food be present, they stay until the advent of the third issue, then travel north also beyond the boundary line. The third begins to pair upon reaching latitude 44° and spreads all over the country up to Manitoba.

Courtship.—Dr. Alexander Wetmore (1920) gives the following interesting account of the courtship of the redhead:

The peculiar mating display of these birds seen on several occasions was observed to advantage on June 4. A party of four males and three females were swimming in open water, two of the birds apparently being mated. Suddenly one of the females began to display, approaching one of the males with her head held high, sometimes jerking it up and down and again holding it erect, and at intervals calling *quek que-e-ek*, the last a peculiar rattling note. The male chosen extended his neck, holding his head erect, frequently whirling quickly to show the female his back, or again sank down with his head drawn in while the female bowed before him. At short intervals she opened her mouth and bit at him gently or, if he was swimming, sprang quickly in front of him with her head erect and back partly submerged. She transferred her attentions from one male to another in turn, even approaching the one who apparently was mated. The males showed considerable jealousy over these favors and drove each other about in fierce rushes. At intervals they called, the note being a curious drawn-out groaning call, resembling the syllables *whee ough* given in a high tone. As it was given the male sometimes raised his breast, elevated his head, and erected his crest. Again he threw his head straight back so that it touched his dorsum above the rump, with the throat up and the bill pointing toward the tail. The bill was then thrown up and head brought again to the erect position as the call was made. The curious actions of the male in calling continued after he was mated, and the strange call note was heard often. Mated males were seen driving savagely at their mates and biting at them while they escaped by diving.

Nesting.—My first experience with the nesting habits of the redhead was gained in North Dakota in 1901 where we found it breeding abundantly in all of the larger sloughs where there was plenty of deep open water in the center, surrounded by extensive areas of cat-tail flags (*Typha latifolia*), bullrushes (*Scirpus lacustris*), and tall reeds (*Phragmites communis*); its nest was most often found among the flags or bullrushes, growing in water a foot or more deep, and least often among the *Phragmites* which usually grew in shallower or drier places. I quote from my own published notes (1902) as follows:

We first met with it on June 3 in a large slough in Nelson County, where the water was not over knee-deep, except in a few scattered open spaces, and where the reeds and flags were somewhat scattered and open. A pair of Canada geese nested in this slough and two pairs of marsh hawks, but it was chiefly tenanted by yellow-headed blackbirds, coots, and long-billed marsh wrens. The blackbirds fairly swarmed in this slough, and the constant din of their voices was almost bewildering, especially whenever one of the marsh hawks sailed over the slough, which sent them all up into the air at once, cackling and squeaking, hovering and circling about for a few moments, and then settling down into the reeds again. Redheads were flying back and forth across the slough, killdeers, willets, and Wilson phalaropes were flying about the shores, and long-billed marsh wrens were singing among the flags on all sides. While wading along a shallow ditch through a small patch of last year's flags, a big brown duck sprang into the air from a clump of tall reeds, and, after a short search, I found my first nest of the redhead, well concealed among the reeds. It was a handsome nest, well made of dead reeds, deeply hollowed and lined with broken pieces of the reeds mingled with considerable white down, especially around the upper rim; it measured 16 inches in diameter outside and 8 inches inside, the upper part of the rim being about 10 inches above the water; it rested on a bulky mass of dead reeds built up out of the shallow water, the whole structure being firmly held in place by the live growing reeds about it. It held 11 handsome eggs, in which incubation had just begun. I could not photograph this nest, as it was raining hard, but I collected the nest and eggs, which are now in my cabinet.

We found the redheads breeding in two large, deep sloughs in Steele County. One of these, in which we found four nests of the redhead, is illustrated in the photograph. In the open part of this slough, shown in the foreground, the water was too deep to wade, but, in the southern end of the slough, shown in the background, the water was seldom deeper than the tops of our hip boots, and in many places quite shallow. The principal growth was the tall slough reeds, quite thick in some places, and often as high as our heads, with numerous thick patches of tall cat-tail flags and several patches of the "queen of the prairie" reeds growing in the drier portions. The redheads' nests were all located in the shallower parts of the slough where the reeds and flags were growing less thickly.

The redheads' nests found here on June 10 contained 6, 10, 14, and 16 eggs, respectively, none of which were collected. The latter of these is shown in the photograph, it was located in the center of a tangled mass of broken-down dead flags, in a nearly dry, open space, near the edge of the slough, well concealed from view by the arching over of the dead flags above it. The bird proved to be a close sitter, as we twice flushed her from the nest. We tested one of the eggs and found it far advanced in incubation.

Mr. J. H. Bowles (1909) gives the following attractive account of the nesting habits of the redhead in Washington:

They are essentially lovers of shoal bodies of fresh water, and in summer resort in considerable numbers to the larger lakes of central Washington for the purpose of rearing their young. One of their favorite breeding grounds may be found at Moses Lake, a beautiful body of water situated in the north central part of the State. At this place, in the summer of 1906, it is certain that at least 150 pairs remained to nest. Paddling our canoe along the margin of the lake, close to its heavy fringe of cat-tails, we would flush a pair or two at intervals of every hundred feet. As is customary with all waterfowl during the nesting season, they were remarkably tame, allowing such a close approach as to give an excellent view of the handsome nuptial plumage of the male.

Leaving the canoe and plunging at random into the sea of rushes, fortune may favor us sufficiently to permit of our happening upon one of their nests. This is a heavy, deep basket of rushes placed in the thickest of the growth, either upon a small muddy island left by the receding water, or built up amongst the flags upon the matted dead stems which cover the surface of the lake in these places. It is a structure of such beauty as to cause the bird student to pause almost breathless upon its discovery.

The mother duck has heard his noisy approach long since and departed, first carefully spreading over the eggs a heavy blanket taken from the lining of the nest. This consists entirely of down of the most delicate shade of white faintly tinged with gray, which the duck plucks from her own breast. A faint glimpse only can be obtained of the 12 or 14 greenish-drab eggs which seem completely to fill the nest, but let the sun be shining brightly with the dense green rushes for a background, and be sure that fatigue, soaked clothing, mosquitoes, and a dozen other discomforts will instantly vanish from remembrance at the sight.

Mr. Robert B. Rockwell (1911) writes of the nesting habits of this species in the Barr Lake region of Colorado:

The redheads' nests, like those of the teal, exhibited a wide variation in structure and location. The first two nests were found June 10, 1906. These, containing five fresh eggs and nine incubated eggs, respectively, were within 2 feet of each other, in burrows in the top of a large muskrat house at the edge of a small lake in a sparse growth of cat-tails. The birds had burrowed in about 18 inches, lined the cavity with down, and deposited the eggs at the end of the cavity. A careful examination of all the muskrat houses seen (and they were so conspicuous that in all probability none was overlooked) during the balance of 1906 and the full nesting seasons of 1907 and 1908, failed to reveal any other similarly located nests of this species.

On May 31, 1907, we found a beautiful set of 11 fresh eggs in a large, bulky nest somewhat resembling an overgrown nest of the coot; but much less compact and not so neatly cupped or lined as the average coot's nest. There was little or no downy lining in the nest which was built in an average growth of cat-tails over about 18 inches of water, and some 20 yards from the open water of the lake. There was no apparent attempt at concealment, and it was very conspicuous owing to its large size. The female flushed widely, with a good deal of noise, when we were fully 40 yards from the nest, thus attracting our attention to it. Eight of these eggs hatched on or about June 20, the remaining three being addled.

The finest nest of this species which came to our attention was found June 15, 1907, in a dense cat-tail swamp between two small rush-encircled lakes. It was a beautifully built structure of dead cat-tail blades, mostly broken into small pieces, well built up above the surface of the water (which at this spot was only a few inches deep), deeply cupped, plentifully lined with down, and well concealed in the dense cat-tail growth.

Eggs.—The redhead incubates on a large set of eggs; my notes record various numbers from 6 to 22, but many of the largest sets contain eggs of other species; probably the redhead itself usually lays from 10 to 15 eggs. The redhead also seems to be careless about laying its eggs in other ducks' nests. In North Dakota we found one of its eggs in a ruddy duck's nest and in three cases we found three to four of its eggs in the nests of the canvasback, on which the latter duck was incubating. These two species seem to have a peculiar habit of building nests in which large numbers of eggs are laid, by both species, but are apparently not incubated; we found two such nests in Saskatchewan, one of which contained 19 eggs; this set is now in my collection and apparently contains eggs of the redhead, canvasback, and mallard; it was evidently a canvasback's nest originally. Messrs. Willett and Jay (1911) mention a nest found at San Jacinto Lake, California, which contained 27 eggs and which "was undoubtedly the product of at least two females, as there were 17 eggs of one type and 10 of another. In fact the 10 eggs may not be redheads' at all, as they resemble very much the eggs of the pintail."

The eggs of the redhead can generally be distinguished from those of other species by their color, size, and texture, but I have seen eggs that were puzzling; the nest, however, is always distinctive; it is built like that of the canvasback, but the down in it, which is usually mixed with the reeds or flags, is *whiter* than that of the canvasback. Although the down is practically white, certain portions of it have a slight grayish tinge. The down in the nests of this and all other species of diving ducks is more closely matted or in less well-defined, fluffy pieces than the down in the nests of the surface feeding ducks.

The shell of the egg is extremely hard and tough, with a smooth, glossy surface; it will dull the cutting edges of an egg drill in a short time. The color varies from "pale olive buff," matching almost exactly certain types of mallard's eggs, to a pale "cream buff" or "cartridge buff." The eggs are larger than mallard's eggs and more glossy and they are very different in color from canvasback's eggs. Taking into account the nest, the down, and the size, color, and texture of the eggs, there should never be any difficulty in recognizing a redhead's nest, even if the bird were not clearly seen. In shape the eggs vary from a somewhat rounded to a considerably elongated elliptical ovate; they are sometimes nearly oval. The measurements of 79 eggs in various collections average 61.2 by 43.4 millimeters; the eggs showing the four extremes measure **66.8** by 43.5, 66.2 by **45.5, 58** by 41.8 and 61.5 by **41.2** millimeters. Incubation is performed by the female alone and lasts for a period of 22, 23, or 24 days.

Plumages.—The downy young is quite different from other ducklings, being more uniformly colored with less contrast between the light and dark areas. The upper parts, including the crown, back, rump, and tail are "light brownish olive," but the deep color of the basal portion of the down is much concealed by the light yellowish tips; the side of the head and neck, including the forehead and a broad stripe above the eye, are "olive-ocher" paling to "colonial buff" on the throat and chin; the remainder of the under parts is "colonial buff" with deeper shadings; there are shadings of "chamois" on the sides of the head and neck, but no conspicuous dark markings; in some specimens there are suffusions of brighter yellow in all of the lighter-colored parts, such as "amber yellow" or "citron yellow"; there is a yellowish spot on each of the scapulars and on each side of the rump. All of the colors become paler and duller as the duckling increases in size.

The plumage develops in the young redhead in the same sequence as in the young canvasback and when 7 or 8 weeks old it is fully feathered, except the wings, and is a little more than half grown.

The scapulars and back are dark gray, edged with brownish, the breast reddish brown, the belly mottled with brown and white, and the head is reddish brown.   This is the juvenal plumage in which the sexes are alike and somewhat resemble the adult female; young birds, however, are more mottled, with less clear brown above and less clear white below.   The contour feathers are fully acquired and the young bird is nearly fully grown before the wings are fairly started.   Young males are generally browner and darker than young females, particularly on the breast and head.   By November the black feathers begin to appear on the breast and neck of the young male, each black feather being tipped with brown, which wears off later; in December the red feathers appear in the head and neck, and the gray vermiculated feathers in the back, scapulars, and flanks are assumed; and by January, or February at the latest, the plumage is practically adult, though the full perfection of the adult plumage is not acquired for at least another year.   The progress toward maturity in the young female is practically the same, though the change is not so conspicuous.   Probably young birds breed during their first spring.

The adult male has a partial eclipse plumage, involving a double molt of much of its plumage; the molt into this plumage begins early in August, the flight feathers are shed about the middle or last of August, and the full winter plumage is complete again in October or November.   In the eclipse plumage the head and neck become browner, the breast and under parts become mottled, as in the breeding female; there are many brown feathers in the back, the rump is largely brownish, and the crissum is veiled with light edgings. The adult female assumes during the nesting season and the summer a more mottled plumage than is worn in the winter; the clear dark brown of the upper parts is veiled with lighter edgings, and the clear white of the under parts is mottled with brownish.

Food.—The favorite feeding grounds of the redhead during the summer are in the open lakes of the interior where it dives in deep water or in shallower places to obtain the roots and bulbs of aquatic plants or almost any green shoots which it can find; it is not at all particular about its food and is a gluttonous feeder.   It also dabbles with the surface-feeding ducks in the muddy shallows where it finds insects, frogs, tadpoles, and even small fishes and water lizards. Audubon (1840) says that "on several occasions" he has "found pretty large acorns and beechnuts in their throats, as well as snails, entire or broken, and fragments of the shells of various small unios, together with much gravel."

Dr. D. G. Elliot (1898) writes:

Redheads feed much at night, especially if the moon is shining, and at such times are exceedingly busy, and the splashing of diving birds the coming and going of others, and the incessant utterings of their hoarse note, are heard from dark to daylight. They also feed by day, if the weather has been stormy, but on quiet, pleasant days they rarely move about much, but remain quietly out in the open water, sleeping, or dressing their feathers, or occasionally taking a turn beneath the surface as though more in an exploring mood, than for the purpose of seeking food.

Among the vegetable food of the redhead, Dr. F. Henry Yorke (1899) has recognized in its food the following genera of plants: *Vallisneria, Limnobium, Zizania, Iris, Nymphaea, Nuphar, Myriophyllum, Callitriche,* and *Utricularia.*

In its winter home on or near the seacoast it frequents the tidal estuaries, as well as the ponds, and feeds in company with the canvasback, the scaup ducks, and the baldpate, diving in deep water for the roots, as well as the stems and buds of the wild celery (*Vallisneria*), on which it becomes very fat and its flesh assumes a flavor almost indistinguishable from that of the canvasback. But it does not wholly confine itself to this food, feeding largely on other aquatic plants and on marine animal life, which detracts from the flavor of its flesh.

Behavior.—Prof. Walter B. Barrows (1912) says of the flight of the redhead:

It travels in V-shaped flocks like geese, and flies with great rapidity, but the common statement that its speed reaches 100 miles per hour is certainly a gross exaggeration. It is safe to say that no species of duck when migrating flies more than 50 or 60 miles per hour—most species hardly more than 40 miles.

Doctor Elliot (1898) writes:

The flocks rarely alight at first, even when there may be numbers of ducks congregated on the water, but traverse the length of the sound or lake as if reconnoitering the entire expanse and trying to select the best feeding ground. After having passed and repassed over the route a few times, the flock begins to lower, and gradually descending, at length the wings are set and the birds sail gradually up to the chosen spot, usually where other ducks are feeding, and drop in their midst with many splashings. But while this is the usual method adopted by newcomers, sometimes the program is changed and the birds, attracted by a large concourse of their relatives, particularly if the day be calm and the sun shining with considerable heat, will suddenly drop from out the sky with a rapid zigzag course, as if one wing of each duck had been broken, and they cross and recross each other in the rapid descent, their fall accompanied by a loud whirring sound, as the air is forced between the primaries. On such occasions the flock is mixed all up together in a most bewildering manner, until, arriving a few feet above the water, the wings become motionless and the birds glide up to and alight by the side of their desired companions.

Early in the morning, and again late in the afternoon, the redhead regularly takes a "constitutional." The flocks that have been massed together during the night or the middle of the day, rise from the water, not all together but in companies of several dozen, and stringing themselves out in long, irregular lines, each bird a little behind and to one side of its leader, fly rapidly up and down, at a considerable height over the water. Sometimes these morning and evening promenades are performed at a great elevation, so that the movement of the wings is hardly perceptible. On such occasions they appear like a dark ribbon against the sky, and the comparison is strengthened by the fact that every movement of the leader elevating or depressing his course is imitated exactly by all those which follow, and so the line has frequent wavy motions like currents passing through it, as when a ribbon is held in the fingers and a flip given to it which causes it to undulate along its whole length.

Doctor Elliot (1898) has well described the note of this species as a "hoarse guttural rolling sound, as if the letter R was uttered in the throat with a vibration of the tongue at the same time. It is easily imitated, and the bird readily responds to the call of its supposed relative." Rev. J. H. Langille (1884) gives an entirely different impression of it; he writes:

Not infrequently the males are quite noisy, loudly uttering their deep-toned *me-ow*, which is the precise imitation of the voice of a large cat. The female, especially if rising from her nest or out of the water, has a loud, clear *squak*, on a higher tone than that of the mallard or dusky duck, and so peculiar as to be readily identified by the ear, even if the bird is not in sight.

Doctor Yorke (1899) confirms this impression, saying: "The redheads' cry whilst floating about in compact bunches resembles the mewing or cry of a cat, but their call is a very modest quack."

Fall.—The fall migration of the redhead follows soon after that of the canvasback and spreads out over much the same route; from its main breeding grounds in the central part of the continent, the prairie regions of southern Canada, and the Northern States, it migrates almost east, through the region of the Great Lakes to the coasts of southern New England, southeast to the Chesapeake Bay region, and south through the Mississippi Valley to the Gulf coast; there is also probably a southwestward migration to the Pacific coast and a southward one to Mexico. Redheads migrate in large flocks by themselves or late in the season they often mingle with scaup ducks. They become very abundant in the fall along the southern coast of New England, especially in the large fresh or brackish ponds on Marthas Vineyard, where several thousand of them are reported as congregating every fall; some of the ponds, which are controlled by sportsmen, are planted with *Vallisneria, Potamogeton,* and other duck foods which have attracted an increased number of redheads and scaup ducks. A party of four men are said to have killed 110 of these two species in five hours' shooting.

Game.—Redheads are abundant on the Chesapeake, where they are shot in large numbers with the canvasbacks from the batteries; when feeding on wild celery their flesh is of fine flavor. They are very popular as game birds on the lakes and sloughs of the Mississippi Valley; they travel about in large flocks and are easily decoyed to wooden decoys set near the hunter's well-concealed blind or sink box. A net set on poles around the gunner's boat or duck float may be rendered quite inconspicuous by weaving branches or grass into it so that it will match its surroundings; the ducks do not seem to notice it and very good shooting may be had from such a blind. Doctor Yorke (1899) writes:

Their playgrounds are in open waters upon large lakes, or some distance from the shore on the coasts, where they float about in rafts or flocks. They are easily lured to shore by tolling, either by a red handkerchief raised and lowered, or by some odd moving object, for they are most inquisitive birds; sometimes a dog is trained to run along the shore and bark at the water's edge, the gunner lying concealed close by; even after being shot at, they soon seem to forget the occurrence and gradually work in again to the object which had previously attracted them. Should, however, a few baldpates be mixed up with them, these soon spoil the game; being more suspicious, the baldpates will keep turning and swimming back without approaching within shot, drawing the redheads with them Even upon a flight, the baldpates lead many flocks of redheads away out of shot by their shying away from any object which they distrust and which the redheads would have unhesitatingly approached.

Winter.—The redhead winters as far north as it can find open water; Mr. Thomas McIlwraith (1894) states that for—

two seasons a flock of 100 or 150 remained in Lake Ontario all winter, about half a mile from shore, opposite the village of Burlington. The birds spent most of their time at one particular place, sometimes diving sometimes sitting at rest on the water, and always close together, as if for greater warmth. When the weather moderated in March they shifted about for a few days, and then went off to the northwest, the direction taken by most waterfowl when leaving this part of Ontario in the spring

Occasionally they linger too long in freezing lakes and some of them perish, but they are usually more hardy and better able to take care of themselves than the canvasbacks. Even in its winter haunts on the seacoasts the redhead prefers to feed in fresh-water ponds, associating with baldpates, scaup ducks, mallards, and shovellers; it also frequents brackish ponds and salt-water estuaries in company with the canvasbacks. It must venture out onto the ocean at times, for it is known to winter in the Bahamas, occurring there in large flocks.

On the coasts of Virginia and North Carolina the redhead is abundant in winter. I saw many large flocks in Back Bay and Currituck Sound, usually flying high in the air. The practiced eye of the experienced gunner can recognize them at a distance by their flight; they seem, to me, to move their wings more rapidly than canvasbacks; they look darker and shorter; and they fly in more irregular formations and more erratically. On pleasant smooth days, especially if they have been shot at in the morning, they may be seen flying out to sea in large flocks to spend the day in safety; they return again toward night to the fresh-water bays to feed on the roots of foxtail grass and wild celery.

Mr. J. A. Munro (1917) says that the redhead is "the commonest duck on Okanogan Lake," British Columbia—

in winter. Late in January, when their feeding grounds at the south end of the lake become frozen, they congregate in enormous flocks in the vicinity of Okanagan Landing. The prevailing winds are southerly and serve to keep the shallow water here free of ice. Several specimens of pondweeds (*Potamogeton*) afford an abundant food supply. By February 15, the flocks have reached their maximum and number several thousand. They remain in these large bands until March, when they move north. A small number remain and breed. Males outnumber females in the proportion of 15 to 1. Courtship commences about the last week in February. This is interesting in view of the fact that they are one of the last ducks to breed.

DISTRIBUTION.

Breeding range.—Central and western North America. Probably breeds in Newfoundland (Sandy River) and has been found breeding in southeastern Maine (Calais). Otherwise east to southeastern Michigan (St. Clair Flats). South to southern Wisconsin (Lake Koshkonong), southern Minnesota (Heron Lake), central western Nebraska (Garden and Morrill Counties), southern Colorado (San Luis Valley), northwestern New Mexico (Lake Burford), southwestern Utah (Rush Lake), central Nevada (Ruby Lake), and southern California (Riverside and Los Angeles Counties). West nearly to the coast in southern California (Ventura County), to the inland valleys farther north (San Joaquin and Sacramento Valleys), central Oregon (Klamath and Malheur Lakes), central Washington (east of the Cascade Mountains), and central southern British Columbia (Swan Lake). North to central British Columbia (Lac la Hache), central Alberta (Edmonton), Great Slave Lake (Fort Resolution rarely), central Saskatchewan (Saskatchewan River), and south central Manitoba (Lake Winnipegosis).

Winter range.—Mainly in the southern United States. East to the Atlantic coast and to the Bahamas. South to the West Indies (Cuba and Jamaica), the Gulf of Mexico, and central western Mexico (Manzanillo). West to the Pacific coast of Mexico and the United States. North to southern British Columbia (Okanogan Lake), southeastern Arizona (San Pedro River), northeastern Colorado (Barr Lake), northern Arkansas (Big Lake), probably southern Illinois, and eastern Maryland (Chesapeake Bay). More rarely north to Lake Erie and Ontario and southern New England.

Spring migration.—Early dates of arrival: Indiana, central, March 6; Ontario, southern, March 14; Iowa, central, March 8; Wisconsin, southern, March 10; Manitoba, southern, April 12; Maine, Scarboro, March 27. Average dates of arrival: Ohio, Oberlin, March 10; Ontario, southern, March 24; Iowa, Keokuk, March 7; Minnesota, Heron Lake, March 26; Manitoba, southern, April 21.

Fall migration.—Early dates of arrival: Ontario, southern, September 10; Virginia, Alexandria, October 5; Iowa, Iowa City, October 6; Missouri, St. Louis, October 16; Texas, San Angelo, October 1. Average dates of arrival: Ontario, southern, September 19; Pennsylvania, Erie, October 7; Virginia, Alexandria, October 12.

Casual records.—Has wandered on migrations northwest to southern Alaska (Kodiak Island).

Egg dates.—California, Colorado, and Utah: Twenty-seven records, April 23 to July 7; fourteen records, May 24 to June 9. Minnesota and North Dakota: Eighteen records, May 18 to June 28; nine records, June 3 to 17. Manitoba and Saskatchewan: Nine records, June 1 to July 6.

AITHYIA FERINA (Linnaeus).

EUROPEAN POCHARD.

HABITS.

The common pochard of the Old World is closely related to our redhead; authorities differ as to whether it is a distinct species or only a subspecies. The American bird is larger, more intensely colored and has black edges on the wing coverts which the European bird lacks; our bird also lacks the black base of the bill, which is conspicuous in the pochard. Audubon evidently regarded the two birds as identical, and Wilson, although expressing some doubt, apparently agreed with him. Nuttall followed their lead.

Dr. Barton W. Evermann (1913) recorded the capture of a specimen of the European pochard on St. Paul Island, in the Pribilof group, on May 4, 1912, which constitutes the first and, so far as I know, the only record of the occurrence of this species in North America. It was probably a straggler from its Asiatic range; it is widely distributed in central Asia, as far east as Lake Baikal in southern Siberia, and perhaps farther east.

In compiling the life history of this species, which is entirely unknown to me, I can not do better than to quote from the excellent and very full account of it written by Mr. John G. Millais (1913). Referring to the haunts of this pochard, he writes:

The home of this pochard is large, fresh-water lakes, or big reed-inclosed swamps with deep water pools in the center, where they can dive for food and remain beyond the reach of the gun. They are not adverse to still tidal estuaries, generally of brackish water, but seem to regard the sea itself merely as a place of refuge when driven from their true homes. Where pochards are most at home are large open stretches of fresh water that contain wide areas that are not of too great a depth. They seem to like lakes with rather muddy bottoms, where vegetation grows on pure sand, in which there is an abundance of water insects and much molluscae. From such a center they travel out at night to smaller ponds, and return at daybreak to their sanctuary. This proves that the pochard is intelligent, and, like all diving ducks, first considers its safety and then its food supply. In migration time, single birds or a few together may be found in quite small pools, but they never stay long in such places, but pass on until they find safety in numbers. As a rule, pochards keep well to the center of a lake or offshore during the day, and are only to be seen diving near reed beds or close to the banks of sluggish rivers, where they receive continuous protection. They are at all times suspicious of man, and at once swim for deep water on the least alarm. Even during gales they like to keep just out of shot of shore on the edge, as it were, of rough water, and take just as much advantage of bank shelter as is compatible with safety. This sense of caution is also evinced in their methods of going ashore to sleep and preen, for they generally chose some gentle shallow or low sloping island over which some members of the flock can see at all times, and on which the rest of the flock can rest at midday. On smaller pools they show an affection for those small green islands round which the muddy bottom produces an abundance of plant growth. Where constantly protected, it is common to see them in close proximity to the rushy banks where alders and willows grow and keep off the winds. Like all ducks, they seem to dislike a draft, and avoid wind-swept areas of water.

Spring.—The pochards that intend to nest within a certain area follow the general rule of all diving ducks and arrive in one flock, generally on the largest sheet of open fresh water in the neigborhood from the 10th to the 15th of March, or even later if the weather is still inclement. The curious groaning wheeze of the male may now be frequently heard, and courtship commences. The finest colored males being those of 2 years or over, are always the first to pair, and drive off the young males which, at a distance, may appear to be adult. The latter remain in a flock apart and seem to be easily discouraged from paying attention to the females. If, however, there is not a preponderance of males, as there usually is in the case of this duck, these young males will often pair with the females, who are quite ready to make love to them.

Courtship.—At the commencement of courtship, generally on the first warm day, several males are to be seen showing off before one duck. In most birds, pairing is generally due to the disposition on the part of the female to accept attentions, and you will usually notice that some particular female is in advance of the rest of her sex in this respect. As I have shown in my drawing, which is done direct from life, four or five males are crowding round one female who, in turn circles round some male, dipping her bill in the water stretching her neck low on the water, and occasionally uttering her coarse cry of *kurr-kurr-kurr.* The males continuously keep up their curious groan, which is somewhat like a man affected with asthma and being told by the doctor to "take a deep breath." In addition to this call, they also utter a soft low whistle, which the spectator must be close at hand to hear. The first attitude of the male consists in throwing the head and neck back until the back of the head touches a point between the shoulders. This is repeated constantly at the commencement of courtship. The more common display is to blow the neck out with air, with the head raised horizontally, and utter the groan as the air is released. During this show a distinct "kink" is to be observed in the lower part of the neck, whilst the center is unusually swollen. The fullest display is usually performed as the male approaches the female. The male then lies very flat on the water and stretches the head and neck to the fullest extent, at the same time blowing out the neck and frequently turning the head on one side so to display its full beauty. Two or three males may thus often be seen together laying themselves out to attract the female's attention, and the effect is somewhat striking. During these moments of intense excitement the pupil of the eye of the male nearly disappears, and the eye itself seems to blaze a very rich lacquer red.

Nesting.—In the breeding season pochards seem to prefer small lakes whose sides are overgrown with dense vegetation or even large reed-beds. Small islands are also very attractive to them, but, if absent, they will seek out nesting sites that run into meadows of sedge and grass, from which flow channels connected with the main pools. They are not at this season averse to the close proximity of man, and the pair of birds keep very close together until the female commences to sit. The nest is usually built just above the level of the water on the edge of a clump of reeds where the soil is firm, or in the center of a small island. As a rule, it is entirely covered with undergrowth and well lined with down. The female usually deposits from 7 to 9 eggs. Leverkühn records one nest of 10 eggs and Saunders and Naumann one each of 13 eggs, whilst Professor Newton had a clutch of 14 eggs sent from Yorkshire, probably the result of two females laying in the same nest. Full clutches are usually to be found in England in the first or second week in May, and the second week in May in Germany, and third week in May in Scotland.

"The female shows great devotion," says Naumann, "during the time of sitting. She approaches the nest with caution, flies past it accompanied by the male without, however, circling round it, lowers herself with the male on to the water at some distance away from it, and both sit there motionless for some time, with very erect necks, until finally the female, swimming in an attitude of diving, or running, huries back to it. The male meanwhile remains on the open water close by and warns her of the approach of any danger with a loud gabbling cry, but is always the first to take to flight, and later on, when the sitting is over, troubles himself no more about her, stays in the daytime far away from her on the open water near, and only comes back to her in the evening if she leaves the nest for a rest."

The female keeps adding down, plucked from her own breast, to her nest as incubation proceeds, until there is a considerable quantity deposited, and with this she covers the eggs carefully if she leaves the nest.

Eggs.—The eggs are a somewhat broad oval with the shell waxy and smooth but not glossy. In color a pale greenish gray, generally tinged with yellow. After being blown they often assume a dull brownish drab color. Average size of 100 eggs, 61.4 by 43.6 millimeters. Maximum, 68 by 45.5 and 64 by 46.5; minimum 57.2 by 43 and 61 by 39.2 (or in inches, 2.42 by 1.72) (F. C. R. J.).

Young.—Observers seem to be agreed that the young do not leave the nest until the day after they are hatched, and they are then tended with the most assiduous care by the mother. At first she keeps them close to the edge of the reeds, especially if there is any wind, and dives for food, which she breaks up and offers to them. Very soon they learn to catch flies and pick up floating seeds, and they may be seen diving of their own accord when only a day or two old. The cry of the young is a gentle *peep*, which they emit until fully fledged and able to fly. From early days the young are expert divers, and soon learn to escape by that method if threatened with danger, but on first alarm they pack closely together, as if for mutual protection.

Before reaching full powers of flight, pochards, as well as other ducks which nest in central Europe, have many enemies to contend with. No doubt large pike kill them in numbers. Rats and others account for a certain number, whilst hen and (on the Continent) marsh harriers account for a few. Magpies, carrion and hooded crows search out the nests and destroy the eggs.

Food.—The principal food in summer and autumn is vegetable and fresh-water mollusks. They eat large quantities of the roots, seeds, leaves, and flowers of aquatic plants, which they take and swallow at the bottom. They are especially fond of the seeds of *Polygonum amphibium*, and, in the autumn, of the seeds of *Potamogeton marinus* and *P. pectinatus*, also the tender parts of *Myriophyllum*. In confinement they refuse many hard foods such as acorns, etc., which surface-feeding ducks will eat with avidity. In summer the young birds eat quantities of floating insects, but the old birds seem to take few of these, although they catch numbers of water beetles, small fish, tadpoles, and small frogs. With their liking for seeds of all kinds, it is not difficult to get pochards to feed on any sort of grain or bird seeds. Like other diving ducks, they swallow a considerable quantity of sand or small stones, to assist digestion. Pochards seldom go on land to feed unless upon some mound of mud and water reeds which drought or a falling lake has exposed. They also seldom tip up the hind part of the body to reach food with the bill. They are not averse, however, to taking floating seeds and insects off the surface of the water.

Behavior.—Generally busy feeding at night, they like to rest and sleep a great part of the day with bill tucked into the shoulder feathers. In this attitude they remain for hours half asleep, but not so soundly that they avoid using their feet to maintain their position in the same spot.

It has often struck me, in watching a flock of pochards, that there is always an unusual preponderance of males, and a party can usually be recognized at a considerable distance by the redheads and shining lead-blue bills of the males. Females and young are always more difficult to distinguish from other ducks owing to their more uniform color. With their feet so far to the rear they walk with a decided roll, keeping the body in a fairly vertical position. But when standing still or taking alarm ashore, they raise the breast and assume a somewhat upright attitude. They never stay long on land, on which they appear to be little at home, but on the water they are expert swimmers and quick in all their movements. They swim deep, with the tail trailing in the water, and when engaged in diving further sink the body, depress the tail under water, and even allow the water to wash over the mantle.

In diving, their leg push is powerful and creates a considerable swirl after the bird passes out of sight. The bird swims rapidly to the bottom and probes in every direction for food, staying under as long as a minute, and then floating quickly to the surface with legs stationary on either side. Generally they come to the top in very nearly the same place at which they have dived. Nearly all their food is swallowed where it is found, but I have seen them bring fish to the surface, where it is passed across the bill several times until rendered soft enough to swallow whole. Certain roots are also treated in the same fashion. Naumann states that pochards can remain under water for "nearly three minutes." This may be possible, but I have never timed one, even in confinement, to stay so long beneath the water.

Their flight is rapid and "scurrying." The wings, not being large, have to be beaten quickly to bear the weight of the body, and the pace is not very swift. It is accompanied by a rushing sound; the birds fly very close together in a somewhat compact mass. When high in the air they often assume a V-formation, as if desirous of being led by some experienced individual, and the whole flock sometimes indulge in a remarkable "header," or plunge from the sky down to some sheet of water where they wish to alight. They can not rise easily from the water unless there is a considerable breeze, and sometimes scurry along the surface for some distance before getting under way. They also alight on the water somewhat clumsily. In the air they are readily recognized by the large head, body, and feet, short stumpy tail, and short wings. Although this duck may be said to be cautious on large sheets of water, it is not a difficult bird to approach even in large flocks, especially in a small sailing boat, and this may be due to its disinclination to fly, especially as it must come *upwind* toward the point of disturbance. I have sailed right in amongst pochards and scaup in October before putting them to flight. On small ponds they show even greater tameness, and, if undisturbed, will often consort with pinioned birds and tame species, and soon become as tame as domestic ducks. There are many instances of wild pochards joining domesticated ducks, and remaining with them for months. I have never found pochards on the sea in Scotland except during hard frosts. A few days of 10° below freezing point and I was certain to find pochards on the Moray Firth, where I shot with the big gun for three seasons, and if the frost continued for more than 10 days the birds left for the winter, most probably for the open water of the southwest, not returning until the lakes were open in March.

Game.—When found on small ponds pochards are by no means shy, and will generally allow a gunner to walk within gunshot if simple precautions are taken, but it is a mistake to shoot these birds in such places if there is a desire on the part of the landowner to establish the species as a resident, for all ducks soon learn the spots where they are protected, and will not tolerate much molestation. If specimens are required, or the needs of the pot are pressing, it is much better to attack the birds on large sheets of water or on the estuaries, which they are not easily made to forsake. In the autumn these large flocks are easily approached by a small sailing boat to within gunshot of an 8-bore, or even a full choke 12-bore, but if numbers are wanted the punt gun will do great execution in their serried ranks. I have seldom fired at pochards on the sea, but one frosty morning in February, 1891, when returning from an unsuccessful raid on the widgeon in Castle Stuart Bay, Moray Firth, I spied a small but dense flock of duck in Campbeltown Bay, not far from the village. These were about 60 pochards driven to the sea by stress of weather from the various Nairnshire lochs. Knowing that they would be tame and had doubtless never seen a punt, I reserved fire until I was within 80 yards, and cut a clear lane right through the flock, killing dead 20 birds, and afterwards recovering 2 winged ones. On the east coast of Scotland such a shot with the big gun is rare, but I have seen occasions on Loch Leven (where, Heaven forbid, a punt gun should ever be used) and the Loch of Strathbeg when a very much larger number of birds could easily have been killed. There are sometimes good opportunities of getting a shot at these ducks at flight, when they leave the estuaries or large lakes, and pass out to feed on smaller sheets of water at dusk. I was once waiting at a point on the Island of Mugdrum, Tay Estuary, when, hearing a rush of wings, I looked up, and had just time to snap two barrels into a flock of duck that passed on my left; the result was six pochards down, but I lost two in the darkness. If it is desired to shoot pochards on a small lake, it is much better to drive them off it, and station the gun or guns away from the water, as this form of shooting does not seem to terrify them nearly so much as stalking them from the shore. They are not more or less difficult to kill than other diving ducks, but require to be hit well forward, as winged birds may give much trouble.

DISTRIBUTION.

Breeding range.—Sub-Arctic portions of Europe and Asia. East to southern Siberia (Lake Baikal). South to eastern Persia, the Caspian Sea, northern Algeria, and southern Spain. West to the British Isles. North to the Sub-Arctic portions of Scandinavia, Finland, and Russia.

Winter range.—From the Mediterranean Basin (Morocco to Egypt) to India, China, and Japan.

Casual records.—Wanders to Iceland, the Faroes, Azores, and Canary Islands. Accidental in the Commander Islands (Bering Island, May 13, 1911) and the Pribilof Islands (St. Paul Island, May 4, 1912).

ARISTONETTA VALISINERIA (Wilson).

CANVASBACK.

HABITS.

The lordly canvasback, the most famous American game bird, from the standpoint of the epicure, is distinctly a Nearctic species and was discovered or, at least, first described by Alexander Wilson. It must have been taken by earlier sportsmen, but it was apparently not recognized as different from its near relative the European pochard, which it superficially resembles.

Spring.—It is a hardy species wintering just below the frost line, and one of our earliest migrants. The first spring flight appears above the frost line before the ice disappears from the ponds, lingers but a short time and passes on northward as fast as the ice breaks up ahead of it. The dates vary greatly in different seasons depending on the breaking up of winter conditions, but the migration often begins in February and is generally well under way by early March. The general direction is northwestward over the Great Lakes, for the birds wintering on the Atlantic coast, northward from the Gulf of Mexico through the Mississippi Valley and northward and eastward from Mexico and the Pacific coast, in converging lines toward its main breeding grounds in the prairie and plains region of central Canada.

Courtship.—All through the spring immense flocks of canvasbacks congregate on the larger lakes on or near their breeding grounds, floating in dense masses far out from shore, playing, feeding, or resting until the time arrives to break up into pairs for the breeding season. This usually occurs before the middle of May, but I have seen them in large flocks as late as the last week in May in southern Saskatchewan. I have never seen their courtship performances and can not find any description of them in print. But Dr. Arthur A. Allen has sent me the following interesting notes on what he has observed under somewhat artificial conditions.

Upon several occasions prior to 1917 I had observed small groups of canvasbacks on Cayuga Lake behaving in a manner which I took to be their courtship performance. Several females would draw together holding their heads up and their neck-stiff until they were almost touching breast to breast, when about an equal number of males would swim rapidly around them. Occasionally the males were seen to throw their heads back toward their tails, or one of the females would dart out at a male that approached more closely. These performances took place at some distance from shore, however, and many of the details were missed.

During February of 1917, however, several pairs of canvasbacks were captured and placed with clipped wings on a small pond within 100 feet of my windows where they could easily be observed. They became quite tame in a remarkably short time and before the summer was over would eat from one's hand. About the middle of April they were first observed going through their courtship performances, and, inasmuch as they paid scant attention to one on the shore of the pond less than 20 feet away, every detail could be watched. First signs of excitement were evidenced by the males which began to call. As the canvasback is normally a very quiet duck this immediately attracted my attention. The call consists of three syllables *ick, ick, cooo,* with a little interval between the second and third. When the first two syllables are being produced the bird opens his bill slightly and then with considerable force appears to inhale quickly, jerking his bill as he does so. It appears as though this sudden inhalation abruptly closes the glottis so as to produce the two rather high-pitched, sharp, quick, *ick, ick* notes. Accompanying these notes the back of the neck swells and the feathers rise as though a gulp of air were being swallowed. Immediately, however, it seems as though exhalation occurred with the bill closed, accompanied by a low *cooo* like a muffled bark or distant moo of a cow and not so very different from the ordinary grunting note of the male bird when alarmed. Accompanying this note the chin swells out for an instant with a curious swelling about the size of an ordinary marble.

Very frequently this note was accompanied by the head-throwing performance, already referred to, the *ick, ick* notes being given when the head was thrown back, and the *cooo* when the head was brought forward again, the swelling on the chin being noticeable as the head assumed the normal position. This head-throwing performance was practically the same as has been described by Doctor Townsend for the golden-eye and has been observed by me frequently while watching redheads and scaup ducks as well.

The calls of the males were answered by the females with a low, guttural *cuk cuk*. The four females then drew together until their breasts nearly touched, jerking their heads and holding their necks stiff and straight as they did so. The males then began swimming about them in circles, sometimes with their heads close to the water after the fashion of the mallards, sometimes calling as already described, and frequently jerking their heads so that the occiput struck the back. Occasionally one of the males would approach a little closer to the females and then one of the females would lower her head and chase him away, returning to her stand in the middle of the circle. This performance was observed many times but there were no further developments, and the birds never paired or selected mates on this pond so far as I could observe.

Nesting.—In the summer of 1901 we found the canvasbacks breeding quite abundantly in Steele County, North Dakota. Even then their breeding grounds were being rapidly encroached upon by advancing civilization which was gradually draining and cultivating the sloughs in which this species nests. Since that time they have largely, if not wholly disappeared from that region, as breeding birds, and their entire breeding range is becoming more and more restricted every year, as the great northwestern plains are being settled and cultivated for wheat and other agricultural products. This and other species of ducks are being driven farther and farther north and must ultimately become exterminated unless large tracts of suitable land can be set apart as breeding reservations, where the birds can find congenial surroundings. As my experience with the nesting habits of the canvasback in North Dakota will serve to illustrate its normal methods, I can not do better than to quote from what I (1902) have already published on the subject, as follows:

The principal object of our visit to the sloughs in Steele County was to study the breeding habits of the canvasbacks; so, soon after our arrival here, late in the afternoon of June 7, we put on our hip-boots and started in to explore the northern end of the big slough shown in the photograph. In the large area of open water we could see several male canvasbacks and a few redheads swimming about, well out of gun range. Wading out through the narrow strip of reeds surrounding the open water, and working along the outer edge of these, we explored first the small isolated patches of reeds shown in the foreground of the picture. The water here was more than knee-deep, and in some places we had to be extremely careful not to go in over the tops of our boots so that progress was quite slow. We had hardly been wading over 10 minutes when, as I approached one of these reed patches, I heard a great splashing, and out rushed a large, light-brown duck which, as she circled past me, showed very plainly the long sloping head and pointed bill of the canvasback.

A short search in the thick clump of tall reeds soon revealed the nest with its 11 eggs, 8 large, dark-colored eggs of the canvasback and 3 smaller and lighter eggs of the redhead. It was a large nest built upon a bulky mass of wet dead reeds, measuring 18 inches by 20 inches in outside diameter, the rim being built up 6 inches above the water, the inner cavity being about 8 inches across by 4 inches deep. It was lined with smaller pieces of dead reeds and a little gray down. The small patch of reeds was completely surrounded by open water about knee-deep, and the nest was so well concealed in the center of it as to be invisible from the outside. The eggs were also collected on that day, and proved to be very much advanced in incubation.

The other nests of the canvasback that we found were located in another slough, about half a mile distant, which was really an arm of a small lake separated from the main body of the lake by an artificial dyke or roadway with a narrow strip of reeds and flags on either side of it. In the large area thus inclosed the water was not much more than knee-deep, except in a few open spaces where it was too deep to wade.

Here among open, scattered reeds, the pied-billed grebes were breeding abundantly. A few pairs of ruddy ducks had their nests well concealed among the tall thick reeds. Coots and yellow-headed blackbirds were there in almost countless numbers. Long-billed marsh wrens were constantly heard among the tall thick flags. Red-winged blackbirds, soras, and Virginia rails were nesting abundantly in the short grass around the edges. Marbled godwits and western willets were frequently seen flying back and forth over the marshes acting as if their nests were not far away and clamorously protesting at our intrusion. Killdeers and Wilson phalaropes hovered about us along the shores. Such is the home of the canvasback, an ornithological paradise; a rich field indeed for the naturalist, fairly teeming with bird life. Our time was well occupied during our visit to this interesting locality, and the days were only too short and too few to study the many interesting phases of bird life before us, but we devoted considerable time to the canvasback, and, after much tiresome wading, succeeded in finding three more nests in this slough. The first of these was found on June 8, while wading through a thick patch of very tall flags, higher than our heads; we flushed the female from her nest and had a good look at her head as she flew out across a little open space. The nest was well concealed among the flags, but not far from the edge. It was well built of dead flags and reeds in water not quite knee-deep, and was sparingly lined with gray down. This nest contained 11 eggs, 7 of the canvasback and 4 of the redhead, which were collected on June 13 and found to be on the point of hatching.

Another nest, found on June 8, was located in a small, isolated clump of reeds, surrounded by water over knee-deep, on the edge of a large pondlike opening in the center of the slough, as is admirably illustrated in the photograph kindly loaned me by Mr. Job. The nest was beautifully made of dead and green reeds firmly interwoven, held in place by the growing reeds about it, and sparingly lined with gray down. It was built up out of the water, and was about 5 inches above the surface of the water; the external diameter was about 14 inches and the inner cavity measured 7 inches across by 4 inches deep. The nest and eggs, now in my collection, were taken on June 11, at which time incubation was only just begun; it contained eight eggs of the canvasback and one of the ruddy duck. All the canvasbacks' nests that we found contained one or more eggs of the ruddy duck or redhead, but we never found the eggs of the canvasback in the nest of any other species. The canvasbacks are close sitters, generally flushing within 10 feet of us, so that we had no difficulty in identifying them by the peculiar shape of the head; in general appearance they resemble the redheads very closely, except that the female canvasback is lighter colored above. The gray down in the nest will also serve to distinguish it from the redhead's nest, which is generally more profusely lined with white down.

In the extensive marshes near the southern end of Lake Winnipegosis and about the Waterhen River in Manitoba we found canvasbacks breeding abundantly in 1913, where we had ample opportunities for studying their nesting habits and the development of the young in captivity, in connection with Mr. Herbert K. Job's extensive experiments in hatching and rearing young ducks. On June 5 we examined seven nests of this species scattered over a wide area of marshy prairie; five of these nests contained 8 eggs each, one held 9 and one held 10 eggs, in various stages of incubation, but mostly well advanced. Most of the nests were in typical situations, more or less well concealed in thick clumps of bulrushes (*Scirpus lacustris*) or flags (*Typha latifolia*), but several were located in open places among short sedges (*Scirpus campestris*) where they were in plain sight. As we approached a small pond hole surrounded by a wide border of these sedges, the brown dead growth of the previous season, our guide pointed out a nest, about halfway from the shore to the open water, on which we could plainly see the duck sitting, only slightly concealed by the low scanty vegetation. The nest was one of the handsomest I have ever seen, a large, well-built structure of dead reeds, flags, and sedges, placed in shallow water and built up 9 inches above it; it measured 18 inches in outside diameter with an inner cavity 4 inches deep and 8 inches in diameter; it was profusely lined with the characteristic gray down which covered the whole interior and upper part of the nest, as if more warmth were necessary in such an exposed situation. In this, and in other similar cases, where incubation was advanced the ducks sat very closely and allowed us to walk up to within a few feet before leaving the nest.

All of the slough-nesting ducks seem to be very careless about laying their eggs in the nests of other species, which may be due to inability to find, or lack of time to reach, their own nests. Occasionally nests are found which are used as common dumping places for several species, where eggs are deposited and perhaps never incubated; we found such a nest at Crane Lake, Saskatchewan, on June 7, 1905, which contained 19 eggs, of at least three different species— canvasback, redhead, and mallard, and possibly others; the nest was partially broken down on one side and some of the eggs had rolled out into the water; it was originally a canvasback's nest, but had apparently been deserted.

The down in the canvasback's nest is large and soft in texture, but not so fluffy as in the surface-feeding ducks. It varies in color from "hair brown" to "drab." The breast feathers in the nest are whitish, but not pure white.

Eggs.—The canvasback usually lays from seven to nine eggs, but the set is often increased, if not usually so, by the addition of several eggs of the redhead, ruddy duck, or other species. The eggs when fresh can be readily distinguished by their color, which is a rich grayish olive or greenish drab of a darker shade than that usually seen in the eggs of other species. They vary in shape from ovate to elliptical ovate and have much less luster than the eggs of the redhead.

The measurements of 88 eggs in various collections average 62.2 by 43.7 millimeters; the eggs showing the four extremes measure **66.8** by 43.2, 63 by **45.8, 56.5** by 40.7, and 57 by **38.8** millimeters.

Incubation lasts for 28 days and is performed entirely by the female. The males desert the females as soon as the eggs are laid and gather into large flocks in the lakes and large open spaces in the sloughs.

Plumages.—The downy young show their aristocratic parentage as soon as they are hatched in the peculiar wedged-shaped bill and head. The color of the upper parts—crown, hind neck, and back— varies from "sepia" to "buffy olive." The under parts are yellow- ish, deepening to "amber yellow" on the cheeks and lores, brighten- ing to "citron yellow" on the breast, fading out to "naphthalene yellow" on the belly and to almost white on the throat. The mark- ings on the side of the head are but faintly indicated; below the broad yellow superciliary stripe is a narrow brown postocular stripe and below that an indistinct auricular stripe of light brown. The yellow scapular patches are quite conspicuous, but the rump spots are hardly noticeable. The colors become duller and browner as the young bird increases in size.

Before the young bird is half grown, or when about 5 weeks old, the first feathers begin to appear on the flanks and scapulars; at about the same time small "russet" feathers appear on the face, and the head soon becomes fully feathered; the breast plumage comes next, then the tail; and the last of the down is replaced by feathers on the neck and rump before the wings are even started. The young bird is fully grown before the wings appear and is 10 or 12 weeks old before it can fly. The sexes are nearly indistinguishable up to this age, but the young male is more clearly "russet" brown on the head than the female; both have light throats and brown backs. The young male, however, makes rapid progress toward maturity and soon begins to acquire the red head and the vermiculated black and white feathers of the back; by November he has assumed a plumage much like the adult, except that all the colors are duller or mixed with juvenal feathers and the back is darker, about the color of an adult male redhead. By the following spring only a few vestiges of the immature plumage are left, a few brown feathers in the back, light edgings in the breast, and less perfection in the wings.

The canvasback has a partial eclipse plumage which it wears for a short time only. The head and neck become mottled with dusky and dull brown; the black chest is mixed with brown and gray feathers; and the belly is more or less mottled. Dr. Arthur A. Allen tells me that molting begins from the first to the middle of August, possibly somewhat earlier, as it is inconspicuous at first. By September 1 they are in full eclipse, such as it is. Breeding plumage begins to show in October and they are in full plumage again by November 1.

Food.—The principal food of the canvasback, or at least the food which has made it most famous as a table delicacy, is the so-called "wild celery" (*Vallisneria spiralis*); it is in no way related to our garden celery and is more commonly known as "eelgrass," "tape grass" or "channel weed"; it grows most abundantly in the Chesapeake Bay region and is supposed to be the chief attraction for the vast number of canvasbacks and other ducks which resort to these waters in winter; but it also grows abundantly all along the Atlantic coast in estuaries and tidal streams, where the current is not too swift, the long slender, ribbonlike leaves floating in or out with the tide in dense masses, often so thick as to impede the progress of boats or seriously interfere with the use of oars. The canvasback prefers to feed on the root of the plant only, which is white and delicate in flavor and said to resemble young celery; it is obtained by diving and uprooting the plant; the roots are bitten off and eaten and the leaves or stems are left to float away in tangled masses. While feeding on *Vallisneria* the canvasback is often accompanied by other species of ducks which appreciate the same food, such as the redhead, baldpate, and scaup duck; the redhead and scaup can dive almost as well as the canvasback and so succeed in pulling up the roots for themselves; but the baldpate has to be content with the parts discarded by the canvasback or with what it can steal by force; the baldpate frequently lies in wait for the canvasback and, as soon as it appears on the surface with a bill full of choice roots, attacks it and attempts to steal what it can; the American coot also persecutes the canvasback in the same way. Audubon (1840) says, writing of its food in Chesapeake Bay, that the *Vallisneria*—

is at times so reduced in quantity that this duck, and several other species which are equally fond of it, are obliged to have recourse to fishes, tadpoles, water lizards, leeches, snails, and mollusca, as well as such seeds as they can meet with, all of which have been in greater or less quantity found in their stomachs.

On the inland lakes, streams, and marshy ponds, along its migration routes and on its breeding grounds, the canvasback lives on a variety of food both vegetable and animal, such as aquatic plants of various kinds, wild oats, water-lily and lotus seeds, small fishes, crustaceans, mollusks, insects and their larvae. Dr. F. Henry Yorke (1899) has added the following list of plants eaten by the canvasback: Teal moss (*Limnobium*), blue flag (*Iris versicolor*), water chinquapin (*Nymphaea lutea*), tuber-bearing water lily (*Nymphaea tuberosa*), yellow pond lily (*Nuphar Kalamanum*), water milfoil (*Myriophyllum*), water starwort (*Callitriche*), bladderwort (*Utricularia*) and a number of other water plants. Grinnell, Bryant, and Storer (1918) say:

In California the canvasback partakes of more animal food, for wild celery does not grow in this State. On the shallow water of the tidelands and marshes it feeds extensively on crustaceans and shellfish, thereby acquiring a "fishy" taste and thus becoming undesirable as a table bird. The stomachs of some canvasbacks collected on San Pablo Bay contained clams (*Mya arenaria*), and snails (*Odostomia*, species); one stomach from Tia Juana Slough, near San Diego, contained periwinkles (*Cerithidea californica*), and another from the same place contained grass blades, stems, and roots. A stomach from Guadalupe, San Luis Obispo County, was filled with barley, there being 22 whole kernels and many hulls; but there is a possibility that this was bait put out by hunters.

In connection with the feeding habits of the canvasback it may be well to call attention to an interesting case of lead poisoning in this species, resulting from feeding on grounds which have been shot over considerably.

Mr. W. L. McAtee (1908) published the following account of it in The Auk:

Conditions similar to those described by Mr. J. H. Bowles for the Nisqually Flats, Puget Sound, exist at Lake Surprise, Texas. To the latter locality canvasbacks resort from November to March. About the 1st of January, each year, many of these ducks are found among the rushes along the shore in various stages of sickness. Some can dive, but can not fly, and all become emaciated. A part of these of course are cripples, but most of them, although free from wounds, are plainly diseased, and according to the belief of those who have had most experience with them, the cause is lead poison from shot in the gizzards. No fewer than 40 shot have been taken from a single gizzard and the shot generally bear evidence of more or less attrition. As the season advances, the diseased ducks gradually disappear; the greater part die, but some it is thought recover. According to the information at hand no other species than the canvasback is thus affected at Lake Surprise.

Ducks secure a great deal of their food by sifting mud through their bills; if shot are abundant in mud, it is not hard to understand how the birds may collect a considerable number in a day. Resisting digestion to a marked degree, as shot do, the quantity in the gizzard is added to day by day, the ducks continuing to feed over the same grounds, until finally the gizzard is clogged with shot, and malnutrition, if not actual poisoning, ensues. Epidemics, such as we now have evidence of on Puget Sound and at Lake Surprise, in all probability will increase in number, adding another to the almost overwhelming array of unfavorable conditions against which our ducks must more and more hopelessly struggle.

Behavior.—The flight of the canvasback, though apparently labored, is really quite rapid, strong, and well sustained. When migrating or when flying to and from their feeding grounds they fly in wedge-like flocks, usually at a considerable height and with more velocity than is apparent. When on the wing the canvasback can be recognised by the long, slender neck and head, carried in a downward curve, by the long pointed bill and by the sharp-pointed wings; it is a longer and more slender bird than the redhead; when sitting on the water it can be distinguished from the redhead or the scaups, almost as far as it can be seen, by the extreme whiteness of the back.

The canvasback is essentially a diving duck and one of the most expert at it; it swims low in the water like a grebe and dives quickly, swimming for long distances under water, using its wings for this purpose; if pursued it comes to the surface only for an instant, diving again promptly and swimming away so far and so swiftly as to distance its pursuer; it hardly pays gunners to chase the crippled birds, as they are tough and hard to kill, as well as skillful divers; well-trained retrievers have been taught that it is useless to attempt to catch them. The canvasback can dive to great depths and is said to be able to obtain its food at a depth of from 20 to 25 feet.

Mr. Thomas McIlwraith (1894) noted a peculiarity in its diving habits, which is decidedly grebelike:

Before going under water it throws itself upward and forward, describing a curve as if seeking to gain impetus in the descent, just as boys sometimes do when taking a header off a point not much above the water level.

The vocal performances of this species are not elaborate nor are they frequently heard. The male has a harsh, guttural croak or "a peeping or growling note. The female canvasback can quack almost as well as the black duck, and also gives voice to screaming *currow* when startled," according to Eaton (1910).

Fall.—The fall migration route of the canvasback from its main breeding grounds on the central plains of Canada is peculiar and interesting, as it has shown some marked changes within recent years; it has always been somewhat fan-shaped, spreading out in three directions; the two main flights have been, in a general way, southeastward to the Atlantic coast from Delaware southward, and southward through the Mississippi Valley to the Gulf of Mexico; a third line of flight of less importance takes a more southwesterly course to Mexico and the Pacific coast. There is also a southward flight along the Pacific coast of birds which have bred in British Columbia and Alaska. During recent years canvasbacks have been increasing in abundance, during the fall migration, in the vicinity of the Great Lakes, in New York and in southern New England, indicating a more northerly range or a more directly eastward migration route to the Atlantic coast. Previous to 1895 records of this species in Massachusetts were exceedingly scarce and it was regarded as very rare or a mere straggler; during the next few years records became more frequent and since 1899 there have been records of canvasbacks taken every year, with increasing frequency, until now the bird has become a regular, if not a common, visitor in certain localities. For a full account of this interesting change in habits, I refer the reader to Mr. S. Prescott Fay's (1910) excellent paper on the subject in The Auk. Such a marked change in a migration route is not easily accounted for, though several causes may have had their effect in bringing it about. I believe that the principal cause has been the increased population of the Mississippi Valley and the Central West, which has brought about the draining and cultivation of many of its former feeding grounds and resting places; this, with the increased persecution by gunners throughout its former migration route, has driven the birds farther northward to a migration route along the Great Lakes. The species seems to be declining in abundance on the Gulf coast, which would seem to support this theory, though it may mean a reduction in the numerical strength of the species as a whole. Its breeding grounds have become more restricted to more northerly localities, which has also tended to give it a more northerly migration route eastward. The theory has been advanced that the wild celery now grows more abundantly farther north than formerly, though I doubt if this can be proven, or if it has had as much effect as the other two causes.

Prof. Walter B. Barrows (1912) says of its fall migration in Michigan:

This duck is seen almost invariably in flocks, these gathering often into large companies of many hundred individuals. Like the redhead this species in Michigan is more common along the Great Lakes than on the ponds and streams of the interior, yet it occurs sparingly in the latter situations.

In the fall it reappears in October and in places where food conditions are favorable may remain until late December. Its favorite food, the "eelgrass," or so-called wild celery (*Vallisneria spiralis*) has been planted in several places during recent years and attracts many kinds of ducks.

Game.—On account of its world-wide fame as a table bird and its prominence as a game bird, it seems worth while to devote some space to the consideration of the canvasback from the standpoint of the sportsman and to give some account of the methods of hunting it. Professor Barrows (1912) says of the methods employed during its migration through Michigan:

Formerly the birds were slaughtered by all sorts of abominable devices, including night floating, punt guns, sail boats, and steam launches, * * * as well as by more legitimate methods of decoys. At present they are sometimes obtained by "sneaking" or drifting down upon flocks in the open water in a boat more or less concealed by rushes, bushes, and similar disguises, but the greater number are shot from blinds or hiding places over painted wooden decoys.

Good shooting used to be found on the inland lakes, on the early spring migration, which occurs while the lakes are still partially covered with ice. The gunner selects a small open water hole, which the ducks have been seen to frequent, where he anchors his decoys in the water, pulls his skiff up onto the ice and builds a blind around it of ice cakes, where he can lie concealed within easy gunshot of the water holes; a decoy, which can be made to dive by pulling a cord, will help to attract passing flocks which are looking for a feeding place. New arrivals will usually decoy readily to such places, but birds which have spent some time in the vicinity soon learn to avoid such dangerous water holes and frequent the places where they can feed in safety.

The coast region of Virginia and North Carolina with its numerous estuaries and tributary streams has always been the most famous winter resorts of canvasbacks, and many other species of wild fowl, in North America. Vast hordes of canvasbacks, redheads, scaup ducks, as well as geese and swans formerly frequented these waters, attracted by the mild climate and the abundance of food. Several generations of gunners, by persistent and constant warfare, have seriously reduced the numbers of these hosts of wildfowl, but the birds are still sufficiently plentiful to attract sportsmen in large numbers and to keep alive the various gunning clubs which now control nearly all of the best shooting grounds. Some of the more destructive methods of killing ducks, such as night shooting and wholesale slaughter with swivel guns, have been prohibited by law. Netting ducks in gill nets sunken a short distance below the surface proved very destructive, but was abandoned as the ducks caught in this way became water soaked and of inferior flavor.

One of the oldest and most sportsmanlike methods of shooting ducks on Chesapeake Bay is known as point shooting. The sportsman lies concealed in a blind, with a retriever to pick up his birds, and waits for passing flocks to come near enough for a shot. The best flight is early in the morning, between dawn and sunrise when the ducks are flying to their feeding grounds; they usually fly around the points rather than over them; but if the wind is favorable, they often come within gunshot. This kind of shooting requires considerable practice and hard shooting guns, for the canvasbacks fly swiftly, often high in the air and are hard to kill, all of which makes it attractive to the true sportsman. Similar shooting is obtained on narrow sand bars where the ducks fly directly overhead; this is even more difficult. Canvasbacks are also shot over decoys at the points, from blinds on the flats, and from water holes in the ice on the rivers.

An interesting ancient method of shooting canvasbacks was by tolling them in with a small dog, especially trained for the purpose. Some quiet place was selected where a large flock of canvasbacks was bedded a short distance offshore and where the hunters could conceal themselves in some suitable ambush near the water. A small dog was kept running up and down the beach after sticks or stones, with a white or red handkerchief fluttering from some part of his body, which would so arouse the curiosity of the ducks that they would raise their heads and swim in toward shore to study the cause of such peculiar actions. Often their discovery of the hidden danger came too late, for as they turned to swim away they would receive a broadside from a battery of guns and large numbers would be killed. Tolling is now prohibited in many places.

The old-fashioned dugout, in which the hunter lay concealed with his boat covered with eelgrass has been entirely replaced by the modern surface boat or battery, an ingenious contrivance from which more canvasbacks are shot than by any other method. It consists of a stout wooden box, just long enough and deep enough to effectually conceal a man while lying down, surrounded by a broad wooden platform, attached to its upper edge; the platform is also surrounded with frames covered with canvas; it is so constructed and ballasted that the platform floats flush with the surface of the water and the box is entirely below it; the platform is constantly awash, but the water is kept out of the box by projecting flanges. The battery is towed out to the shooting grounds and anchored with 200 or more wooden decoys anchored around it. The gunner is entirely out of sight, except from overhead, as he lies flat in the bottom of the box until the birds are near enough, when he rises and shoots. An assistant is needed with a sailboat, launch, or skiff to pick up the birds.

When the canvasbacks first come in the fall, they gather in large numbers in the salt waters of Chesapeake Bay. During November they come down into the fresh waters of Back Bay, Virginia, and Currituck Sound, North Carolina, their favorite winter resorts. Here they feed on the roots and seeds of the foxtail grass which grows abundantly in these bays, but will not grow in salt water. The growth of this excellent duck food, on which the numerous duck clubs largely depend for their good shooting, is being much injured by carp and by the increasing abundance of swans; both of these species root up or trample down this grass so extensively that the feeding grounds for ducks are seriously injured. An open season on swans might reduce their numbers and improve the duck shooting. The canvasbacks, like the redheads, will feed in the bays all day, if not disturbed, but usually large flocks, or flocks of flocks, may be seen flying out to sea in the morning and back again at night.

Winter.—The canvasback is a late migrant and often lingers in the vicinity of the Great Lakes until driven farther south by the freezing of its favorite lakes and ponds, which sometimes proves disastrous. Mr. Elon H. Eaton (1910) says, of its occurrence in the central lake region of New York:

The winters of 1897–98 and the three following winters were remarkable for the large flocks of canvasbacks which appeared about the 1st of December on these waters and remained until early in March. On Canandaigua Lake a flock of nearly 1,000 canvasbacks passed a large part of the winter, and on Keuka Lake flocks of 200 birds were frequently seen. In February, 1899, many of these ducks were killed on Canandaigua Lake about the air holes which remained open. Most of those killed were in poor flesh and some were picked up on the ice in a starving condition.

The freezing of Cayuga Lake in February, 1912, caused the death of many canvasbacks and other ducks by starvation; I quote from Mr. Alvin R. Cahn's (1912) interesting paper on the subject as follows:

These ducks suffered, to all appearances, as much as any species on the lake. A flock of 22 was approached to within 30 feet one afternoon before they gave any heed, but finally they rose heavily and flew low over the ice a distance of 60 yards, where they lit, and immediately assumed a resting posture. Two of these ducks were captured alive, both being taken almost as easily as one would take an apple from the ground. The first made one feeble flight when approached, but that was all. He was followed and picked up off the ice without a struggle. The second was taken from the ice without having made any attempt to fly. The condition of both of these birds was pitiful, to say the least. Hardly able to stand erect, and too feeble to mind what was going on around them, they sat on the ice in a more or less dazed condition. The feathers were unpreened, and those of the breast and belly were yellow and matted with grease. Both of these birds were found on the ice of Fall Creek. There are records of 22 canvasbacks that were found dead within this area.

DISTRIBUTION.

Breeding range.—Western North America. East to the eastern edge of the prairie region in central Manitoba (Lake Winnipegosis, Lake Manitoba, and Shoal Lake), rarely in southern Minnesota (Heron Lake), and casually in southern Wisconsin (Lake Koshkonong). South to central western Nebraska (Garden and Morrill Counties), northern New Mexico (Cimarron), northern Utah (Box Elder and Davis Counties), and western Nevada (Pyramid Lake). West probably to eastern or central Oregon and Washington and to southern British Columbia (Lumby and Grand Forks), and central British Columbia (Lac la Hache). North to central Alaska (Fort Yukon), northern Mackenzie (Anderson River), and Great Slave Lake (Fort Rae and Fort Resolution).

Winter range.—Southern North America. East to the Atlantic coast of United States. South to Florida, the Gulf coasts of Louisiana and Texas, central Mexico (valley of Mexico and Mazatlan); rarely to Cuba and Guatemala. West to the Pacific coasts of northern Mexico and United States. North to southern British Columbia (Puget Sound region and Okanogan Lake), northwestern Montana (Flathead Lake, until frozen), northern Colorado (sparingly), northeastern Arkansas (Big Lake), southern Illinois, and eastern Maryland (Chesapeake Bay); rarely as far north as Lakes Erie and Ontario, and eastern Massachusetts (Boston).

Spring migration.—Average dates of arrival: Iowa, Keokuk, March 12; Minnesota, Heron Lake, March 28; Nebraska, central, March 14: North Dakota, northern, April 18; Manitoba, southern, April 21; Pennsylvania, Erie, March 13 to 26; Ohio, Oberlin, March 17; New York, Cayuga Lake, April 1.

Fall migration.—Early dates of arrival: Maine, Pittston, about October 8; Long Island, Mastic, October 11; Virginia, Alexandria, October 15; California, southern, October 20. Late dates of departure: Maine, Falmouth, November 14; Rhode Island, Middletown,

November 18; Minnesota, Heron Lake, November 27; Pennsylvania, Erie, December 21.

Casual records.—Accidental in Bermuda (October 30, 1851). Rare or accidental on migrations east to New Brunswick and Nova Scotia.

Egg dates.—Manitoba, Saskatchewan, and Alberta: Eighteen records, May 26 to 27; nine records, June 1 to 11. Minnesota and North Dakota: Twelve records, May 9 to June 25; six records, May 31 to June 11. Colorado and Utah: Four records, May 23 to June 20.

FULIGULA FULIGULA (Linnaeus).

TUFTED DUCK.

HABITS.

This widely distributed Palaearctic species is closely related to our ring-necked duck and might be said to replace it throughout its extensive breeding range from western Europe to extreme eastern Asia. Audubon (1840) says, referring to the ring-necked duck:

We are indebted for the discovery of this species to my friend the Prince of Musignano, who first pointed out the difference between it and the tufted duck of Europe. The distinctions that exist in the two species he ascertained about the time of my first acquaintance with him at Philadelphia in 1824, when he was much pleased on seeing my drawing of a male and a female, which I had made at Louisville, in Kentucky, previous to Wilson's visit to me there. Wilson supposed it identical with the European species.

Mr. Ned Hollister (1919) has also referred to this relationship.

There is, so far as I know, but one record of the capture of a tufted duck in North American territory, for which we are indebted to Dr. Barton W. Evermann (1913) who reported the capture of a female on St. Paul Island, Alaska, on May 9, 1911. "The bird was accompanied by the male which escaped."

I have never seen this species in life, but fortunately Mr. J. G. Millais (1913) has written a very full and satisfactory life history of it from which I shall quote, as follows:

Throughout its range the tufted duck is essentially an inhabitant of open sheets of fresh water, preferring those of moderate size that have a considerable depth in the center, and whose shallows are overgrown with reeds and other aquatic plants. They also like lakes with numerous islands and backwater, not too narrow, where they can sit and preen in the shallows in non-feeding hours, and whose vegetation gives them protection from the wind. In fact, all ducks that frequent open lakes of fresh water dislike drafts and take full advantage of the cover that grows along the 'banks, either sitting under the lee, or resting and diving at such a distance from shore that some protection is afforded. It is only in still weather or moderate breezes that they assemble in numbers on the open and deep parts of a lake, or when subject to frequent disturbance.

Spring.—On large lakes, like Loch Leven, where tufteds intend to breed, most of the adult birds are paired off by the end of March, and keep closely together during the early part of the breeding season. There are, however, many small lakes and ponds where tufteds breed, which are not frequented by the birds in winter, owing probably to the fact that they have been frozen. On these the tufteds arrive in one small flock late in February or early in March, and at once commence courtship and pairing. As soon as they are paired they become very tame, and it adds much to the charm of a day's spring fishing on Loch Leven to see these charming, birds, with hosts of other ducks, circling round the boat, and taking but little notice of intruders in the sanctuary.

Courtship.—Early in March the large flocks of tufted ducks split up into smaller companies; and if the weather is fine, and they are observed with the glass, it will be seen that a constant commotion is taking place amongst their ranks. Two or three males are sometimes to be seen "showing off" before a duck, and vice versa, some drakes seem to have a decided attraction for the females, which swim rapidly round and alongside them, dipping their bills frequently, and uttering their harsh call. The courtship of the male tufted duck is probably the most undemonstrative of any of the *Anatidae*. I myself, and such good observers as Mr. Gerald Legge and Mr. Hugh Wormald, who have these birds constantly under notice both in a wild and domesticated condition, have never seen any show on the part of the male except the following: The male swims rapidly past the female but without turning his head in her direction, and extends the neck to its full length. At the same time the bill is raised to an angle of 45°, and retained in this position for some seconds, whilst it utters frequently a low gentle whistle something like the word *hoi*, and well-nigh impossible to express onomatopoetically. In many cases in which male birds are furnished with ornaments of exceptional beauty, we notice that these parts are displayed in courtship, but in the case of the tufted drake, the bird seems to be incapable of displaying his long crest in any fashion, for at the period of courtship it hangs limp at the back of the head as at other seasons. In moments of excitement the pupil of the eye almost disappears, as it does in so many birds, and the golden iris seems to blaze with unusual fire.

Nesting.—Considering the fact that tufted ducks pair early, it is somewhat curious that they are not by any means early breeders. It is not long before they seek out a suitable nesting site, but it is generally well into May before the females think of nesting. The site chosen is generally only a few yards from the water, often amid dense herbage or the top of a sloping bank on some island, tongue of land, or embankment. At Patshull 8 or 10 nests are annually placed on a low dike separating two small lakes. The distance of each nest from the water would be 2 to 5 yards and the site hidden in rushes and coarse grass where a few stunted willows grow. I have found them with little covering but a few grass blades, and at other times, some 20 yards from the water, in a thicket of willows, Scots fir, bramble, and reeds. A favorite position is in center of a tuft of rushes, only slightly raised above the level of the lake. Naumann says they will travel as far as 80 to 100 paces from the water to make the nest in a clump of sedge or osier, rushes or tufts of grass, in places once wet and now dry.

Eggs.—Eggs rather coarse in texture, with some gloss; color yellowish brown. Sometimes with greenish tinge. Shape frequently a much elongated ellipse or long oval. Average size of 150 eggs, 59.0 by 40.9 millimeters; maximum, 65.9 by 46.3 and 63.9 by 47.2; minimum, 53 by 38 (or in inches, 2.32 by 1.61). (F. C. R. J.)

The number of eggs in a clutch usually ranges from 8 or 9 to 10, but at times much larger numbers are found. R. J. Ussher has recorded a nest with 14 from Ireland, and in May, 1899, Mr. Malloch, of Perth, sent me a photo of a nest high up on the old castle at Loch Leven which also contained 14 eggs. The Rev. F. C. R. Jourdain has met with clutches of 11, 12, 15, 16, and 18, but the last seemed certainly to be the produce of two ducks. The late T. E. Buckley found clutches of 16 and 17 in Caithness. Newton in the *Ootheca* mentions 21 eggs as found in one nest, and Jourdain found a duck sitting on a pile of 28 eggs at Osmaston, Derbyshire, which, of course, she was quite unable to cover, but in this case about five ducks were laying in one nest. Stuart Baker speaks of 40 eggs as found in one nest, but gives no details.

Young.—The incubation period is somewhat variable, lasting from 23 to 28 days. A female sits very close, and only deserts her nest in presence of urgent danger. She plucks her breast of the dark-gray down, and surrounds her eggs with it, as well as covering them with it when necessity compels her to obtain some food. When the young escape from the eggs, they follow the mother at once to the water, and crowd very closely round her as she swims. If disturbed by man, she will fly a short distance and dive, when the young, even if very small, at once imitate her movements. In a very few days the young are expert divers. During the first days of life the young are largely fed by the mother, or, to speak more correctly, have food placed before them by the parent, who obtains it from the bottom and then breaks it up, when it is at once swallowed by the hungry brood. All the time she is so engaged the latter are busy catching flies and *diptera* on the surface as they swim along.

Young tufted ducks begin to dive very soon after they enter the water. Mr. Wormald allows his young birds to enter a pond and seek for food as soon as they are hatched. As instancing their lack of knowledge in the art of diving and their quick acceptance of this method of gaining their food, Mr. J. Whitaker tells me the following interesting fact, which he noticed at Rainworth in the summer of 1912. A female tufted duck led her bunch of young ones which had just been hatched, to the middle of the pond. She then dived immediately; the young rushed in every direction on the surface of the water, evidently under the impression that they had lost their mother. She reappeared in a minute, however and all the brood hurried to her side.

At the next dive they did not appear to be so frightened, but looked about waiting for her reappearance. The third time she dived two of the young ones copied her movements, and in a very short period the whole of the family were diving with their mother in quite professional fashion. This little incident shows how quickly education may be completed in birds whose instincts naturally trend in certain directions.

Food.—When diving for its food the tufted duck makes a full semicircle with the head and neck, and, giving a vigorous kick, passes quickly out of sight, leaving a boil on the troubled waters. It remains below the water from a few seconds to half a minute, and finds most of its food on the bottom. Like most of the fresh-water diving ducks, it will take quantities of food on the surface such as flies, *diptera*, and duckweed, of which it is especially fond. Even when quietly preening on shore I have seen a tufted duck dash at and swallow a small frog that incautiously sprang into a shallow beside it. Most authorities speak of the food as being entirely animal, but this is not the case. Dresser, however, does not make this mistake, and correctly states that it will eat roots, seeds, and the buds of aquatic plants. I have never seen the tufted duck actually feed on land, for we must not regard habits developed in confinement as natural. Its chief food consists of aquatic animals of various kinds, fresh-water mussels and snails, insects, frogs, and tadpoles. Various tufted ducks that I have kept in confinement caught quantities of flies, water beetles, small fish, and ate large quantities of pondweed. They can, however, be easily "fed off" on to grain. "In the stomachs of some killed in Bavaria," says Naumann, "Jackel found fish spawn, a grass frog (*Rana esculenta*), mussels (*Pisidum fontinale*), the larvae of *Phryganea* and *Ephemera*, and the seeds of *Polygonum amphibium*, *persicaria* and *Lapathifolium*, *Rumex*, and *Potamogeton*." The stomachs of tufted ducks generally contain a quantity of sand, fine shell, or small stones.

Behavior.—Unless the sun is shining, when the snow-white flanks of the males appear bright and glistening, the appearance of this duck is very black. It swims low in the water, with the head well sunk between the shoulders. The tail is usually carried just above the water, but when alarmed, wounded, or bent on feeding, it can sink the body and depress the tail below, or even just under the water. At a short distance the golden eye and blue bill are very noticeable, and in the spring the eye of the male is very bright, the pupil being almost indistinguishable, especially when courtship is proceeding. If alarmed near the shore, the tufted is very quick to apprehend danger. It raises the head, stiffens the neck, sinks the body slightly, and at once commences to swim to deep water at a considerable pace. Should it consider that it is not possible to gain a point beyond gun range by swimming, it rises at once with considerable splashing and some noise, especially if the day is calm or the wind offshore, and then quickly rising with rapid beating wings, it passes away. Before leaving a lake tufted ducks always circle over the water many times sometimes rising to a height of 20 or 30 yards, and sometimes diving through the air toward the water again, which they always seem loath to abandon. The flight is rapid and very straight once the birds have decided on their course. They have a very black-and-white appearance in the air, and if the sun is upon them, even a glistening or "twinkling," which can be recognized from a great distance.

During the day the companies of tufted ducks spend most of the time in resting, preening, and feeding, but as evening comes on they become restive and keep much on the wing. Like many other ducks, most of their journeys are performed at night, which fact is proved by their frequent disappearance from certain lakes and appearance in the morning on others. When traveling over short distances the flight is generally performed within gunshot of the land or water, but when making longer journeys they mount to a considerable height in the air like the golden-eye. They generally fly in very close irregular companies in a swift arrowlike manner, swinging and swaying to take advantage of any wind breaks, hills, woods, etc. In summer single pairs of birds will resort to any pond that is quiet and undisturbed, preferring those that are well lined with sedge, rushes, grassy banks, embankments, and heather islands. In autumn the immatures may be found even in pools in fields, wide drains, and large sewage tanks. I have killed several in an unsavory bog right in the heart of the town of Glasgow known as the Postle Marsh, but they do not seem partial to swift-flowing rivers and if found there it is certain that they are only on passage. As a rule they come ashore on long low tongues of land or small islands from which a good view may be obtained, and are very quick to take alarm given by the cries of other birds or the use of their own eyes. During gales of wind they are very clever to take advantage of the shelter of islands or headlands, and yet remain on the water just out of shot of any spot that may hide a gunner. Resting on the water they appear to be asleep, yet their little feet are in motion all the time so as to prevent the wind from drifting them too far into rough water Thus they will maintain one position for hours at a time.

On shore they walk slowly and clumsily, with a decided roll. On the water they are expert divers and, when feeding, keep in close companies. When feeding they dive all together or very quickly one after the other, remaining below from a few seconds to a minute (generally 50 seconds). They are very buoyant and rise to the surface with a "jump" at different points, when they at once reassemble and commence diving again. In this manner they spend a great part of the day. When at the bottom, in clear water, they probe the mud, sand, or pebbles in search of food, and, like the golden-eye, I have seen them turn over stones of considerable size with the bill. The food is swallowed under water as a rule, but if a fish, frog, or large piece of succulent root or vegetable matter is found it is brought to the surface and crushed or broken up before being swallowed. The usual cry, uttered by both sexes, but somewhat louder and harsher in the case of the female, is *korr, korr, korr*, or *ka-ka-ka, karr*. They emit this when rising, quarreling, or suddenly alighting, or on being scared. The call of the male in spring is a low gentle whistle, and the onlooker must be at close range to hear it at all.

Game.—On the whole the tufted duck is not very easy to shoot on large sheets of water. But even in such places they may be stalked from behind banks or through woods, and watched when swimming within shot of the shore. When the flock is found on feed the gunner can then run in and obtain his chance as the birds rise to the surface. When little disturbed it is possible to sail within gunshot of a flock on the open water, but the old birds are usually difficult to obtain in this way unless they are "cornered" in some backwater or arm of the lake, when they will not fly overhead but pass within shot to the open waters of the lake. I have killed many by lying hidden on small islands in Loch Leven. There they will pass at close range on stormy days, but always keep well out of shot of the larger islands. Winged birds shot from the shore are seldom recovered unless shot again at once before they commence to dive, but from a boat winged birds may be tired out and killed more easily than pochard or scaup, since they neither possess the constitution nor vitality of these ducks. On small lakes or ponds tufted ducks are easily shot, as there is always some corner or point of land where the gunner can stand in bushes and hide himself to intercept them as they leave the place. It is merely necessary to find this spot and send a man round to drive the birds and they will come straight to the gunner. Moreover, in leaving small sheets of water tufted ducks do not rise high, and so offer an easy mark.

Winter.—Speaking of the winter habits of the species in Germany, Naumann says: "Although they seem fairly unsusceptible to cold, as long as ice does not entirely close the pieces of water to them, yet for all colder lands they remain birds of passage. From September or the beginning of October onward they assemble in small companies on larger sheets of water, and these flocks grow bigger in proportion as the year advances until finally in November or December they have become flocks of many thousands; at the approach of frosts they endeavor to prevent the complete freezing of certain places on the water by continued movement, and all at first start on their journey together if they can no longer succeed in doing this and the water is altogether covered with ice. They wander off in great flocks in search [of water] from which only a few occasionally through some mishap become separated, for afterwards on still open places on the rivers you seldom come across heron-duck (tufteds), and these will soon follow after, so that in the middle of the winter (unless it is quite a mild one) there are none to be seen in our country. Whilst those assembled in the north and east of Germany desert us in order, some of them winter in southern lands in Switzerland, Italy, and Hungary, on large inland lakes, or on the sea coasts."

Some remain in the sea or the north and east coast of Germany, but generally about the tidal estuaries. Their appearance on the open sea Naumann very rightly regards as exceptional.

DISTRIBUTION.

Breeding range.—Palaearctic region. From the Faroes, the British Isles, and Norway entirely across Europe and Asia to eastern Siberia and Kamchatka. North to about 70° N. and south to about 50° N.

Winter range.—Southern Europe, northern Africa to Abyssinia and southern Asia (India, China, and Japan), and Formosa.

Casual records.—Wanders to Madeira, Liberia, the Seychelle, Pelew, Marianne and Philippine Islands, and Borneo; also the Kurile and Pribilof Islands (St. Paul Island, May 9, 1911).

FULIX MARILA NEARCTICA (Stejneger).

AMERICAN SCAUP DUCK.

HABITS.

The scaup duck of Europe, which is closely related to or perhaps identical with our own bird, was so named, according to some of the earlier writers, on account of its habit of feeding on the beds of broken shellfish which are locally called scaup; but it is equally fair to assume that its name may have been chosen from its resemblance to one of its characteristic notes. The two American species of scaup ducks resemble each other in general appearance and almost intergrade in size and color, so much so that they have often been confused; still intelligent gunners have long recognized two species of "bluebills," one larger and one smaller. The subject of this sketch is known to the gunners as the "big bluebill," "big blackhead," and a variety of other names. It is very distinct from the lesser scaup in its distribution and habits; it breeds much farther north over a much wider area, which is practically circumpolar; its migration routes are quite different; and its winter home is mainly on our more northern seacoasts, where it is more of a salt-water duck than its smaller relative.

Spring.—From its principal winter home on the Atlantic coast the spring migration is decidedly northwestward, through the Great Lake region to the interior of Canada and Alaska; there is also a northward migration up the Mississippi Valley and another northward, and perhaps northeastward, from the Pacific coast. The species breeds abundantly in northern Alaska, but we do not know positively whether all of these birds have migrated from Pacific coast winter resorts or not. Dr. F. Henry Yorke (1899) says of the spring migration in the central valleys:

The first issue stays but a short time, soon passing northward as fast as the ice disappears, for they rarely leave the frost line until the ice has departed, working up in the interior, through the lakes and overflowed bottoms below St. Louis, following behind the ringbills. Some years they arrive in great numbers, while at other seasons they are very few. They prefer still to running water; naturally, large ponds and lakes, bayous, bays, and inlets are their favorite resorts.

Courtship.—The courtship of the scaup duck is described by Mr. John G. Millais (1913) as follows:

The male scaup anxious to pair approaches the female with head and neck held up to their fullest extent, the bill being raised in the air to an angle of 50° to 60°. If the female responds to this she also lifts the neck stiffly, at the same time uttering a crooning sort of note like the words *Tuc-tuc-turra-tuc*. If alarmed, or pretending to be so, she swims away quickly with powerful strokes, uttering her quacking cry, *Scaar-scaar*. When paired the female often comes up to the male and bows her head gently several times. The actual show of the male is a quick throw up of the head and neck, which is greatly swollen with air as it extends. At the summit of extension the bird utters a gentle cry like the words *Pa-whoo*, only uttered once. As he makes this show, the female sometimes swims round him, lowering the head and dipping the bill in the surface of the water and making a gentle call, *Chup-chup*, or *Chup-chup-cherr-err*. Quite as frequently the cry of the male is uttered after the head is raised and slightly lowered. The male also utters a very low whistle. Except the harsh loud cry of the female, all these calls of pairing scaup are very low in tone, and the spectator must be within a few yards of the birds to hear them.

Nesting.—The best known, and probably the most populous, breeding grounds of the greater scaup duck are in northern Alaska. Dr. Joseph Grinnell (1900) describes three nests, which he found in the Kotzebue Sound region, as follows:

In the Kowak Delta this species was quite common in June, and on the 14th of that month I took a set of 11 fresh eggs, also securing the female as she flushed from the nest. This nest was on a high, dry hummock, about 10 yards from the edge of a lake. It was almost hidden from view by tall, dead grass of the previous year's growth. The eggs rested on a bed of finely broken grass stems, while the rim of the nest was indicated by a narrow margin of down. A second set of 10 fresh eggs was taken on the same day and the nest was similar in construction, but was out on the tundra between two lakes, and fully a quarter of a mile from either. A set of seven fresh eggs taken on the 15th was quite differently situated. The nest was almost without feathers or down, and consisted of a neat saucer of matted dry grass blades, supported among standing marsh grass and about 4 inches above the water. It was in a broad, marshy swale about 30 feet from a small pond of open water. The swale was drained into the main river channel by a slough, so that in this case there was little danger of a rise in the water of more than an inch or two.

Mr. Hersey collected four sets for me in the vicinity of St. Michael in 1915. Three of the nests were more or less concealed in tufts of grass close to the shores of small ponds; the nest cavities were lined with fine, dry grass and in one case a well-formed nest of this material was made; no down was found in the newest nests containing fresh eggs, but, as incubation increased, considerable down became mixed with the grass. One nest, found June 19, was in a clump of dead flags in a pond 3 feet out from the shore and surrounded by water; the nest was made of bits of broken flags mixed with dark gray down and a few white breast feathers of the duck; it contained eight eggs.

Mr. Chase Littlejohn (1899) has published the following notes on a nesting colony of this species which he found on an island near the end of the Alaska Peninsula. He writes:

The island contains about 4 acres, one-half of which is about 50 feet above sea level; but on both the east and west ends there is quite an area only a few feet above water. These gravel points are covered for the most part with a species of salt weed less than 1 foot in height, common to the seashore of that country. Among these weeds on the west end there is a colony of about 50 pairs of scaups which have, to my knowledge, bred there for several years; while on the east end not a single nest can be found, although the conditions are practically the same. Furthermore, there is quite an area on the west end well suited to their wants; but they prefer to occupy a narrow strip along the edge of the weeds and place their nests close together, some of them not over 2 feet apart, others 10 at the most, showing that they prefer to be neighbors, I can not remember one isolated nest of *Fuligula marila*, and I have found many.

At the eastern extremity of its American breeding range the scaup duck has been found breeding, at least twice, in the Magdalen Islands, Quebec. Mr. Herbert K. Job found a nest, on June 29, 1900, in a small grassy islet, one of a series of small islets known as the "egg nubbles," in the great pond near East Cape; the nest was a bed of down in the thick grass and held nine fresh eggs. I have explored this pond several times since, but have never succeeded in even seeing a scaup duck. Rev. C. J. Young also found them breeding here in 1897 and sent Prof. John Macoun (1909) the following note, received from one of his correspondents:

I found a bluebill's nest in a strange place, after you left me. It was in a bunch of rushes at the head of the bay, growing in water that took me up to my middle to reach them.

The greater scaup may breed in North Dakota, Manitoba, and Saskatchewan, but during our various explorations in these regions we found no positive evidence to prove it. There is, however, a positive nesting record of the species farther south. Mr. W. H. Collins (1880) reported finding a nest at St. Clair Flats, Michigan, in 1879, which he identified by shooting the female.

The nest was built in a tuft of flags and composed of rushes and wild rice lined with some down and feathers. It was situated similarly to the redheads' nest resting in the water, and being held in place by the tuft of flags in which it was built.

The down in the nest of the scaup duck is small, soft, and compact in texture and "clove brown" or "bone brown" in color, with small inconspicuous, lighter centers. The breast feathers mixed with it are small and white or grayish white.

Eggs.—The scaup duck lays ordinarily from 7 to 10 eggs; sometimes only 5 or 6 constitute a full set and as many as 19 and 22 have been reported; probably these larger sets are the product of two females. The eggs can usually be recognized by their size and color. The color is about the same as in eggs of the lesser scaup and ring-necked ducks, a much darker olive buff than in other ducks' eggs.

It varies from "deep olive buff" or "olive buff" to "yellowish glaucous." The shell is smooth, but not glossy when fresh. The shape is usually elliptical ovate. The measurements of 180 eggs in various collections average 62.4 by 43.7 millimeters; the eggs showing the four extremes measure **68.5** by 44, 59 by **48**, **54.5** by 41.5 and 66.3 by **40.7** millimeters.

Young.—The period of incubation is said to be about four weeks, but probably it is nearer three and one-half weeks. This duty is performed by the female alone, as she is deserted by her mate as soon as incubation begins. Mr. Hersey's Alaska notes state that "there is little doubt that the males of this species go out to sea as soon as the females have laid their eggs and there molt into the eclipse plumage. I never saw any about the tundra ponds after incubation had begun." The female evidently assumes full care of the young also, leading them about in the ponds and marshes and teaching them to catch flies and other insects. Mr. Hersey came upon a female with a brood of nine young in a large pond out on the tundra. She did not fly or dive, but, calling her young about her, swam to the farther side of the pond. As he walked around the shore she kept at a distance and would not allow him to come nearer. When he withdrew she came ashore with her brood and led them away.

Plumages.—The downy young scaup duck is a swarthy duckling, deeply and richly colored with dark brown on the upper parts. The crown, hind neck, and entire back are a deep rich "raw umber," darker than any color in Ridgway's standards, with glossy reflections of bright "argus brown"; this color invades the lores and cheeks and shades off gradually on the neck and sides into the color of the under parts; the sides of the head and neck are "old gold" or "olive ocher," shading off to "colonial buff" on the throat and to "cream buff" and "cartridge buff" on the belly; an area of darker color, approaching that of the upper parts, encircles the lower neck and fore breast and invades the posterior under parts, restricting the light-colored belly. There are no light-colored spots on the scapulars and rump, as seen in the surface-feeding ducks. All the colors become duller with increasing age. The white plumage of the breast and belly is the first to develop, then the brown scapulars, the tail, the head and the back; the young bird is fully grown before the last of the faded down disappears from the neck; and the wings are the last of all to be developed. This flapper stage lasts all through August and into September, while the adults are also flightless and in the partial eclipse plumage.

In the first fall plumage in September young birds of both sexes are much alike and resemble the adult female superficially, but the white face is confined to the lores and chin, instead of including the forehead, as in the adult female, and it is more or less mottled with brown; the head, neck, and chest are paler brown than in the old female. At this age males can generally be distinguished from females by having the lesser wing coverts somewhat vermiculated with grayish white. In October young birds begin to assume a plumage more like the adults in both sexes and a steady progress toward maturity continues through the winter and spring. During October the greenish-black feathers begin to appear in the head and neck; the brown feathers of the back are replaced gradually by gray vermiculated feathers; and the first white feathers vermiculated with black appear in the scapulars. In November the first black feathers appear in the chest. By February the head is practically adult, and the remainder of the plumage closely resembles that of the adult, before summer. Young birds of both sexes do not breed the first spring but remain in flocks by themselves completing the molt.

A complete molt of both old and young birds occurs in summer, but it produces only a partial eclipse plumage, and after this molt young birds become practically indistinguishable from adults, although the full perfection of plumage is not acquired until a year later. Mr. Millais (1913) says of the eclipse plumage:

The adult male is somewhat late in assuming its eclipse dress, and seems to require to be in good health to attain it, for both pochards and scaup which I have kept in confinement have not fully changed as the wild birds do. About the middle or end of July the adult male passes into a fairly complete eclipse. The whole of the wings, scapulars, back, rump, tail and chest are at once molted direct to the winter dress, a feature of the chest feathers being a broad band of white on the edge of every feather. But an intermediate or temporary plumage for July, August, and September is furnished in a large number of eclipse feathers for parts of the head, neck, mantle, and flanks. The head becomes a dull brownish black, showing light gray on the cheeks (due to the old winter feathers reaching the extremity of their length). A few white feathers come into the lores (showing a distinct affinity to female plumage), the neck assumes a gray collar, and the nape and mantle, instead of being black, are filled with new gray and black vermiculated feathers similar to those on the back. The flanks instead of being white as in spring and winter, are now filled with white feathers finely vermiculated with brown. All of these new eclipse feathers are again molted gradually. From the end of September, when the bird is still in eclipse dress, till the end of October new winter feathers are constantly coming in and displacing the old ones; and the full winter plumage is not assumed until November.

Dr. Arthur A. Allen tells me that in both of the scaup ducks molting begins about mid-August and the birds are in full eclipse by mid-September. Breeding plumage begins to show again in mid-October, but the full plumage may not be attained until the following April, though some birds, probably the oldest, are practically in full plumage by December 1st.

Food.—The feeding grounds of the scaup duck are mainly in fairly deep water at a safe distance from the shore where their food is obtained by diving; they are expert at this and can remain under water for 50 or 60 seconds. Where food is plentiful they often feed in large companies, diving separately, indiscriminately, or all in unison; they show no particular system in their manner of diving and are not very careful about posting sentinels to watch for dangers; sometimes the whole flock will be below the surface at the same time, so that an approach is fairly easy. In their summer homes in fresh-water lakes and ponds they more often feed on or near the surface, where they live on fish fry, tadpoles, small fishes, small snails and other mollusks, flies, and water insects; they also eat some vegetable food, such as the buds, stems, roots, and seeds of floating and submerged water plants. Dr. F. Henry Yorke (1899) has identified

212   BULLETIN 126, UNITED STATES NATIONAL MUSEUM.

the following genera of water plants in the food of this duck: *Vallisneria, Lymnobium, Zizania, Piper, Elymus, Iris, Nuphar, Nymphaea, Myriophyllum, Callitriche,* and *Utricularia.*

During the winter on the seacoast its food consists of surface-swimming crustaceans, crabs, starfish, and various mollusks; small mussels, obtained by diving in the mussel beds, form the principal part of its animal food at this season; but it also eats considerable vegetable food, such as the buds and root-stocks of wild celery (*Vallisneria*), and the seeds and succulent shoots of *Zostera marina*. In the Chesapeake Bay region the scaup ducks feed on the roots of the wild celery with the canvasbacks and redheads, where they are quite as expert as any of the diving ducks in obtaining these succulent roots; consequently they become very fat and their flesh, which is ordinarily undesirable, acquires an excellent flavor.

Mr. Arthur H. Norton (1909) found the stomach of a scaup duck, killed on the coast of Maine in winter, filled with shells of *Macoma balthica*. Dr. J. C. Phillips (1911) reported that the stomachs of scaup ducks, killed on Wenham Lake, Massachusetts, in the fall, "held animal and vegetable matter in equal proportions, the items being bur reed, pondweed, and bivalves (*Gemma gemma*)."

Behavior.—Mr. Millais (1913) gives the best description of the flight of this species that I have seen; I can not do better than to quote his words, as follows:

In flight they proceed at a rapid pace in a somewhat compact formation. The birds fly very close together, and the sound produced by their wings is somewhat loud and rustling. On rising to fly the neck is straightened out, and the bird runs along the surface of the water with considerable splashing for a few yards, but the distance traveled on the surface of the water is coincidental with the amount of head wind. In calm weather, if not much disturbed, they are always liable to take to wing, and if the boat does not press them they will swim away for a long time before turning around and facing up wind. When sitting on the sea, scaup often keep in one long unbroken line parallel to the coast, and when rising the first bird at one end takes wing and is followed in order right across the flock. When flying they keep at a moderate elevation, but if the wind is offshore and they are desirous of coming in to some estuary, they nearly always strike the sands or part of the coast line which they desire to cross at exactly the same spot every day and at a considerable height. As they approach the waters of the estuary and feeding grounds the leading birds then often make a dive downward, their movements being followed in line by the rest of the flock, so that if the line of birds is a long one it often has a curious *waving* appearance. Doubtless this rising high as they approach the coast line is dictated by common sense, for it is on the sands and rocky shore they are most often shot at, and they learn caution from bitter experience. When on migration by day I have seen scaup circling at a great height, but when leaving the sea or open water for the feeding grounds at night scaup as a rule do not fly much above 30 feet above the land or water. I have, when waiting for duck on the mussel beds at dawn and sunset, occasionally obtained shots at flight at scaup, and the sound of their rushing wings has often foretold their approach, when, if they could be seen in time, I have occasionally made successful shots. When in small parties scaup may sometimes be seen flying in oblique formation like other ducks, but when in large companies they generally hold together in a solid phalanx, or in one long unbroken line massed in several places.

The scaup duck is a bold, strong swimmer, making the best of speed even in rough weather; it is a hardy sea duck, unexcelled in its powers of swimming and diving. It is not particularly shy and can usually be approached with a little caution; but it must be hard hit to be secured, as it is tough, has a thick coat of feathers, and is such a powerful swimmer and diver that it is useless to pursue a wounded bird unless it is shot over at once. It dives quickly and swims rapidly away under water with its wings tightly closed, as many of the best divers do. I have seen scaup ducks, which I had dropped as if killed, sit up and shake themselves, dive before I could shoot them over, and never show themselves again; if the sea is at all rough, they can easily escape without showing enough to be seen. Mr. Charles E. Alford (1920) has published some interesting notes on the diving habits of this species, which seems to dive with extreme regularity for definite periods; the dives varied in duration from 25 to 29 seconds and the periods between the dives varied from 11 to 19 seconds.

Except during the mating season, as described above, scaup ducks are usually silent. Their most characteristic note is a harsh, discordant "scaup, scaup," from which their name may have been derived. They also occasionally utter soft guttural or purring sounds.

Fall.—The fall migration is the reverse of the spring route, southeastward through the Great Lakes to the Atlantic seacoast and southward through the Mississippi Valley. Professor Cooke (1906) says that these two routes are clearly revealed in the fall, "when this species scarcely occurs in Indiana, though common both to the east and west of that State." The first flights come fairly early in the fall, with the first frosts, probably made up of the more tender birds which have bred farther south and hatched out earlier. The later flights consist of hardier birds from the far north, which come rushing down ahead of the wintry storms and cold weather, probably driven out by ice and snow. They frequent the lakes, larger ponds, and rivers, feeding and resting on open water, even in rough weather; they often gather in large flocks, which has given them the names, "raft duck" and "flocking fowl"; a dense pack of these unsuspicious birds resting on a sand bank in a river or floating on the surface of a pond often offers a tempting shot to the unscrupulous gunner.

Game.—The true sportsman, however, finds excellent sport in shooting these swift-winged ducks over decoys. They decoy readily to the painted wooden decoys used by the bay men of Long Island and the Chesapeake and large numbers are killed from the floating batteries, such as are used for shooting canvasbacks and redheads.

This method is well described by Mr. Dwight W. Huntington (1903) as follows:

Just before daybreak we reached the place determined upon and found it unoccupied. The battery was placed in the water, the decoys were arranged about it within close range, and my gunner sailed away to leave me lying below the surface of the bay in the box with its wide rim floating on the water. As the first light came in the east I could see the ducks, mostly scaups and redheads, flying swiftly across the dim gray light. Soon there was a rush of wings quite close to my head as a flock of blackheads swung in to the decoys. Sitting up I fired two barrels at the shadowy forms, but nothing struck the water, and the noisy whistling of wings was soon lost in the darkness. As the sun came up the ducks came rapidly, sometimes one or two, more often a flock. I shot at every one, with but poor success. The cramped position, the hasty shot from a sitting position, were new to me and strange, and it was some time before I began to kill the ducks. A single bird coming head on was about to settle to the decoys, when I fired at him at close range, and he struck the water dead. Shortly afterwards I made a double from a flock, and with growing confidence my shooting improved. I soon had a goodly lot of scaups showing black and white upon the waves as they drifted with the breeze. Meanwhile the bay man, who had been cruising far enough away not to alarm the ducks, approached and gathered in the slain.

Mr. Walter H. Rich (1907) says that on the coast of Maine—

most of the bluebills are killed from the "gunning float," the gunner clad in a white suit and the little craft itself "dressed down" to the water's edge with snow and ice to represent a floating ice cake. It is no wonder that the poor victims are "deluded" for it needs sharp eyes and close attention to make out anything dangerous in an object so harmless in appearance. There is commonly little trouble in approaching within easy range of a flock if the gunner is skilled in handling his craft, but to get within shot reach is not all, for any duck which can last out the New England winter will carry off a good load of shot, as the bird must have an abundance of vitality and an extra-heavy suit of underwear to endure the climate. Both of these our hero has.

Winter.—On the Pacific coast there is a southward migration route from the Alaskan breeding grounds and probably a southwestward flight from the interior. Mr. W. L. Dawson (1909) writes of their arrival on the Washington coast:

At Semiahmoo Spit, upon our northern boundary, the bluebills begin to arrive from the north about the 20th of September, and their numbers are augmented for at least a month thereafter. The earlier arrivals come in small flocks of from a dozen to 25 individuals, borne upon the wings of a northwest breeze, and as they pass the narrow promontory of sand, the waiting gunners exact toll of those which enter the harbor. Upon the waters of the inner bay, Drayton Harbor, the incoming birds assemble in a great raft, five or ten thousand strong, and if undisturbed, deploy to dive in shallow water, feeding not only upon the eelgrass itself, but upon the varied forms of life which shelter in its green fastnesses.

About half an hour before sunset, as though by some preconcerted signal, a grand exodus takes place. Flock joins flock as the birds rise steadily against the wind. Mindful of their former experience, the ducks attain a height of two or three times that at which they entered the harbor and, strong in the added confidence of numbers, the serried host, some 40 companies abreast, sweeps over the spit in unison—a beautiful and impressive sight. Some five minutes later a second movement of a similar nature is organized by half as many birds remaining; while a third wave, containing only a hundred or so of laggards, leaves the harbor destitute of scaups.

On the way to their winter homes on the seacoast these ducks often linger in the lakes until driven out by the ice and often many perish in the freezing lakes. Mr. Alvin R. Cahn (1912) describes such a catastrophe on Cayuga Lake, New York, as follows:

The largest flock seen was just off Portland Point. This flock was discovered at rest upon the ice, and so close together were they, and so numerous, that the birds gave the appearance of a solid black line, and it was not until one had approached to within 100 yards of them that one could be sure that it was indeed a flock of ducks. The birds were quite indifferent to being approached, and it was not until one was within 200 feet of them that they showed any signs of uneasiness. When within 100 feet, they rose slowly and flew some little distance down the lake, where they settled once more into their compact formation. It was not until they rose that one realized that there were easily over 400 ducks in the flock. It was all but impossible for these birds to rise clear of the ice. The indifference shown toward unguarded approach, the reluctance with which they rose, the short distance which they flew, in fact, their every action bespoke exhaustion and weakness. In a small piece of open, rapidly flowing water in Fall Creek, a female of this species was caught by hand without difficulty. The bird, too exhausted even to try to fly, could make no headway against the current, and was therefore easily captured. It was too weak to eat, and died within 24 hours. Two peculiar incidents with regard to bluebills have been brought to my notice. One specimen was found while still alive, in which over half the webbing of both feet had been frozen and dropped off. Another was found frozen in a cake of ice, nothing but the head and about half the neck protruding from the mass. The duck, still alive, was chopped out, when it was found that the ice had in some way frozen over the duck, leaving water next to the body. This was undoubtedly kept from freezing by the action of the legs and the body heat. The bird was uninjured, and after being fed, seemed little the worse for its experience.

Large numbers of scaup ducks spend the winter on the New England coast and they are especially abundant in the Vineyard Sound region, south of Cape Cod, and on the ocean side of Long Island. Here they may be seen in large flocks, sometimes numbering several hundred, riding at ease on the rough or choppy sea. Their movements are largely governed by the condition of the mussel beds on which they feed. From Chesapeake Bay to Currituck Sound, North Carolina, they are also abundant and are regarded as one of the desirable species of game birds. On the coast of Louisiana, according to Beyer, Allison, and Kopman (1906), "the occurrence of this species is confined chiefly to the colder parts of the winter. This species is seldom found away from the coast, and occurs more frequently on the open Gulf waters than any other species."

DISTRIBUTION.

Breeding range.—The North American form breeds east to the west coast of Hudson Bay (Churchill), southwestern Ungava (Great Whale River), and casually to the Gulf of St. Lawrence (Magdalen Islands and northern New Brunswick).   South to southeastern Michigan (St. Clair Flats, casually), formerly northern Iowa (Clear Lake), rarely, if at all, now in Minnesota or North Dakota, but probably still in southern or central Manitoba (Lake Winnipeg), central Saskatchewan (Prince Albert), central Alberta (Buffalo Lake), and central British Columbia (east of the Cascades).   West to the Aleutian Islands (Atka and Agattu) and the Bering Sea coast of Alaska. North to the Arctic coasts of Alaska and Canada.   In the eastern and southern portions of its breeding range it is rare or casual.   The European form breeds in Iceland, the Faroe Islands and northern Europe and Asia, from Scotland to Bering Sea, and from about 70° north latitude southward.

Winter range.—North America mainly on the seacoasts of the United States.   On the Atlantic coast from Maine to Florida, most commonly from southern New England and Long Island to North Carolina.   On the Gulf coasts of Louisiana and Texas nearly to the Mexican boundary.   On the Pacific coast from the Aleutian Islands to southern California (San Diego).   A few winter in the Great Lakes, and a few in the southwestern interior (Colorado, Utah, Nevada, Arizona, and New Mexico).   The European form winters south to the Mediterranean Sea, northern Africa (Algeria, Tunis and Egypt), the Black and Caspian Seas, the Persian Gulf, northern India (rarely), China, Japan, and Formosa (Taiwan).

Spring migration.—Early dates of arrival: Indiana, central, March 1; Ohio, Oberlin, March 9; Illinois, northern, March 6; Manitoba, southern, March 31; Yukon, Fort Reliance, May 1; Alaska, St. Michael, May 8, and Kowak River, June 1.   Average dates of arrival: Illinois, northern, March 23; Ohio, Oberlin, March 24; Ontario, southern, March 30; Iowa, central, March 16; Minnesota, Heron Lake, April 2; Manitoba, southern, April 16; Mackenzie, Fort Simpson, May 24.   Late dates of departure: Florida, Pinellas County, March 4; Long Island, Mastic, May 30; Massachusetts, Nantucket, May 1.

Fall migration.—Early dates of arrival: Labrador, Ticoralak, October 11; New York, Long Island, September 26; Virginia, Alexandria, October 18; South Carolina, Mount Pleasant, October 31.   Late dates of departure: Alaska, St. Michael, October 15; Quebec, Montreal, November 14; Minnesota, Heron Lake, November 27.

Casual records.—Rare or casual on both coasts of Greenland (Neanortalik, Godhavn, and Stormkap, June 21, 1907), eastern Labrador and Newfoundland.

Egg dates.—Alaska and Arctic America: Eleven records, June 14 to July 5; six records, June 15 to 20.   Manitoba, Saskatchewan, and Alberta: Twelve records, May 25 to July 6; six records, June 12 to 19.   Iceland: Eight records, May 30 to July 10.

FULIX AFFINIS (Eyton).

LESSER SCAUP DUCK.

HABITS.

Unlike the larger scaup duck, this species is distinctly an American duck, but of wider distribution on this continent. It is more essentially an inland species, showing a decided preference for the smaller lakes, ponds, marshes, and streams, whereas its larger relative seems to prefer the larger lakes in the interior and the seacoast in winter. Its breeding range is more extensive and its center of abundance during the breeding season is much farther south, its chief breeding grounds being in the prairie regions of central Canada and the Northern States. Though differing in distribution and in their haunts, the two species are closely related and much alike in appearance, so much so that so good an observer as Audubon failed to distinguish them; nearly all that he wrote about them evidently referred to the lesser scaup, with which he was most familiar, and he criticised Wilson for some of his remarks which evidently referred to the greater scaup. Adult males of the two species are, of course, easily recognized, but the females and young birds are so much alike and vary so much in size that they are often confused. Rev. W. F. Henninger writes me that a series of *Fulix affinis* which he has examined measure up to the minimum measurements given for *Fulix marila* and that the males show both purple and green reflections on the head; this suggests the possibility of intergradation between the two species.

Spring.—The lesser scaup duck is not one of the earliest migrants, but it begins to move northward from its winter home soon after the melting ice and snow begin to indicate the coming of spring. On its migration it follows the courses of the larger streams and rivers, but when it settles down to feed it soon spreads out into the sloughs, marshes and shallow ponds. Prof. Lynds Jones (1909) says that, in Ohio, "it literally swarms in the marshes during late March and the most of April, where feeding companies cover large areas of the open waters of the marshes." Where spring shooting is allowed it flies wildly about, seeking refuge on the open lakes beyond range, but on certain reservoirs where it is not molested it appreciates the security and becomes very tame. In such places a few birds linger well into the summer and some apparently remain to breed.

Courtship.—Very little has been published about the courtship of this species, but Audubon (1840) makes the following brief reference to it:

At the approach of spring the drakes pay their addresses to the females, before they set out on their journey. At that period the males become more active and lively, bowing their heads, opening their broad bills, and uttering a kind of quack, which to the listener seems produced by wind in their stomach, but notwithstanding appears to delight their chosen females.

Dr. Alexander Wetmore (1920) gives the following account of it:

A pair rested in open water in front of me when suddenly the female began to swim back and forth with the head erect, frequently jerking the tip of her bill up while the male drew his head in on his breast and lowered his crest, giving his crown a curious flattened appearance. The female turned alternately toward and away from the male, sometimes biting gently at him, while occasionally he responded by nipping at her with open mouth. At short intervals she dove toward him, barely sliding under his breast, and emerged at once only a few feet away, or at times advanced toward him brushing against him and then turning away. A second male that tried to approach was driven away by quick rushes though the female paid no attention to him. She continued her diving and finally at intervals the male began to dive with her, both emerging at once. As the display continued he joined her under the water more and more frequently and finally both remained below the surface for over 30 seconds where copulation apparently took place. When they emerged the female swam away for a short distance with the male following her. Frequently during these displays the female gave a peculiar rattling, purring call like *kwuh-h-h-h-h* while the males whistled in a low tone.

Nesting.—Although they arrive on their breeding grounds fairly early, they are very deliberate about nesting preparations and are among the later breeders. All through the extensive western prairies these little ducks may be seen, throughout May and the first half of June, swimming about in pairs in the little marshy creeks, sloughs, and small ponds; they are apparently mated when they arrive and seem to enjoy a protracted honeymoon. In the Devils Lake region in North Dakota we found the lesser scaup duck nesting abundantly in 1901 and examined a large number of nests. On the small islands in Stump Lake, now set apart as a reservation, we found 16 nests of this species in one day, June 15, and all of the eggs proved to be fresh or nearly so. The nests were almost invariably concealed in the taller prairie grass, but some nests were located under small rosebushes and one was placed against the side of a small rock surrounded by tall grass, but in a rather open situation. The nest consisted of a hollow scooped in the ground, profusely lined with very dark down mingled with a little dry grass and occasionally a white feather from the breast of the bird. The females seemed to be very close sitters; we always flushed the bird within 10 feet of us or less; but when once flushed they seemed to show no further interest in our proceedings. The males apparently desert the females after incubation has begun and flock by themselves in the sloughs or small ponds. Lesser scaup ducks occasionally lay in other duck's nests; we found one of their eggs in a gadwall's nest and one in a white-winged scoter's nest; but we found no evidence that other ducks ever lay in the scaup's nests.

In southwestern Saskatchewan the lesser scaup duck was not so abundant as in North Dakota, but still quite common; we found 6 nests in situations similar to those described above; three of these were on that wonderful island in Crane Lake, more fully described under the gadwall.   In Manitoba, about Lake Winnipegosis, we found a few nests one of which was in a different situation from any other we had seen; it was built like a canvasback's nest in the water near the edge of a clump of bulrushes (*Scirpus lacustris*), but it contained the dark down and the characteristic eggs of the lesser scaup.   Nests have been reported by other observers in such situations, but the nest is usually placed on dry ground.   MacFarlane (1891) found over a dozen nests of this species near the northern limit of the wooded country on the east side of the Anderson River, of which he says:

They were usually found in the midst of a swamp—a mere hole or depression in the center of a tuft of turf or tussock of grass, lined with more or less down, feathers, and hay.

Dr. Joseph Grinnell (1909) refers to a nest found by Mr. Littlejohn on an island in Glacier Bay, Alaska; it was at the edge of a small pond "placed within a heavy growth of grass about a foot from the water's edge, and consisted of grass stems lined with a little down from the parent's breast."

The down in the lesser scaup duck's nest is indistinguishable from that of its largest relative, "clove brown" or "bone brown" in color, with inconspicuous lighter centers.   The small breast feathers in it are white or grayish white.

Eggs.—The lesser scaup duck lays from 6 to 15 eggs, but the commonest numbers run from 9 to 12.   The eggs are like those of the larger scaup duck but they are decidedly smaller.   The shape varies from elliptical ovate to nearly oval.   The shell is smooth and slightly glossy.   The color varies from "ecru olive" or "dark olive buff" in the darkest eggs to "deep olive buff" in the lightest eggs.   When seen in the field the deep *café au lait* color of all the scaup duck's eggs is distinctive and unmistakable, but in cabinet sets it has usually faded more or less and is not so conspicuous.   It is always much easier to identify ducks' eggs in the field than in collections, for there is usually something about the eggs, the nest, or the bird which is distinctive.   The measurements of 88 eggs, in the United States National Museum and the writer's collections, average 57.1 by 39.7 millimeters; the eggs showing the four extremes measure 61.5 by 38, 59 by 42.5, and 50 by 35.5 millimeters.

Young.—Incubation is performed by the female alone and probably lasts from three to four weeks.   When the young are hatched she leads them to the nearest water, which is usually not far distant, and teaches them how to escape from their numerous enemies and how to catch their insect food.   While engaged in rearing and studying young ducks in Manitoba we employed Indians to catch the small young of this and some other species which we did not succeed in hatching in the incubators. Mr. Hersey's notes describe their methods as follows:

A brood of lesser scaup found on a small pond acted very differently from young golden-eyes. Instead of separating they drew close together and swam back and forth a few feet. The Indians plunge into the pond, clothes and all and drive the brood toward our end. They swim along quietly and as they near the end of the pond the Indians close in until they are within a few feet of the birds. Then suddenly the birds begin to dive, each one swimming under water past the men and coming up well out toward the middle of the pond. If the water is clear the Indians will watch the young bird swimming and catch them under water, but if muddy, they all get safely by and then the whole performance is repeated.

It often happens when a brood dives in this way, that one or more birds get separated from the rest. The single birds are picked out and captured first, while the rest of the brood wait at the other end of the pond. It is no easy matter to catch one of these youngsters. When he realizes he is being chased he makes every effort to get back to his brothers and sisters, pattering along the surface much faster than a man can move through the water. However, they usually head him off and he then returns to diving. After a while he gets tired and diving once more swims under water until close to the shore when he crawls into the grass. Once on land he loses no time but pushes his way rapidly through the grass. Unless his pursuer is quick he will yet make his escape, but the Indians aware of this habit watch the tops of the grass closely, following his movements by the slight waving and soon overtake him.

Plumages.—The downy young is darkly and richly colored. The upper parts are dark, lustrous "mummy brown" or "sepia," shaded with "brownish olive"; these colors are darkest and most lustrous on the posterior half of the back and lightest on the shoulders; the dark colors cover the upper half of the head and neck, the back and the flanks, fading off gradually into a dusky band around the lower neck and encroaching on the ventral region posteriorly. The color of the under parts, which covers the lower half of the head, throat, breast, and belly, varies in different individuals; in some it runs from "olive ocher" to "primrose yellow," but in most specimens from "chamois" to "cream buff"; these colors are brightest and richest on the cheeks and on the breast. The markings on the head are usually indistinct, but a superciliary buff stripe, a loral dusky stripe and a postocular dusky stripe are discernible in the majority of a series of 11 specimens in my collection. There is also an indistinct yellowish spot on each scapular region, but none on the rump. The colors become duller and lighter as the duckling grows older.

So far as I can judge from the study of available material, the sequence of plumages to maturity and the seasonal molts of adults are practically the same as in the greater scaup duck. Young birds do not breed during their first spring and become practically adult in plumage after their first complete summer molt, or when from 14 to 16 months old. The eclipse plumage of the adult male is only partial and not conspicuous. The adult female seems to have a distinct breeding plumage, which is much browner than the winter plumage and in which the white face wholly or partially disappears.

Food.—The feeding habits of the lesser scaup are much like those of the greater scaup, except that the smaller species is confined almost wholly to fresh water. Mr. Vernon Bailey (1902) writes:

Like all of the genus, the lesser scaups are great divers and keep much in the open lakes, often in large flocks, where they dive for food, or sleep and rest on the water in comparative safety. They can not resist the temptation of the rice lakes, however, and swarm into them by thousands to fatten on the delicious grain, which they glean from the mud bottoms after it has been threshed out by the wind and the wings of myriads of coots and rails. While they eat, the hunters lie hidden in the tall rice and on the ridges which they must pass in going from lake to lake, and in spite of their bullet-like flight the sadly thinned flocks show the penalty they have paid for leaving the open water.

Their animal food consists of small fry and fish spawn, tadpoles, pond snails and other small mollusks, worms, crawfish, water insects, and larvae. They also consume a variety of vegetable food among which Dr. F. Henry Yorke (1899) has identified all the plants mentioned under the preceding species. The stomachs of this species, taken by Dr. J. C. Phillips (1911) in Massachusetts, contained "seeds of burreed, bayberry, and saw grass (*Cladium effusum*), and snails (*Lunatia heros*) and ants."

Behavior.—The lesser scaup like its larger relative, is an expert diver and can remain under water for a long time, grubbing on the bottom for its food. Like many of the best divers, its large and powerful feet enable it to swim rapidly beneath the surface without the use of its wings, which are held tightly closed. It swims away so rapidly under water when wounded that it is useless to pursue it; it is said by gunners to cling to the weeds or rocks on the bottom until dead; it seems more likely that in most cases it swims away to some place where it can hide or that it skulks away with only its bill above water; Mr. W. E. Clyde Todd (1904), however has published the following note from Mr. Samuel E. Bacon:

I once wounded a duck of this species in shallow water and, wading out to where I saw it last, I found it holding to a strong weed by its bill, 2 or 3 feet below the surface, stone dead.

Audubon (1840) writes:

The scaup duck seems to float less lightly than it really does, its body being comparatively flat. It moves fast, frequently sipping the water, as if to ascertain whether its favorite food be in it. Then turning its head and glancing on either side to assure itself of security, down it dives with all the agility of a merganser, and remains a considerable time below. On emerging, it shakes its head, raises the hind part of its body, opens its short and rather curved wings, after a few flaps replaces them, and again dives in search of food. Should any person appear when it emerges, it swims off a considerable distance, watches every movement of the intruder, and finally either returns to its former place or flies away.

On the wing, as well as on the water, the lesser scaup duck is a very lively, nervous, and restless bird; its flight is very swift and often erratic with frequent unexpected twists and turns, which make it a difficult bird for the sportsman to hit; but it often flies in large flocks, closely bunched, or with a broad front of many birds abreast, which gives the gunner an opportunity for an effective raking shot. When much disturbed by shooting these ducks fly about from one lake to another high in the air twisting and turning in a most erratic manner and finally darting down almost vertically, making the air whistle with their wings.   Audubon (1840) says:

When these birds are traveling, their flight is steady, rather laborious, but greatly protracted. The whistling of their wings is heard at a considerable distance when they are passing over head. At this time they usually move in a broad front, sometimes in a continuous line. When disturbed, they fly straight forward for a while, with less velocity than when traveling, and, if within proper distance, are easily shot. At times their notes are shrill, but at others hoarse and guttural. They are, however, rarely heard during the day, and indeed like many other species, these birds are partly nocturnal.

Fall.—The fall migration starts rather late with this species, as it is one of the last to leave its northern breeding grounds, and it proceeds southward in a leisurely manner in advance of the frost line. The migration route is practically a reversal of the route traversed in the spring, mainly in the interior, over the sloughs, marshs, lakes, and rivers most frequented by gunners. Constant persecution by sportsmen keeps these little ducks on the move and they have little time to rest and feed, except at night on the larger lakes. They decoy readily and many are shot over live or wooden decoys from blinds made in the rushes near their feeding grounds; they are killed by ambushed gunners on their fly ways between the marshes and the lakes where they roost; and they are hunted out of the cover where they feed in the sloughs of the North and the rice fields of the South. They are safe only in the center of some large body of water.

Mr. Todd (1904) quotes another interesting note from Mr. Bacon, describing the departure of these ducks from Lake Erie as follows:

On one occasion I saw, as I believed, all the lesser scaups in this neighborhood start for the south. The bay had frozen over a few nights before, and on this particular afternoon a large flock of these ducks kept circling over the lake, sometimes high in the air, again dropping swiftly to the surface and skimming along for a mile or so. Finally having evidently gathered into one flock all the birds of the vicinity, they rose to a great height and, starting southward, were soon lost to view.

Winter.—They are very abundant all winter throughout the southern half of the United States, where they find some safe havens of rest. Large numbers winter on the Indian River in Florida and on the lakes in the interior of that State; on Lake Worth they are very abundant and so tame that they have learned to feed almost out of the hands of the winter tourists. On the Louisiana coast they are the commonest ducks and they soon learn to appreciate the security which they find on the protected reservations. It must be a relief to them to find such a wild fowl paradise after running the gauntlet of shooting grounds and sportsmen's clubs.

DISTRIBUTION.

Breeding range.—Northern interior of North America. East to the west coast of Hudson Bay (Churchill) and southeastern Ontario (Lake Temiskaming). South to northern Ohio (Lake, Lorain, and Sandusky Counties), southern Wisconsin (Lake Koshkonong), southeastern Iowa (Keokuk), possibly northern Nebraska (Cherry County), and northeastern Colorado (Barr Lake, few). West to northwestern Montana (Teton County), central British Columbia (Quesnelle Lake and Lac la Hache), and the coast of southern Alaska (Glacier Bay). Has bred casually near San Francisco, California. North to central Alaska (Yukon River) and the northern limit of timber in northern Canada (Mackenzie and Anderson River regions). Breeding records east of Hudson Bay probably refer to *marila*, and perhaps some of the northern records do.

Winter range.—Southern North and Central America. East to the Atlantic coast of United States, the Bahamas, and the Lesser Antilles (St. Thomas, St. Lucia, Trinidad, etc.). South to Panama. West to the Pacific coast of Central America and United States. North to southern British Columbia (Vancouver and Okanogan Lake), southeastern Arizona (San Pedro River), northeastern Colorado (Barr Lake), northeastern Arkansas (Big Lake), southern Illinois, and eastern Maryland (Chesapeake Bay); rarely as far north as Long Island and Massachusetts (Boston).

Spring migration.—Early dates of arrival: Iowa, Keokuk, February 21; Minnesota, Heron Lake, March 5; Michigan, southern, March 11; Alberta, Stony Plain, April 19; Mackenzie, Fort Simpson, May 24. Average dates of arrival: Colorado, Loveland, March 12; Iowa, central, March 21; Minnesota, Heron Lake, March 22; Ohio, Oberlin, March 24; New York, Cayuga Lake, April 1; Illinois, Chicago, April 6; Ontario, Ottawa, April 26; Manitoba, Raeburn, April 9. Late dates of departure: Panama, March 25; Lower California, San Martin, April 23; Porto Rico, Culebra Island, April 21; Florida, Wakulla County, May 23; New York, Cayuga Lake, June 24.

Fall migration.—Early dates of arrival: Ontario, Ottawa, October 12; New York, Cayuga Lake, October 1; Virginia, Alexandria, September 25; Florida, Wakulla County, October 18; Panama, November 25. Average dates of departure: Ontario, Ottawa, November 11; New York, Cayuga Lake, November 15; Manitoba, southern, November 18; Minnesota, southern, November 13; Iowa, Keokuk, December 2. Late dates of departure: Quebec, Montreal, November 12; Ontario, Ottawa, November 21; Illinois, Chicago, December 22.

Casual records.—Accidental in Bermuda (December 19, 1846, January 8, 1849, and February 25, 1876). Rare in migration to Newfoundland and Nova Scotia. Accidental in Greenland (Egedesminde).

Egg dates.—Alaska and Arctic America: Eighteen records, June 17 to July 18; nine records, June 22 to July 5. Manitoba, Saskatchewan and Alberta: Thirteen records, May 20 to July 14; seven records, June 10 to July 3. Minnesota and North Dakota; Eighteen records, May 1 to July 10; nine records, June 12 to 25.

PERISSONETTA COLLARIS (Donovan).

RING-NECKED DUCK.

HABITS.

Although usually classed with the scaup ducks and resembling them in general appearance, this species seems to be more closely related to the European tufted duck than to any American species. Wilson figured and described it under the name, " tufted duck," supposing it to be identical with that species. Its gray speculum separates it from the scaups and its black head and conspicuous crest make it seem very distinct from the redhead, though female ringnecks and redheads look very much alike. I am interested to note that, since I wrote the above, Mr. Ned Hollister (1919) has very ably advanced a similar theory. Its distribution is similar to that of the lesser scaup. It is essentially a fresh-water duck of the interior. It prefers the marshes and sloughs to the open lakes and streams and is less gregarious than the scaups.

Spring.—It is not an especially early migrant in the spring but usually appears soon after the breaking up of the ice, coming along with the scaup ducks and frequenting much the same resorts, but flocking by itself in small scattered groups around the marshy edges of the ponds and in the sloughs.

Courtship.—Audubon (1840) refers briefly to the courtship of this species as follows:

They have an almost constant practice of raising the head in a curved manner, partially erecting the occipital feathers, and emitting a note resembling the sound produced by a person blowing through a tube. At the approach of spring the males are observed repeating this action every now and then while near the females, none of which seem to pay the least attention to their civilities.

Nesting.—The first account we have of the breeding habits of the ring-necked duck was funished by Dr. T. S. Roberts (1880); he found a nest on May 27, 1876, near Minneapolis, Minnesota, and on June 1 he shot the female and collected the eggs. He described the nest, as follows:

The situation chosen for the nest was in a narrow strip of marsh bordering a large shallow pond or slough. About halfway between the shore and the edge of the open water was a mass of sunken débris, probably the remnants of an old muskrat house, which reached nearly or quite to the surface of the water, here about 8 inches deep. On this foundation was the nest, a rather compact, bulky structure, built mainly of fine grass with a little moss intermingled. Outside, the grass is long and circularly disposed, while the bottom, inside, is composed of short broken pieces and the inside rim of fine grass bent and loosely tangled together with considerable down among it. Measurements were not taken before removing the nest, but in its present condition the walls and base are 2½ inches thick, the diameter inside 6 inches, and the depth of the cavity 3 inches. The clutch was nine eggs, which contained small embryos.

My own experiences with the nesting habits of this species have been rather unsatisfactory, but I will give them for what they are worth. On June 12, 1901, while exploring some extensive wet meadows about the sources of a branch of the Goose River in Steele County, North Dakota, I flushed a strange duck from her nest; she circled past me two or three times within gunshot, so that I had a fairly good look at her, but I had no gun with me at the time; I judged from her appearance and gait that she was a scaup, but she lacked the white speculum of the other two species. I made two subsequent visits to the nest alone and on the following day Doctor Bishop and Mr. Job went with me; she proved too shy for us to shoot, but we all concluded that she was a ring-necked duck, as the eggs were unmistakably scaup's and if she had been one of the two other species we would certainly have seen the conspicuous white speculum. The nest was well concealed in thick grass in a rather open place in the meadow about 10 yards from the river; it was made of bits of dry grass and thickly lined with very dark gray down; it contained 10 eggs, nearly fresh. This set is now in the collection of Herbert Massey, Esq., in England.

I found another doubtful nest in the Crane Lake slough in Saskatchewan on June 23, 1906; while hunting through the bullrushes (*Scirpus lacustris*) for canvasback's nests, I flushed a small duck from her nest and shot at her as she went fluttering off over the water, but lost her in the bullrushes; the eggs were evidently scaup ducks' and I felt certain that she had a gray speculum. The nest was in a thick clump of dead bullrushes, made of dry bullrushes and lined with very dark down; it measured 10 by 12 inches in diameter, the inner cavity was 7 inches across and 3 inches deep; the rim was built up about 3 inches above the water, and the eggs were wet and partly in the water; there were eight eggs, one of which was a redhead's.

Mr. Herbert K. Job (1899) found a nest in the Turtle Mountain region of North Dakota, on June 14, 1898, which he felt sure was a nest of the ring-neck duck. He writes:

It was in a reedy, boggy bayou, or arm of a lake, which was full of bitterns, black terns, and bronzed, red-winged and yellow-headed blackbirds. I was on my way out to photograph a bittern's nest already found, and was struggling along more than up to my knees in mud and water, when a smallish duck flushed almost at my feet from some thick, dead rushes, disclosing 12 buffy eggs, nearly fresh. The clear view within a yard of the pearly gray speculum and the total absence of white on the wing told the story. She alighted near by in open water, and gave me and my companion such fine opportunity to study her with the glass and note every detail of her plumage, both as she sat and as she flew back and forth before us, that it was not necessary to sacrifice her for identification. Nothing was seen of the male.

Maj. Allan Brooks (1903) records a nest which he found in the Cariboo District of British Columbia on June 27, as follows:

The nest was in a tussock of grass, in eight inches of water: it was composed of coarse green grass and arched over with the drooping blades of the tussock. The nine eggs contained small embryos.

From this and the foregoing records it would seem that the ring-necked duck habitually builds its nest in wet situations and not on dry ground, as is usually the case with the lesser scaup duck. The down in the ring-necked duck's nest is smaller and a little lighter in color than that of the lesser scaup duck; it is "warm sepia" or "bister," with lighter centers; the breast feathers in it are pure white or pale grayish, tipped with white.

Eggs.—The set seems to consist of from 8 to 12 eggs, which are practically indistinguishable from those of the lesser scaup; the shape, texture of shell, and color are all the same; further description would be useless repetition. The measurements of 75 eggs, in various collections, average 57.5 by 39.8 millimeters; the eggs showing the four extremes measure **60.5** by 41.9, 58 by **42.2**, **53.5** by 38.5 and 54.9 by **38** millimeters.

Young.—I have no data on the period of incubation. The female alone performs this duty, as the males desert the females during this period and do not assist in the care of the young. I have never seen a brood of young ringnecks and can not find anything in print about their behavior or their development.

Plumages.—I have never seen a small downy young of this species, which was positively identified, but Maj. Allan Brooks (1903) says:

The young in down are very light colored, resembling the young of the canvas-back and redhead, and quite different from the dusky, unspotted young of the lesser scaup.

His excellent plate shows this to advantage. A larger downy young bird, measuring 8 inches long in the skin, collected by Major Brooks and now in the United States National Museum, I should describe as follows: The whole head, except the posterior half of the crown, is yellowish, shading from "chamois" or "cream buff" on the cheeks and auriculars to "colonial buff" on the throat; the posterior half of the crown and the occiput are "bister," nearly separated by points of yellow from a broad band of "bister" which extends down the hind neck to the back; narrow dusky postocular streaks are faintly suggested; the dark color of the back changes gradually from "sepia" anteriorly to "bister" posteriorly; the under parts are "ivory yellow" tinged with "cream buff;" there are two large scapular patches, two narrow wing stripes and two small rump patches of "cream buff"; there is also a narrow streak of the same color in the center of the upper back. All of these colors would probably be richer and brighter in a younger bird.

In the juvenal plumage, during the early fall the two sexes are very much alike. The upper parts are dull, dark brown, or blackish brown, with lighter edgings; the under parts are mottled with dull, light brown, and whitish; the wings are similar to those of the adult female, the secondaries in the speculum being dull gray, subterminally dusky and only very narrowly, if at all, tipped with white; the sides of the head and neck are mottled with brown and whitish; the crown is deep brownish black, mottled with brown; and the chin is broadly white.

During September and October the sexes differentiate rapidly. New black feathers appear in the head and neck of the young male; new black feathers with a greenish gloss gradually replace the brown feathers of the mantle; and the under parts become whiter, with white vermiculated feathers appearing among the brown feathers of the flanks. By the last of December the young male is in nearly full plumage with the two rings on the bill in evidence with only a few brown feathers left in the back; the brown neck ring is hardly noticeable, the wing is still immature and all the colors are duller. Further progress is made toward maturity during the winter and spring, but it is not until after the new wings are acquired, at the complete molt the next summer, that young birds become indistinguishable from adults, when about 14 months old.

I have never seen the eclipse plumage of the ring-necked duck but it probably has a partial eclipse plumage, or a prolonged double molt in August, very much like what takes place in the tufted duck, to which it is closely related.

Food.—Dr. F. Henry Yorke (1899) says:

The feeding grounds are more inshore than those of the bluebills, and they feed more upon seeds such as frog bit, duck and pond weed, being very fond of bulbs of the nonscented water lily, upon which they will gorge themselves and get exceedingly fat; at that time they are counted a delicacy for the table. The playgrounds are in open pieces of water surrounded by weeds and lily pads, in buck brush, willows, and wild rice. The roosting grounds are in buck brush, the edges of timber, down smartweed, and flags.

In his food chart for the ring-necked duck he gives the same list of foods as given for the scaup duck. It is also said to feed on minnows, small frogs, tadpoles, crawfish, snails, insects, aquatic roots, various seeds, and even grains.   Audubon (1840) says:

> Whilst in ponds, they feed by diving and dabbling with their bills in the mud amongst the roots of grasses, of which they eat the seeds also, as well as snails and all kinds of aquatic insects.   When on rivers, their usual food consists of small fish and crays, the latter of which they procure at the bottom.   A male which I shot near Louisville, in the beginning of May, exhibited a protuberance of the neck so very remarkable as to induce me to cut the skin, when I found a frog, the body of which was nearly 2 inches long, and which had almost choked the bird, as it allowed me to go up within a dozen or 15 paces before I took aim.

Mr. Arthur H. Howell (1911) writes:

> The food of the ringneck consists mainly of the seeds and stems of pond weed, hornwort, and other aquatic plants, with many nymphs and larvae of water insects.

Behavior.—Although the ring-necked duck feeds largely in the shallow water of the marshes, it is nevertheless a good diver and can, if necessary, dive in deep water.   Its feet are large and powerful, it dives with its wings tightly closed and swims below the surface very rapidly by the use of feet alone.   It swims lightly and rapidly on the surface and rises readily from the water, making a whistling sound as it does so.   Its flight is swift and vigorous and it is as lively as the other scaup ducks in all its movements.   It flies mostly in small flocks of open formation, rather than in close bunches or lines, so that it does not offer such tempting shots as the other bluebills.   While on its feeding grounds it is also usually more scattered and more often flushed singly or in pairs.   It should be easily recognized in flight by its general resemblance to the scaups and by the absence of the white speculum; the males are conspicuously marked and can be easily recognized by the black back and crested head and by the ringed bill, if near enough; the small white chin does not seem to be very conspicuous in life.   Mr. Horace W. Wright (1910) has called attention to another good field mark; he says:

> A conspicuous feature of his plumage as he sits on the water, even at some distance, is a white band on the side of the breast in front of the wing when closed, having the appearance of a bar, but continuous with the white under the wing when the wing is spread.   With closed wing as the bird sits on the water the upper portion of this white bar lies between the black of the breast and the black of the wing; the lower portion between the black of the breast and the finely barred side.

The female is not so easily recognized, as it closely resembles the female redhead, but, if near enough, the white cheeks, faint white eye-ring and ringed bill may be seen.

Fall.—The fall migrations of these ducks come along slightly in advance of the scaup ducks, southward through the Mississippi valley and southeastward to the South Atlantic States.   They frequent the marshes and small ponds on the way and become very abundant in the rice fields and bayous of the Southern States, where they remain all winter and furnish excellent sport for the gunners. They are generally very fat at that season, when they have been feeding on vegetable food, seeds, and grain, and their flesh becomes excellent in flavor.

Game.—The methods employed for shooting the "ringbills" as they are called, are the same as for the "bluebills." Blinds are set in their fly ways or passes, to and from their feeding grounds, where they decoy well to wooden decoys and where large numbers are killed. Although not so universally abundant as some other species and not so well known, this is one of the most abundant ducks of the South Atlantic and Gulf States in winter. On the coast of Louisiana these ducks spend the night out on the Gulf, but come into the ponds to feed at daybreak. They come in small flocks of from 3 or 4 to 10 or 12, flying with great speed, and drop at once without circling, into the pond they have selected. They seem to have certain favorite feeding ponds, for while one pond will yield excellent sport, the gunner in an adjoining pond may not get a shot. They are naturally not shy and are not easily driven from their favorite feeding grounds. Mr. Arthur H. Howell (1911) writes that on Big Lake, Arkansas, "in November and December it is often the most abundant duck, and gunners there frequently kill as many as 50 birds in a few hours. A few remain all winter."

DISTRIBUTION.

Breeding range.—Central North America. East to northern Saskatchewan (Athabaska Lake region), western Ontario (Lac la Seul), and southeastern Wisconsin (Lakes Koshkonong and Pewaukee). Has been known to breed in southeastern Maine (Calais). South to northern Illinois (formerly at least), northern Iowa (Clear Lake), northern Nebraska (Cherry County), and northern Utah (Salt Lake County). West to northeastern California (Lassen County), central southern Oregon (Klamath Lake), and central southern British Columbia (Chilliwack and Cariboo district). North to the central Mackenzie Valley (Fort Simpson) and Athabasca Lake (Fort Chipewyan). Breeding records from farther north are open to question.

Winter range.—Southern North America. East to the Atlantic coast of United States, the Bahama Islands and rarely to Porto Rico and Cuba. South throughout Mexico to Guatemala. West to the Pacific coast of Mexico and United States. North to southern British Columbia (Okanogan Lake), probably Nevada, New Mexico, and northern Texas, to northeastern Arkansas (Big Lake), southern Illinois (Ohio Valley), and eastern Maryland (Chesapeake Bay). Casual in winter as far north as eastern Massachusetts (Boston).

Spring migration.—Early dates of arrivals: Pennsylvania, Erie, March 15; New York, Niagara Falls, March 10; Massachusetts,

Essex County, April 7; Indiana, English Lake, February 27; Iowa, Keokuk, March 4; Minnesota, Heron Lake, March 15; Alberta, Stony Plain, April 19.   Average dates of arrival: Pennsylvania, Erie, April 16, Indiana, English Lake, March 11; Iowa, Keokuk, March 14; Minnesota, Heron Lake, March 27.   Late dates of departure: Florida, Leon County, March 24; Indiana, northern, May 11; Kansas, May 24.

Fall migration.—Early dates of arrival: Louisiana, Gulf coast, about September 15; Valley of Mexico, September 28; Virginia, Alexandria, October 6.   Late dates of departure: Ontario, Ottawa, November 21; Massachusetts, November 23; New York, Brockport, December 10; Pennsylvania, Erie, December 3; Michigan, Hillsdale, November 26; Indiana, English Lake, November 22.

Casual records.—Accidental in Bermuda (November 13, 1850). Only record for Great Britain is questioned by latest authorities. Has wandered east to Nova Scotia (Sable Island, 1901).

Egg dates.—Manitoba, Saskatchewan, and Alberta: Thirteen records, May 31 to July 6; seven records, June 7 to 19.   Minnesota and North Dakota: Five records, June 1 to 18.

REFERENCES TO BIBLIOGRAPHY.

ALFORD, CHARLES E.
 1920—Some Notes on Diving Ducks. British Birds, vol. 14, pp. 106–110.
ALLEN, CHARLES SLOVER.
 1893—The Nesting of the Black Duck on Plum Island. The Auk, vol. 10, pp.
  53–59.
ALLEN, GLOVER MORRILL.
 1905—Summer Birds in the Bahamas. The Auk, vol. 22, pp. 113–133.
ANDRUS, FRED H.
 1896—Unusual Nesting of American Merganser. The Nidiologist, vol. 3, pp.
  72–73.
ARNOLD, EDWARD.
 1894—My '94 Outing Trip in Northwest Canada. Nidiologist, vol. 1, p. 168; vol.
  2, pp. 11–13 and 23–25.
ARTHUR, STANLEY CLISBY.
 1920—A note on the "Southern Teal." The Auk, vol. 37, pp. 126–127.
AUDUBON, JOHN JAMES.
 1840—The Birds of America, 1840–44.
BAILEY, VERNON.
 1902—Notes in Handbook of Birds of the Western United States, by Florence
  Merriam Bailey.
BAIRD, SPENCER FULLERTON; BREWER, THOMAS MAYO; and RIDGWAY, ROBERT.
 1884—The Water Birds of North America.
BAKER, EDWARD CHARLES STUART.
 1908—The Indian Ducks and Their Allies.
BAKER, FRANK COLLINS.
 1889—Contents of the Stomachs of Certain Birds Collected in Brevard County
  Florida, Between Jan. 5 and April 15, 1889. Ornithologist and Oologist,
  vol. 14, pp. 139–140.
BANGS, OUTRAM.
 1918—Notes on the Species and Subspecies of Poecilonetta Eyton. Proceedings
  of the New England Zoological Club, vol. 6, pp. 87–89.
BARROWS, WALTER BRADFORD.
 1912—Michigan Bird Life.
BEAUPRÉ, EDWIN.
 1906—Unusual Nesting Site of the Black Duck (*Anas obscura*). The Auk, vol.
  23, pp. 218–219.
BENNERS, GEORGE B.
 1887—A Collecting Trip in Texas. Ornithologist and Oologist, vol. 12, pp. 49–52,
  65–69, and 81–84.
BENT, ARTHUR CLEVELAND.
 1902—Nesting Habits of the Anatidae in North Dakota. The Auk, vol. 18, pp.
  328–336, vol. 19, pp. 1–12 and 165–174.
 1912—Notes on Birds Observed During a Brief Visit to the Aleutian Islands and
  Bering Sea in 1911. Smithsonian Miscellaneous Collections, vol. 56,
  No. 32.

BEWICK, THOMAS.
    1847—A History of British Birds.
BEYER, GEORGE EUGENE; ALLISON, ANDREW; and KOPMAN, HENRY HAZLITT.
    1906—List of the Birds of Louisiana. The Auk, vol. 23, pp. 1–15, and 275–282;
        vol. 24, pp. 314–321; vol. 25, pp. 173–180 and 439–448.
BLANCHAN, NELTJE.
    1898–1908—Birds that Hunt and are Hunted.
BOWLES, JOHN HOOPER.
    1908—Lead Poisoning in Ducks. The Auk, vol. 25, pp. 312–313.
    ·1909—The Birds of Washington.
BREWSTER, WILLIAM.
    1900—Notes on the Breeding Habits of the American Golden-eyed Duck or
        Whistler. The Auk, vol. 17, pp. 207–216.
    1902—An undescribed form of the Black Duck (*Anas obscura*). The Auk, vol.
        19, pp. 183–188.
    1909—Something More about Black Ducks. The Auk, vol. 26, pp. 175–179.
BROOKS, ALLAN.
    1903—Notes on the Birds of the Cariboo District, British Columbia. The Auk,
        vol. 20, pp. 277–284.
BROOKS, WINTHROP SPRAGUE.
    1913—An addition to the A. O. U. Check List. The Auk, vol. 30, pp. 110–111.
BRYANT, HAROLD CHILD.
    1914—A Survey of the Breeding Grounds of Ducks in California in 1914. The
        Condor, vol. 16, pp. 217–239.
BUCK, HENRY ROBERT.
    1893—Wood Ducks and Bobwhite. The Nidiologist, vol. 1, p. 54.
BUTLER, AMOS WILLIAM.
    1897—The Birds of Indiana. Department of Geology and Natural Resources.
        Twenty-second Annual Report.
CAHN, ALVIN R.
    1912—The Freezing of Cayuga Lake in its Relation to Bird Life. The Auk, vol.
        29, pp. 437–444.
CALL, AUBREY BRENDON.
    1894—An American Merganser's Nest. The Nidiologist, vol. 1, p. 101.
CARROLL, JAMES JUDSON.
    1900—Notes on the Birds of Refugio County, Texas. The Auk, vol. 17, pp.
        337–348.
CLARK, JOHN NATHANIEL.
    1882—Woodcock and Black Duck. Ornithologist and Oologist, vol. 7. p. 144.
CLARKE, WILLIAM EAGLE.
    1895—On the Ornithology of the Delta of the Rhone. The Ibis, 1895, pp.
        173–211.
COLLINS, WILLIAM H.
    1880—Notes on the Breeding Habits of Some of the Water-Birds of St. Clair
        Flats, Michigan. Bulletin of the Nuttall Ornithological Club, vol. 5,
        pp. 61–62.
COOKE, WELLS WOODBRIDGE.
    1906—Distribution and Migration of North American Ducks, Geese, and Swans.
        United States Department of Agriculture, Biological Survey, Bulletin
        No. 26.
CORDEAUX, JOHN.
    1898—British Birds with Their Nests and Eggs. Order Anseres, vol. 4, pp.
        52–203.

CORY, CHARLES BARNEY.
1880—Birds of the Bahama Islands.
COUES, ELLIOTT.
1874—Birds of the North-West.
CURRIER, EDMONDE SAMUEL.
1902—Winter Water Fowl of the Des Moines Rapids. The Osprey, vol. 6, pp. 71–75.
DAVID, ARMAND, and OUSTALET, EMILE.
1877—Les Oiseaux de la Chine.
DAWSON, WILLIAM LEON.
1903—The Birds of Ohio.
1909—The Birds of Washington.
DUTCHER, WILLIAM.
1907—The Wood Duck. Bird-Lore, vol. 9, p. 189–193.
DWIGHT, JONATHAN, Jr.
1909—The Singular Case of the Black Duck of North America. The Auk, vol. 26, pp. 422–426.
DYBOWSKI, BENEDICT NALENTSCH, and PARREX, A.
1868—Verzeichniss der wahrend der Jahre 1866 und 1867 im Gebiete der Mineralwasser von Darasun in Daurien beobachteten Vögel. Journal für Ornithologie, vol. 16, pp. 330–339.
EASTMAN, A. B.
1915—The Wood Duck. The Oologist, vol. 32, p. 95.
EATON, ELON HOWARD.
1910—Birds of New York.
ELLIOT, DANIEL GIRAUD.
1898—The Wild Fowl of North America.
EMERSON, WILLIAM OTTO.
1901—Nesting of *Spatula clypeata*. The Condor, vol. 3, p. 116.
EVANS, WILLIAM.
1891—On the Periods Occupied by Birds in the Incubation of Their Eggs. The Ibis, 1891, pp. 52–93.
EVERMANN, BARTON WARREN.
1913—Eighteen Species of Birds New to the Pribilof Islands, Including Four New to North America. The Auk, vol. 30, pp. 15–18.
FAY, SAMUEL PRESCOTT.
1910—The Canvasback in Massachusetts. The Auk, vol. 27, pp. 369–381.
FINN, FRANK.
1909—The Waterfowl of India and Asia.
1915—Display of Female *Eunetta falcata*. Zoologist, series 4, vol. 19, p. 36.
FISHER, ALBERT KENRICK.
1901—Two Vanishing Game Birds: The Woodcock and the Wood Duck. Yearbook of Department of Agriculture for 1901, pp. 447–458.
FORBUSH, EDWARD HOWE.
1909—The Mallard. Bird-Lore, vol. 11, pp. 40–47.
FRAZAR, MARTIN ABBOTT.
1887—An Ornithologist's Summer in Labrador. Ornithologist and Oologist, vol. 12, pp. 1–3, 17–20, and 33–35.
GHIDINI, ANGELO.
1911—Les canards du Yang-tsze en Europe. Revue Française d'Ornithologie vol. 2, p. 78.
GIBBS, MORRIS.
1885—Catalogue of the Birds of Kalamazoo County, Michigan. Ornithologist and Oologist, vol. 10, pp. 166–167.

GRINNELL, JOSEPH.
　　1900—Birds of the Kotzebue Sound Region. Pacific Coast Avifauna, No. 1.
　　1909—Birds and Mammals of the 1907 Alexander Expedition to Southeastern
　　　　　Alaska. University of California, Publications in Zoology, vol. 5, pp.
　　　　　171–264.
GRINNELL, JOSEPH; BRYANT, HAROLD CHILD; and STORER, TRACY IRWIN.
　　1918—The Game Birds of California.
HAGERUP, ANDREAS THOMSEN.
　　1891—The Birds of Greenland.
HANNA, G. DALLAS.
　　1916—Records of Birds New to the Pribilof Islands Including Two New to North
　　　　　America. The Auk, vol. 33, pp. 400–403.
　　1920—Additions to the Avifauna of the Pribilof Islands, Alaska, Including Four
　　　　　Species New to North America. The Auk, vol 37, pp. 248–254.
HARTERT, ERNST JOHANN OTTO.
　　1920—Die Vögel der Paläarktischen Fauna.
HATCH, PHILO LUOIS.
　　1892—Notes on the Birds of Minnesota.
HEINROTH, OTTO.
　　1911—Beiträge zur Biologie, namentlich Ethologie und Psychologie der Anati-
　　　　　den. Verhandl. 5th Intern. Ornith. Kongr. Berlin, 1910, p. 589–702.
HENSHAW, HENRY WETHERBEE.
　　1875—Report upon the Ornithological Collections made in Portions of Nevada,
　　　　　Utah, California, Colorado, New Mexico, and Arizona during the years
　　　　　1871, 1872, 1873, and 1874. Report upon Geographical and Geological
　　　　　Explorations and Surveys West of the One Hundredth Meridian.
HOLLISTER, NED.
　　1919—The Systematic Position of the Ring-necked Duck. The Auk, vol. 36,
　　　　　pp. 460–463.
HOPWOOD, CYRIL.
　　1912—A List of Birds from Arakan. Journ. Bombay Nat. Hist. Soc., vol. 21,
　　　　　pp. 1196–1221. Anseres, p. 1220.
HOWELL, ARTHUR HOLMES.
　　1911—Birds of Arkansas. United States Department of Agriculture, Biological
　　　　　Survey, Bulletin No. 38.
HUDSON, WILLIAM HENRY.
　　1920—Birds of La Plata.
HUNTINGTON, DWIGHT WILLIAMS.
　　1903—Our Feathered Game.
JOB, HERBERT KEIGHTLEY.
　　1898—The Enchanted Isles. The Osprey, vol. 3, pp. 37–41.
　　1899—Some Observations on the Anatidae of North Dakota. The Auk, vol. 16,
　　　　　pp. 161–165.
JONES, LYNDS.
　　1909—The Birds of Cedar Point and Vicinity. The Wilson Bulletin, No. 67,
　　　　　vol. 21, pp. 55–76, 115–131, and 187–204; vol. 22, pp. 25–41, 97–115,
　　　　　and 172–182.
KENNARD, FREDERIC HEDGE.
　　1913—The Black Duck Controversy Again. The Auk, vol. 30, p. 106.
　　1919—Notes on a New Subspecies of Blue-winged Teal. The Auk, vol. 36, pp.
　　　　　455–460.

KINGSFORD, E. G.
1917—Wood Duck Removing Young from the Nest. The Auk, vol. 34, pp. 335–336.
KNIGHT, ORA WILLIS.
1908—The Birds of Maine.
LANGILLE, JAMES HIBBERT.
1884—Our Birds and Their Haunts.
LITTLEJOHN, CHASE.
1899—On the Nesting of Ducks. The Osprey, vol. 3, p. 78–79.
MABBOTT, DOUGLAS CLIFFORD.
1920—Food Habits of Seven Species of American Shoal-water Ducks. United States Department of Agriculture, Bulletin, No. 862.
MACFARLANE, RODERICK ROSS.
1891—Notes on and List of Birds and Eggs Collected in Arctic America, 1861–1866. Proceedings of the United States National Museum, vol. 14, pp. 413–446.
1908—List of Birds and Eggs Observed and Collected in the Northwest Territories of Canada, between 1880 and 1894. In Through the Mackenzie Basin by Charles Mair.
MACGILLIVRAY, WILLIAM.
1852—A History of British Birds.
MACOUN, JOHN.
1909—Catalogue of Canadian Birds. Second Edition.
MAYNARD, CHARLES JOHNSON.
1896—The Birds of Eastern North America.
MCATEE, WALDO LEE.
1908—Lead poisoning in Ducks. The Auk, vol. 25, p. 472.
1918—Food Habits of the Mallard Ducks of the United States. United States Department of Agriculture, Bulletin No. 720.
MCGREGOR, RICHARD CRITTENDEN.
1906—Birds Observed in the Krenitzin Islands, Alaska. The Condor, vol. 8, pp. 114–122.
MCILWRAITH, THOMAS.
1894—The Birds of Ontario.
MERRILL, JAMES CUSHING.
1888—Notes on the Birds of Fort Klamath, Oregon. The Auk, vol. 5, pp. 139–146.
MIDDENDORFF, ALEXANDER THEODOROVICH VON.
1853—Reise in den Aussersten Norden und Osten Sibiriens wahrend der Jahre 1843 und 1844 mit Allerhochster Genehmigung auf Veranstaltung der Kaiserlichen Akademie der Wissenschaften zu St. Petersburg ausgeführt und in Verbindung mit vielen Gehehrten herausgegeben. Vol. 2, Zoologie, pt. 2, Wirbelthiere.
MILLAIS, JOHN GUILLE.
1902—The Natural History of the British Surface-Feeding Ducks.
1913—British Diving Ducks.
MOORE, ROBERT THOMAS.
1908—Three Finds in South Jersey. Cassinia, Proceedings of the Delaware Valley Ornithological Club, No. 12, pp. 29–40.
MORRIS, FRANCIS ORPEN.
1903—A History of British Birds. Fifth edition.

MORSE, ALBERT PITTS.
1921—A Sheld Duck (Tadorna tadorna L.) from Essex County, Mass. Bulletin of the Essex County Ornithological Club of Massachusetts, December 1921, p. 68.

MUNRO, JAMES ALEXANDER.
1917—Notes on the Winter Birds of the Okanogan Valley. The Ottawa Naturalist, vol. 31, pp. 81–89.

NELSON, EDWARD WILLIAM.
1887—Report upon Natural History Collections Made in Alaska.

NOBLE, FRANK T.
1906—Why Wounded Ducks Disappear. The Journal of the Maine Ornithological Society, vol. 8, pp. 60–61.

NORTON, ARTHUR HERBERT.
1909—The Food of Several Maine Water-Birds. The Auk, vol. 26, pp. 438–440.

NUTTALL, THOMAS.
1834—A Manual of the Ornithology of the United States and Canada, Water Birds.

OSGOOD, WILFRED HUDSON.
1904—A Biological Reconnaissance of the Base of the Alaska Peninsula. North American Fauna, No. 24.

PEARSON, THOMAS GILBERT.
1891—The Wood Duck. Ornithologist and Oologist, vol. 16, pp. 134–135.
1916—The Shoveller. Bird-Lore, vol. 18, pp. 56–59.

PEARSON, THOMAS GILBERT; BRIMLEY, CLEMENT SAMUEL; and BRIMLEY, HERBERT HUTCHINSON.
1919—Birds of North Carolina. North Carolina Geological and Economic Survey, vol. 4.

PECK, GEORGE DELRANE.
1896—Melanism in Eggs of the Hooded Merganser. The Oregon Naturalist, vol. 3, p. 84.
1911—Reminiscences of the Wood Duck. The Oologist, vol. 28, pp. 35–36.

PHILLIPS, JOHN CHARLES.
1911—Ten Years of Observation on the Migration of Anatidae at Wenham Lake, Massachusetts. The Auk, vol. 28, pp. 188–200.
1911a—A case of the Migration and Return of the European Teal in Massachusetts. The Auk, vol. 28, pp. 366–367.
1912—The European Teal (*Nettion crecca*) again returning to Wenham, Mass. The Auk, vol. 29, p. 535.
1916—A Note on the Mottled Duck. The Auk, vol. 33, pp. 432–433.
1920—Habits of the Two Black Ducks, *Anas rubripes rubripes* and *Anas rubripes tristis*. The Auk, vol. 37, pp. 289–291.

POST, WILLIAM S.
1914—Nesting of the Merganser (*Mergus americanus*) in 1913. The Oriole, vol. 2, pp. 18–23.

PREBLE, EDWARD ALEXANDER.
1908—A Biological Investigation of the Athabaska-Mackenzie Region. North American Fauna, No. 27.

PRESTON, JUNIUS WALLACE.
1892—Notes on Bird Flight. Ornithologist and Oologist, vol. 17, pp. 41–42.

PRJEVALSKY, NICOLAS MICHAELOVICH.
1878—The Birds of Mongolia, the Tangut Country, and the Solitudes of Northern Thibet. Rowley's Ornith. Miscell., vol. 3, pp. 87–110, 145–162.

RADDE, GUSTAV FERDINAND RICHARD VON.
1863—Reisen im Süden von Ost-Sibirien, vol. 2, Die Festlands-Ornis des Südöstlichen Sibiriens.

Rich, Walter Herbert.
    1907—Feathered Game of the Northeast.
Ridgway, Robert.
    1881—On a Duck New to the North American Fauna. Proceedings of United
        States National Museum, vol. 4, pp. 22–24.
    1887—A Manual of North American Birds.
Rinker, Glen.
    1899—Peculiar Nesting of the Hooded Merganser. The Osprey, vol. 4, p. 19–20.
Roberts, Thomas Sadler.
    1880—Breeding of Fuligula Collaris in Southeastern Minnesota, and a Description
        of Its Nest and Eggs. Bulletin of the Nuttall Ornithological Club, vol.
        5, p. 61.
    1919—Water Birds of Minnesota Past and Present. Biennial Report of the State
        Game and Fish Commission of Minnesota, for the Biennal Period Ending
        July 31, 1918.
Rockwell, Robert Blanchard.
    1911—Nesting Notes on the Ducks of the Barr Lake Region, Colorado. The
        Condor, vol. 13, pp. 121–128 and 186–195.
Rolfe, Eugene Strong.
    1898—Notes from the Devils Lake Region. The Osprey, vol. 2, pp. 125–128.
Sage, John Hall.
    1881—Notes from Moosehead Lake, Me. Ornithologist and Oologist, vol. 6
        pp. 50–51.
Sampson, Walter Behrnard.
    1901—An Exceptional Set of Eggs of the Wood Duck. The Condor, vol. 3, p. 95.
Samuels, Edward Augustus.
    1883—Our Northern and Eastern Birds.
Sanford, Leonard Cutler; Bishop, Louis Bennett; and Van Dyke, Theodore
        Strong.
    1903—The Waterfowl Family.
Sawyer, Edmund Joseph.
    1909—The Courtship of Black Ducks. Bird-Lore, vol. 11, pp. 195–196.
Schneider, Frederick Alexander, Jr.
    1893—Nesting of the Cinnamon Teal. The Nidiologist, vol. 1, pp. 20–22.
Schrenck, Leopold von.
    1860—Reisen und Forschungen im Amur-lande in den Jahren 1854–1856, vol. 1,
        pt. 2, Vögel des Amur-Landes.
Sennett, George Burritt.
    1889—A New Species of Duck from Texas. The Auk, vol. 6, pp. 263–265.
Sheldon, Harry Hargrave.
    1907—A Collecting Trip by Wagon to Eagle Lake, Sierra Nevada Mountains.
        The Condor, vol. 9, pp. 185–191.
Simmons, George Finlay.
    1915—On the Nesting of Certain Birds in Texas. The Auk, vol. 32, pp. 317–331
Smith, William G.
    1887—Hybrid Ducks. Ornithologist and Oologist, vol. 12, p. 169.
Stejneger, Leonhard.
    1885—Results of Ornithological Explorations in the Commander Islands and in
        Kamtschatka. Bulletin of the United States National Museum, No. 29.
Strong, Reuben Myron.
    1912—Some Observations on the Life History of the Red-breasted Merganser.
        The Auk, vol. 29, pp. 479–488.

SWARTH, HARRY SCHELWALDT.
  1911—Birds and Mammals of the 1909 Alexander Alaska Expedition. University of California Publications in Zoology, vol. 7, pp. 9–172.
TACZANOWSKI, LADISLAS.
  1873—Bericht über die ornithologischen Untersuchungen des Dr. Dybowski in Ost-Sibirien. Journal für Ornithologie, vol. 21, pp. 81–119.
THOMPSON, ERNEST EVAN.
  1890—The Birds of Manitoba. Proceedings of the United States National Museum, vol. 13, pp. 457–643.
THOMPSON, ERNEST SETON.
  1901—The Mother Teal and the Overland Route. Lives of the Hunted, pp. 195–209.
TODD, WALTER EDMOND CLYDE.
  1904—The Birds of Erie and Presque Isle, Erie County, Pennsylvania. Annals of the Carnegie Museum, vol. 2, pp. 481–596.
  1911—A Contribution to the Ornithology of the Bahama Islands. Annals of the Carnegie Museum, vol. 7, pp. 388–464.
TOWNSEND, CHARLES WENDELL.
  1905—The Birds of Essex County, Massachusetts. Memoirs of the Nuttall Ornithological Club, No. 3.
  1909—The Use of the Wings and Feet by Diving Birds. The Auk, vol. 26, pp. 234–248.
  1911—The Courtship and Migration of the Red-breasted Merganser (Mergus serrator). The Auk, vol. 28, pp. 341–345.
  1912—The Validity of the Red-legged Subspecies of Black Duck. The Auk, vol. 29, pp. 176–179.
  1916—The Courtship of the Merganser, Mallard, Black Duck, Baldpate, Wood Duck, and Bufflehead. The Auk, vol. 33, pp. 9–17.
TURNER, LUCIEN MCSHAN.
  1886—Contributions to the Natural History of Alaska.
VAN KAMMEN, I. T.
  1915—Odd Nesting of the American Merganser. The Oologist, vol. 32, pp 166–168.
WALTON, HERBERT JAMES.
  1903—Notes on the Birds of Peking. The Ibis, 1903, pp. 19–35.
WARREN, BENJAMIN HARRY.
  1890—Report on the Birds of Pennsylvania.
WAYNE, ARTHUR TREZEVANT.
  1910—Birds of South Carolina. Contributions from the Charleston Museum, No. 1.
WETMORE, ALEXANDER.
  1915—Mortality among Waterfowl around Great Salt Lake, Utah. United States Department of Agriculture, Bulletin No. 217.
  1916—Birds of Porto Rico. United States Department of Agriculture, Bulletin No. 326.
  1918—The Duck Sickness in Utah. United States Department of Agriculture, Bulletin No. 672.
  1919—Lead Poisoning in Waterfowl. United States Department of Agriculture, Bulletin No. 793.
  1920—Observations on the Habits of Birds at Lake Burford, New Mexico. The Auk, vol. 37, pp. 221–247 and 393–412.
WILLETT, GEORGE, and JAY, ANTONIN.
  1911—May Notes from San Jacinto Lake. The Condor, vol. 13, pp. 156–160.

WILSON, ALEXANDER.
  1832—American Ornithology.
WITHERBY, HARRY FORBES, and OTHERS.
  1920—A Practical Handbook of British Birds.
WORMALD, HUGH.
  1910—The Courtship of the Mallard and Other Ducks. British Birds, vol. 4,
      pp. 2–7.
WRIGHT, HORACE WINSLOW.
  1910—Some Rare Wild Ducks Wintering at Boston, Massachusetts, 1909–1910.
      The Auk, vol. 27, pp. 390–408.
  1911—The Birds of the Jefferson Region in the White Mountains New Hamp-
      shire. Proceedings of the Manchester Institute of Arts and Sciences,
      vol. 5, pt. 1.
YARRELL, WILLIAM.
  1871—History of British Birds. Fourth Edition, 1871–85. Revised and enlarged
      by Alfred Newton and Howard Saunders.
YORKE, F. HENRY.
  1891—Green-Wing Teal Shooting. The American Field, vol. 35, No. 22, pp.
      533–535.
  1899—Our Ducks.

INDEX.

241

PLATES

PLATE 1. PINTAIL. A flock of pintails on the Ward-McIlhenny Reservation, Louisiana, January 1, 1910, presented by Mr. Herbert K. Job.

PLATE 2. AMERICAN MERGANSER. *Upper:* Nesting site of American mergansers, Lake Winnipegosis, Manitoba, June 19, 1913, referred to on page 3. *Lower:* Nest and eggs in above locality, rocks removed to expose it, June 16, 1903, presented by Mr. Walter Raine.

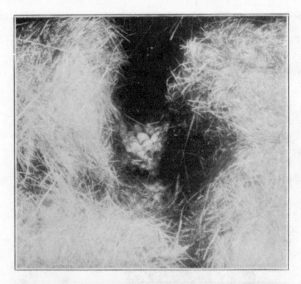

PLATE 3. AMERICAN MERGANSER. *Upper:* Nest and eggs of American merganser, among bales of hay in an old ice house, Lake Winnipegosis, Manitoba, June 18, 1913, bales removed to expose the nest, referred to on page 4. *Lower:* Nesting stub of American merganser, Grand Lake, Maine, June 3, 1920, presented by Mr. Frederic H. Kennard.

PLATE 4. AMERICAN MERGANSER. *Upper:* Female American
merganser on her nest, Four Brothers Islands, Lake Champlain, New
York, July 8, 1910, presented by Mr. Francis Harper. *Lower:* Young
American mergansers, Eagle Lake, California, May 27, 1921, presented
by Mr. Jules Labarthe, Sr., and Mr. Milton S. Ray.

PLATE 5. RED-BREASTED MERGANSER. *Upper:* Nest and eggs of red-breasted merganser, under balsam firs, branches cut away to expose the nest, Magdalen Islands, Quebec, June 21, 1904, referred to on page 15. *Lower:* Another nest of same in above locality, in thick grass, June 20, 1904, referred to on page 15.

PLATE 6. HOODED MERGANSER. Nesting box occupied by hooded merganser, near Tacoma, Washington, April 21, 1907, presented by Mr. J. Hooper Bowles, referred to on page 24.

PLATE 7. MALLARD. *Upper:* Nest and eggs of mallard, among "niggerheads," St. Michael, Alaska, June 9, 1915, from a negative taken by Mr. F. Seymour Hersey for the author. *Lower:* Nest and eggs of mallard, in a clump of reeds, North Dakota, presented by Mr. Herbert K. Job.

PLATE 8. MALLARD. *Left:* Mallard on her nest, 10 feet up on the limb of a maple, among ferns, near Tacoma, Washington, March 17, 1913, presented by Mr. J. Hooper Bowles, referred to on page 37. *Right:* Nest and eggs of mallard, under a rosebush, Crane Lake, Saskatchewan, June 21, 1905, presented by Mr. Herbert K. Job.

PLATE 9. MALLARD. *Upper:* Immature male mallard assuming first winter plumage, in captivity at Ithaca, New York, presented by Dr. Arthur A. Allen. *Lower:* Young mallard, 5 weeks old, Lake Winnepegosis, Manitoba, August 1, 1913, presented by Mr. Herbert K. Job.

PLATE 10. MALLARD. *Upper:* Adult male mallard in full plumage in September. *Lower:* Adult male mallard in eclipse plumage. Both photographs of birds in captivity, Ithaca, New York, presented by Dr. Arthur A. Allen.

PLATE 11. NEW MEXICAN DUCK. New Mexican duck, young male, Las Cruces, New Mexico, July 27, 1920, presented by Mr. Wharton Huber.

PLATE 12. BLACK DUCK. *Upper:* Nesting site of black duck, Magdalen Islands, Quebec, June 21, 1904, referred to on page 52. *Lower:* Nest and eggs of the same in above locality.

PLATE 13. BLACK DUCK. *Upper:* Nest and eggs of black duck, Magdalen Islands, Quebec, presented by Mr. Herbert K. Job. *Lower:* Black duck on her nest, a photograph purchased from Mr. Bonnycastle Dale.

PLATE 14. GADWALL. *Upper:* Nest and eggs of gadwall, Teton County, Montana, May 30, 1918, presented by Mr. A. D. DuBois. *Lower:* Nest and eggs of gadwall, in a clump of reeds, Stump Lake, North Dakota, June 15, 1901, referred to on page 80.

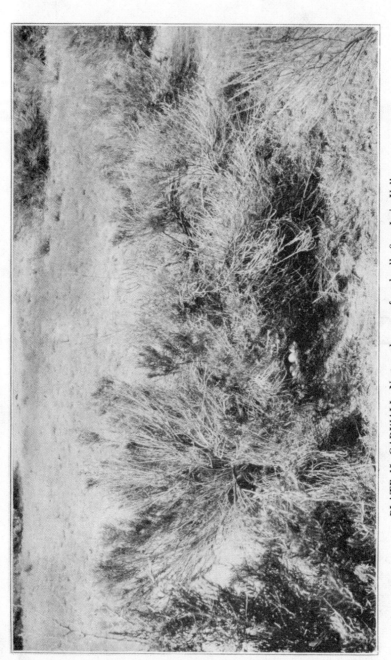

PLATE 15. GADWALL. Nest and eggs of gadwall, San Luis Valley, Colorado, presented by the Colorado Museum of Natural History.

PLATE 16. GADWALL. *Upper:* Young gadwall, 2 weeks old, Lake Winnipegosis, Manitoba, July 28, 1913, presented by Mr. Herbert K. Job.
Lower: Nest and eggs of gadwall, Shoal Lake, Manitoba, June 28, 1912,
presented by Mr. Herbert K. Job.

PLATE 17. BALDPATE. *Upper:* Nest and eggs of baldpate, in a clump of nettles, Stump Lake, North Dakota, May 31, 1901. *Lower:* Another nest of same, in thick, tall grass, same locality and date. Both referred to on page 90.

PLATE 18. EUROPEAN TEAL. *Upper:* Nesting site of European teal, Unalaska, Alaska, June 8, 1911, referred to on page 99. *Lower:* Nest and eggs of same, above locality and date.

PLATE 19. GREEN-WINGED TEAL. *Upper:* Nest and eggs of green-winged teal, Crane Lake, Saskatchewan, June 21, 1905, presented by Mr. Herbert K. Job. *Lower:* Another nest of same, Alberta, June 16, 1906, presented by Mr. Walter Raine.

PLATE 20. GREEN-WINGED TEAL. Nest and eggs of green-winged teal, Alberta, a photograph purchased from Mr. S. S. S. Stansell.

PLATE 21. GREEN-WINGED TEAL. *Upper:* Green-winged teal in full plumage, adult male. *Lower:* Adult male of same in nearly full eclipse plumage. Both photographs of captive birds, Ithaca, New York, presented by Dr. Arthur A. Allen.

PLATE 22. BLUE-WINGED TEAL. *Upper:* Nest of blue-winged teal, covered, Cherry County, Nebraska, July 25, 1911. *Lower:* Same nest, uncovered to show the eggs. Both photographs presented by Mr. Frank H. Shoemaker.

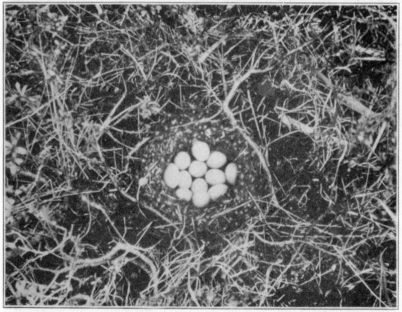

PLATE 23. BLUE-WINGED TEAL. *Upper:* Brood of young blue-winged teal, a few days old. *Lower:* Nest and eggs of blue-winged teal, Magdalen Islands, Quebec, June 16, 1900. Both photographs presented by Mr. Herbert K. Job.

PLATE 24. CINNAMON TEAL. Nest and eggs of cinnamon teal,
Barr Lake, Colorado, presented by the Colorado Museum of Natural
History.

PLATE 25. CINNAMON TEAL. *Upper:* Nest of cinnamon teal, concealed, Barr Lake, Colorado. *Lower:* Same nest, uncovered to show the eggs. Both photographs presented by Mr. Robert B. Rockwell.

PLATE 26. SHOVELLER. *Upper:* Nest and eggs of shoveller, Nelson County, North Dakota, June 7, 1901, referred to on page 137. *Lower:* Nest and eggs of shoveller, Crane Lake, Saskatchewan, June 5, 1905.

PLATE 27. SHOVELLER. *Upper:* Young shovellers, 1 and 2 weeks old, Lake Winnipegosis, Manitoba, July 28, 1913. *Lower:* Young shovellers, nearly fully grown, St. Marks, Manitoba, July, 1912. Both photographs presented by Mr. Herbert K. Job.

PLATE 28. PINTAIL. *Upper:* Nest and eggs of pintail, under a rose bush, Crane Lake, Saskatchewan, June 4, 1905, referred to on page 146. *Lower:* Nest and eggs of pintail, on edge of a cultivated field, Steele County, North Dakota, June 10, 1901, referred to on page 146.

PLATE 29. PINTAIL. *Upper:* Nest and eggs of pintail, Crane Lake, Saskatchewan, June 2, 1905. *Lower:* Nest and eggs of pintail, Lake Manitoba, Manitoba, July 3, 1912. Both photographs presented by Mr. Herbert K. Job.

PLATE 30. PINTAIL. *Upper:* Female pintail, in captivity, Ithaca, New York, presented by Dr. Arthur A. Allen. *Lower:* Young pintail, Klamath River, Oregon, presented by Mr. Wm. L. Finley and Mr. H. T. Bohlman.

PLATE 31. PINTAIL. *Upper:* Adult male pintail in full plumage. *Lower:* Adult male pintail in partial eclipse plumage. Both photographs of birds in captivity, Ithaca, New York, presented by Dr. Arthur A. Allen.

PLATE 32. WOOD DUCK. *Left*: Nesting site of wood duck, in a dead elm stub, about 30 feet up, nest contained 23 eggs, Yates County, New York, May 12, 1907. *Right*: Nesting site of wood duck, in a living elm, about 50 feet up, May 13, 1906, Yates County, New York. Both photographs presented by Mr. Clarence F. Stone.

PLATE 33. WOOD DUCK. *Upper:* Nesting site of wood duck, in a low stub, Yates County, New York, presented by Mr. Clarence F. Stone. *Lower:* Brood of young wood ducks, with mallard foster mother, Amston, Connecticut, July 23, 1920, presented by Mr. Herbert K. Job.

PLATE 34. WOOD DUCK. *Upper:* Adult male wood duck, in full plumage. *Lower:* Adult male wood duck, in eclipse plumage. Both photographs of captive birds, Ithaca, New York, presented by Dr. Arthur A. Allen.

PLATE 35. REDHEAD. *Upper:* Nest and eggs of redhead, Steele County, North Dakota, June 10, 1901, presented by Mr. Herbert K. Job. *Lower:* Nearer view of same nest, referred to on page 177.

PLATE 36. REDHEAD. *Left:* Nest and eggs of redhead, San Luis Valley, Colorado. *Right:* Nearer view of same nest. Both photographs presented by the Colorado Museum of Natural History.

PLATE 37. REDHEAD. *Upper:* Nest and eggs of redhead, Sweetwater Lake, North Dakota, July 15, 1915, presented by Mr. Frank M. Woodruff. *Lower:* Young redheads, St. Marks, Manitoba, July, 1912, presented by Mr. Herbert K. Job.

PLATE 38. CANVASBACK. *Upper:* Nesting site of canvasbacks, Steele County, North Dakota, June 10, 1901, referred to on pages 177 and 191. *Lower:* Nest and eggs of canvasback, above locality and date.

PLATE 39. CANVASBACK. *Upper:* Nest and eggs of canvasback, Steele County, North Dakota, June 11, 1901, referred to on page 192. *Lower:* Nest and eggs of canvasback, in open situation in low sedges, Lake Winnipegosis, Manitoba, June 9, 1913, referred to on page 193. Both photographs presented by Mr. Herbert K. Job.

PLATE 40. CANVASBACK. *Upper:* Young canvasbacks, presented by Mr. Herbert K. Job. *Lower:* Male canvasbacks, in different stages of eclipse plumage, captive birds, Ithaca, New York, presented by Dr. Arthur A. Allen.

PLATE 41. SCAUP DUCK. *Upper:* Nest and eggs of scaup duck, St. Michael, Alaska, June 5, 1915, referred to on page 208. *Lower:* Nearer view of same nest. Both photographs from negatives taken by Mr. F. Seymour Hersey for the author.

PLATE 42. SCAUP DUCK. *Upper:* Nesting site of scaup duck, Magdalen Islands, Quebec. *Lower:* Nest and eggs of scaup duck, in above locality, referred to on page 209. Both photographs presented by Mr. Herbert K. Job.

PLATE 43. LESSER SCAUP DUCK. *Upper:* Nest and eggs of lesser
scaup duck, Crane Lake, Saskatchewan, June 17, 1905. *Lower:* Another
nest of same, Stump Lake, North Dakota, June 15, 1901, referred to on
page 218.

PLATE 44. LESSER SCAUP DUCK. *Upper:* Nest and eggs of lesser
scaup duck, Lake Winnipegosis, Manitoba, June, 1913, referred to on
page 219. *Lower:* Another nest of same, Lake Manitoba, Manitoba,
July 3, 1912. Both photographs presented by Mr. Herbert K. Job.

PLATE 45. LESSER SCAUP DUCK. *Upper:* Young lesser scaup duck, 1 month old, Lake Winnipegosis, Manitoba, August, 1913, presented by Mr. Herbert K. Job. *Lower:* Adult male lesser scaup duck in eclipse plumage, captive bird, Ithaca, New York, presented by Dr. Arthur A. Allen.

PLATE 46. RING-NECKED DUCK. *Upper:* Nest and eggs of ring-
necked duck, Carver County, Minnesota, June 16, 1898, presented by
Dr. Thomas S. Roberts. *Lower:* Nest and eggs of ring-necked duck,
Steele County, North Dakota, June 13, 1901, referred to on page 225.

A CATALOGUE OF SELECTED DOVER BOOKS
IN ALL FIELDS OF INTEREST

A CATALOGUE OF SELECTED DOVER BOOKS
IN ALL FIELDS OF INTEREST

AMERICA'S OLD MASTERS, James T. Flexner. Four men emerged unexpectedly from provincial 18th century America to leadership in European art: Benjamin West, J. S. Copley, C. R. Peale, Gilbert Stuart. Brilliant coverage of lives and contributions. Revised, 1967 edition. 69 plates. 365pp. of text.

21806-6 Paperbound $3.00

FIRST FLOWERS OF OUR WILDERNESS: AMERICAN PAINTING, THE COLONIAL PERIOD, James T. Flexner. Painters, and regional painting traditions from earliest Colonial times up to the emergence of Copley, West and Peale Sr., Foster, Gustavus Hesselius, Feke, John Smibert and many anonymous painters in the primitive manner. Engaging presentation, with 162 illustrations. xxii + 368pp.

22180-6 Paperbound $3.50

THE LIGHT OF DISTANT SKIES: AMERICAN PAINTING, 1760-1835, James T. Flexner. The great generation of early American painters goes to Europe to learn and to teach: West, Copley, Gilbert Stuart and others. Allston, Trumbull, Morse; also contemporary American painters—primitives, derivatives, academics—who remained in America. 102 illustrations. xiii + 306pp.

22179-2 Paperbound $3.00

A HISTORY OF THE RISE AND PROGRESS OF THE ARTS OF DESIGN IN THE UNITED STATES, William Dunlap. Much the richest mine of information on early American painters, sculptors, architects, engravers, miniaturists, etc. The only source of information for scores of artists, the major primary source for many others. Unabridged reprint of rare original 1834 edition, with new introduction by James T. Flexner, and 394 new illustrations. Edited by Rita Weiss. 6⅝ x 9⅝.

21695-0, 21696-9, 21697-7 Three volumes, Paperbound $13.50

EPOCHS OF CHINESE AND JAPANESE ART, Ernest F. Fenollosa. From primitive Chinese art to the 20th century, thorough history, explanation of every important art period and form, including Japanese woodcuts; main stress on China and Japan, but Tibet, Korea also included. Still unexcelled for its detailed, rich coverage of cultural background, aesthetic elements, diffusion studies, particularly of the historical period. 2nd, 1913 edition. 242 illustrations. lii + 439pp. of text.

20364-6, 20365-4 Two volumes, Paperbound $6.00

THE GENTLE ART OF MAKING ENEMIES, James A. M. Whistler. Greatest wit of his day deflates Oscar Wilde, Ruskin, Swinburne; strikes back at inane critics, exhibitions, art journalism; aesthetics of impressionist revolution in most striking form. Highly readable classic by great painter. Reproduction of edition designed by Whistler. Introduction by Alfred Werner. xxxvi + 334pp.

21875-9 Paperbound $2.50

INCIDENTS OF TRAVEL IN YUCATAN, John L. Stephens. Classic (1843) exploration of jungles of Yucatan, looking for evidences of Maya civilization. Stephens found many ruins; comments on travel adventures, Mexican and Indian culture. 127 striking illustrations by F. Catherwood. Total of 669 pp.

20926-1, 20927-X Two volumes, Paperbound $5.00

INCIDENTS OF TRAVEL IN CENTRAL AMERICA, CHIAPAS, AND YUCATAN, John L. Stephens. An exciting travel journal and an important classic of archeology. Narrative relates his almost single-handed discovery of the Mayan culture, and exploration of the ruined cities of Copan, Palenque, Utatlan and others; the monuments they dug from the earth, the temples buried in the jungle, the customs of poverty-stricken Indians living a stone's throw from the ruined palaces. 115 drawings by F. Catherwood. Portrait of Stephens. xii + 812pp.

22404-X, 22405-8 Two volumes, Paperbound $6.00

A NEW VOYAGE ROUND THE WORLD, William Dampier. Late 17-century naturalist joined the pirates of the Spanish Main to gather information; remarkably vivid account of buccaneers, pirates; detailed, accurate account of botany, zoology, ethnography of lands visited. Probably the most important early English voyage, enormous implications for British exploration, trade, colonial policy. Also most interesting reading. Argonaut edition, introduction by Sir Albert Gray. New introduction by Percy Adams. 6 plates, 7 illustrations. xlvii + 376pp. 6½ x 9¼.

21900-3 Paperbound $3.00

INTERNATIONAL AIRLINE PHRASE BOOK IN SIX LANGUAGES, Joseph W. Bátor. Important phrases and sentences in English paralleled with French, German, Portuguese, Italian, Spanish equivalents, covering all possible airport-travel situations; created for airline personnel as well as tourist by Language Chief, Pan American Airlines. xiv + 204pp. 22017-6 Paperbound $2.00

STAGE COACH AND TAVERN DAYS, Alice Morse Earle. Detailed, lively account of the early days of taverns; their uses and importance in the social, political and military life; furnishings and decorations; locations; food and drink; tavern signs, etc. Second half covers every aspect of early travel; the roads, coaches, drivers, etc. Nostalgic, charming, packed with fascinating material. 157 illustrations, mostly photographs. xiv + 449pp. 22518-6 Paperbound $4.00

NORSE DISCOVERIES AND EXPLORATIONS IN NORTH AMERICA, Hjalmar R. Holand. The perplexing Kensington Stone, found in Minnesota at the end of the 19th century. Is it a record of a Scandinavian expedition to North America in the 14th century? Or is it one of the most successful hoaxes in history? A scientific detective investigation. Formerly *Westward from Vinland*. 31 photographs, 17 figures. x + 354pp. 22014-1 Paperbound $2.75

A BOOK OF OLD MAPS, compiled and edited by Emerson D. Fite and Archibald Freeman. 74 old maps offer an unusual survey of the discovery, settlement and growth of America down to the close of the Revolutionary war: maps showing Norse settlements in Greenland, the explorations of Columbus, Verrazano, Cabot, Champlain, Joliet, Drake, Hudson, etc., campaigns of Revolutionary war battles, and much more. Each map is accompanied by a brief historical essay. xvi + 299pp. 11 x 13¾. 22084-2 Paperbound $6.00

How to Know the Wild Flowers, Mrs. William Starr Dana. This is the classical book of American wildflowers (of the Eastern and Central United States), used by hundreds of thousands. Covers over 500 species, arranged in extremely easy to use color and season groups. Full descriptions, much plant lore. This Dover edition is the fullest ever compiled, with tables of nomenclature changes. 174 full-page plates by M. Satterlee. xii + 418pp. 20332-8 Paperbound $2.75

Our Plant Friends and Foes, William Atherton DuPuy. History, economic importance, essential botanical information and peculiarities of 25 common forms of plant life are provided in this book in an entertaining and charming style. Covers food plants (potatoes, apples, beans, wheat, almonds, bananas, etc.), flowers (lily, tulip, etc.), trees (pine, oak, elm, etc.), weeds, poisonous mushrooms and vines, gourds, citrus fruits, cotton, the cactus family, and much more. 108 illustrations. xiv + 290pp. 22272-1 Paperbound $2.50

How to Know the Ferns, Frances T. Parsons. Classic survey of Eastern and Central ferns, arranged according to clear, simple identification key. Excellent introduction to greatly neglected nature area. 57 illustrations and 42 plates. xvi + 215pp. 20740-4 Paperbound $2.00

Manual of the Trees of North America, Charles S. Sargent. America's foremost dendrologist provides the definitive coverage of North American trees and tree-like shrubs. 717 species fully described and illustrated: exact distribution, down to township; full botanical description; economic importance; description of subspecies and races; habitat, growth data; similar material. Necessary to every serious student of tree-life. Nomenclature revised to present. Over 100 locating keys. 783 illustrations. lii + 934pp. 20277-1, 20278-X Two volumes, Paperbound $6.00

Our Northern Shrubs, Harriet L. Keeler. Fine non-technical reference work identifying more than 225 important shrubs of Eastern and Central United States and Canada. Full text covering botanical description, habitat, plant lore, is paralleled with 205 full-page photographs of flowering or fruiting plants. Nomenclature revised by Edward G. Voss. One of few works concerned with shrubs. 205 plates, 35 drawings. xxviii + 521pp. 21989-5 Paperbound $3.75

The Mushroom Handbook, Louis C. C. Krieger. Still the best popular handbook: full descriptions of 259 species, cross references to another 200. Extremely thorough text enables you to identify, know all about any mushroom you are likely to meet in eastern and central U. S. A.: habitat, luminescence, poisonous qualities, use, folklore, etc. 32 color plates show over 50 mushrooms, also 126 other illustrations. Finding keys. vii + 560pp. 21861-9 Paperbound $3.95

Handbook of Birds of Eastern North America, Frank M. Chapman. Still much the best single-volume guide to the birds of Eastern and Central United States. Very full coverage of 675 species, with descriptions, life habits, distribution, similar data. All descriptions keyed to two-page color chart. With this single volume the average birdwatcher needs no other books. 1931 revised edition. 195 illustrations. xxxvi + 581pp. 21489-3 Paperbound $5.00

VISUAL ILLUSIONS: THEIR CAUSES, CHARACTERISTICS, AND APPLICATIONS, Matthew Luckiesh. Thorough description and discussion of optical illusion, geometric and perspective, particularly; size and shape distortions, illusions of color, of motion; natural illusions; use of illusion in art and magic, industry, etc. Most useful today with op art, also for classical art. Scores of effects illustrated. Introduction by William H. Ittleson. 100 illustrations. xxi + 252pp.

21530-X Paperbound $2.00

A HANDBOOK OF ANATOMY FOR ART STUDENTS, Arthur Thomson. Thorough, virtually exhaustive coverage of skeletal structure, musculature, etc. Full text, supplemented by anatomical diagrams and drawings and by photographs of undraped figures. Unique in its comparison of male and female forms, pointing out differences of contour, texture, form. 211 figures, 40 drawings, 86 photographs. xx + 459pp. 5⅜ x 8⅜.

21163-0 Paperbound $3.50

150 MASTERPIECES OF DRAWING, Selected by Anthony Toney. Full page reproductions of drawings from the early 16th to the end of the 18th century, all beautifully reproduced: Rembrandt, Michelangelo, Dürer, Fragonard, Urs, Graf, Wouwerman, many others. First-rate browsing book, model book for artists. xviii + 150pp. 8⅜ x 11¼.

21032-4 Paperbound $2.50

THE LATER WORK OF AUBREY BEARDSLEY, Aubrey Beardsley. Exotic, erotic, ironic masterpieces in full maturity: Comedy Ballet, Venus and Tannhauser, Pierrot, Lysistrata, Rape of the Lock, Savoy material, Ali Baba, Volpone, etc. This material revolutionized the art world, and is still powerful, fresh, brilliant. With *The Early Work,* all Beardsley's finest work. 174 plates, 2 in color. xiv + 176pp. 8⅛ x 11.

21817-1 Paperbound $3.00

DRAWINGS OF REMBRANDT, Rembrandt van Rijn. Complete reproduction of fabulously rare edition by Lippmann and Hofstede de Groot, completely reedited, updated, improved by Prof. Seymour Slive, Fogg Museum. Portraits, Biblical sketches, landscapes, Oriental types, nudes, episodes from classical mythology—All Rembrandt's fertile genius. Also selection of drawings by his pupils and followers. "Stunning volumes," *Saturday Review.* 550 illustrations. lxxviii + 552pp. 9⅛ x 12¼.

21485-0, 21486-9 Two volumes, Paperbound $10.00

THE DISASTERS OF WAR, Francisco Goya. One of the masterpieces of Western civilization—83 etchings that record Goya's shattering, bitter reaction to the Napoleonic war that swept through Spain after the insurrection of 1808 and to war in general. Reprint of the first edition, with three additional plates from Boston's Museum of Fine Arts. All plates facsimile size. Introduction by Philip Hofer, Fogg Museum. v + 97pp. 9⅜ x 8¼.

21872-4 Paperbound $2.00

GRAPHIC WORKS OF ODILON REDON. Largest collection of Redon's graphic works ever assembled: 172 lithographs, 28 etchings and engravings, 9 drawings. These include some of his most famous works. All the plates from *Odilon Redon: oeuvre graphique complet,* plus additional plates. New introduction and caption translations by Alfred Werner. 209 illustrations. xxvii + 209pp. 9⅛ x 12¼.

21966-8 Paperbound $4.00

Jim Whitewolf: The Life of a Kiowa Apache Indian, Charles S. Brant, editor. Spans transition between native life and acculturation period, 1880 on. Kiowa culture, personal life pattern, religion and the supernatural, the Ghost Dance, breakdown in the White Man's world, similar material. 1 map. xii + 144pp.
22015-X Paperbound $1.75

The Native Tribes of Central Australia, Baldwin Spencer and F. J. Gillen. Basic book in anthropology, devoted to full coverage of the Arunta and Warramunga tribes; the source for knowledge about kinship systems, material and social culture, religion, etc. Still unsurpassed. 121 photographs, 89 drawings. xviii + 669pp.
21775-2 Paperbound $5.00

Malay Magic, Walter W. Skeat. Classic (1900); still the definitive work on the folklore and popular religion of the Malay peninsula. Describes marriage rites, birth spirits and ceremonies, medicine, dances, games, war and weapons, etc. Extensive quotes from original sources, many magic charms translated into English. 35 illustrations. Preface by Charles Otto Blagden. xxiv + 685pp.
21760-4 Paperbound $4.00

Heavens on Earth: Utopian Communities in America, 1680-1880, Mark Holloway. The finest nontechnical account of American utopias, from the early Woman in the Wilderness, Ephrata, Rappites to the enormous mid 19th-century efflorescence; Shakers, New Harmony, Equity Stores, Fourier's Phalanxes, Oneida, Amana, Fruitlands, etc. "Entertaining and very instructive." *Times Literary Supplement*. 15 illustrations. 246pp.
21593-8 Paperbound $2.00

London Labour and the London Poor, Henry Mayhew. Earliest (c. 1850) sociological study in English, describing myriad subcultures of London poor. Particularly remarkable for the thousands of pages of direct testimony taken from the lips of London prostitutes, thieves, beggars, street sellers, chimney-sweepers, street-musicians, "mudlarks," "pure-finders," rag-gatherers, "running-patterers," dock laborers, cab-men, and hundreds of others, quoted directly in this massive work. An extraordinarily vital picture of London emerges. 110 illustrations. Total of lxxvi + 1951pp. 6⅝ x 10.
21934-8, 21935-6, 21936-4, 21937-2 Four volumes, Paperbound $16.00

History of the Later Roman Empire, J. B. Bury. Eloquent, detailed reconstruction of Western and Byzantine Roman Empire by a major historian, from the death of Theodosius I (395 A.D.) to the death of Justinian (565). Extensive quotations from contemporary sources; full coverage of important Roman and foreign figures of the time. xxxiv + 965pp. 20398-0, 20399-9 Two volumes, Paperbound $7.00

An Intellectual and Cultural History of the Western World, Harry Elmer Barnes. Monumental study, tracing the development of the accomplishments that make up human culture. Every aspect of man's achievement surveyed from its origins in the Paleolithic to the present day (1964); social structures, ideas, economic systems, art, literature, technology, mathematics, the sciences, medicine, religion, jurisprudence, etc. Evaluations of the contributions of scores of great men. 1964 edition, revised and edited by scholars in the many fields represented. Total of xxix + 1381pp. 21275-0, 21276-9, 21277-7 Three volumes, Paperbound $10.50

AMERICAN FOOD AND GAME FISHES, David S. Jordan and Barton W. Evermann. Definitive source of information, detailed and accurate enough to enable the sportsman and nature lover to identify conclusively some 1,000 species and sub-species of North American fish, sought for food or sport. Coverage of range, physiology, habits, life history, food value. Best methods of capture, interest to the angler, advice on bait, fly-fishing, etc. 338 drawings and photographs. 1 + 574pp. 6⅝ x 9⅜.

22383-1 Paperbound $4.50

THE FROG BOOK, Mary C. Dickerson. Complete with extensive finding keys, over 300 photographs, and an introduction to the general biology of frogs and toads, this is the classic non-technical study of Northeastern and Central species. 58 species; 290 photographs and 16 color plates. xvii + 253pp.

21973-9 Paperbound $4.00

THE MOTH BOOK: A GUIDE TO THE MOTHS OF NORTH AMERICA, William J. Holland. Classical study, eagerly sought after and used for the past 60 years. Clear identification manual to more than 2,000 different moths, largest manual in existence. General information about moths, capturing, mounting, classifying, etc., followed by species by species descriptions. 263 illustrations plus 48 color plates show almost every species, full size. 1968 edition, preface, nomenclature changes by A. E. Brower. xxiv + 479pp. of text. 6½ x 9¼.

21948-8 Paperbound $5.00

THE SEA-BEACH AT EBB-TIDE, Augusta Foote Arnold. Interested amateur can identify hundreds of marine plants and animals on coasts of North America; marine algae; seaweeds; squids; hermit crabs; horse shoe crabs; shrimps; corals; sea anemones; etc. Species descriptions cover: structure; food; reproductive cycle; size; shape; color; habitat; etc. Over 600 drawings. 85 plates. xii + 490pp.

21949-6 Paperbound $3.50

COMMON BIRD SONGS, Donald J. Borror. 33⅓ 12-inch record presents songs of 60 important birds of the eastern United States. A thorough, serious record which provides several examples for each bird, showing different types of song, individual variations, etc. Inestimable identification aid for birdwatcher. 32-page booklet gives text about birds and songs, with illustration for each bird.

21829-5 Record, book, album. Monaural. $2.75

FADS AND FALLACIES IN THE NAME OF SCIENCE, Martin Gardner. Fair, witty appraisal of cranks and quacks of science: Atlantis, Lemuria, hollow earth, flat earth, Velikovsky, orgone energy, Dianetics, flying saucers, Bridey Murphy, food fads, medical fads, perpetual motion, etc. Formerly "In the Name of Science." x + 363pp.

20394-8 Paperbound $2.00

HOAXES, Curtis D. MacDougall. Exhaustive, unbelievably rich account of great hoaxes: Locke's moon hoax, Shakespearean forgeries, sea serpents, Loch Ness monster, Cardiff giant, John Wilkes Booth's mummy, Disumbrationist school of art, dozens more; also journalism, psychology of hoaxing. 54 illustrations. xi + 338pp.

20465-0 Paperbound $2.75

THE RED FAIRY BOOK, Andrew Lang. Lang's color fairy books have long been children's favorites. This volume includes Rapunzel, Jack and the Bean-stalk and 35 other stories, familiar and unfamiliar. 4 plates, 93 illustrations x + 367pp.
21673-X Paperbound $2.50

THE BLUE FAIRY BOOK, Andrew Lang. Lang's tales come from all countries and all times. Here are 37 tales from Grimm, the Arabian Nights, Greek Mythology, and other fascinating sources. 8 plates, 130 illustrations. xi + 390pp.
21437-0 Paperbound $2.50

HOUSEHOLD STORIES BY THE BROTHERS GRIMM. Classic English-language edition of the well-known tales — Rumpelstiltskin, Snow White, Hansel and Gretel, The Twelve Brothers, Faithful John, Rapunzel, Tom Thumb (52 stories in all). Translated into simple, straightforward English by Lucy Crane. Ornamented with headpieces, vignettes, elaborate decorative initials and a dozen full-page illustrations by Walter Crane. x + 269pp.
21080-4 Paperbound $2.50

THE MERRY ADVENTURES OF ROBIN HOOD, Howard Pyle. The finest modern versions of the traditional ballads and tales about the great English outlaw. Howard Pyle's complete prose version, with every word, every illustration of the first edition. Do not confuse this facsimile of the original (1883) with modern editions that change text or illustrations. 23 plates plus many page decorations. xxii + 296pp.
22043-5 Paperbound $2.50

THE STORY OF KING ARTHUR AND HIS KNIGHTS, Howard Pyle. The finest children's version of the life of King Arthur; brilliantly retold by Pyle, with 48 of his most imaginative illustrations. xviii + 313pp. 6⅛ x 9¼.
21445-1 Paperbound $2.50

THE WONDERFUL WIZARD OF OZ, L. Frank Baum. America's finest children's book in facsimile of first edition with all Denslow illustrations in full color. The edition a child should have. Introduction by Martin Gardner. 23 color plates, scores of drawings. iv + 267pp.
20691-2 Paperbound $2.50

THE MARVELOUS LAND OF OZ, L. Frank Baum. The second Oz book, every bit as imaginative as the Wizard. The hero is a boy named Tip, but the Scarecrow and the Tin Woodman are back, as is the Oz magic. 16 color plates, 120 drawings by John R. Neill. 287pp.
20692-0 Paperbound $2.50

THE MAGICAL MONARCH OF MO, L. Frank Baum. Remarkable adventures in a land even stranger than Oz. The best of Baum's books not in the Oz series. 15 color plates and dozens of drawings by Frank Verbeck. xviii + 237pp.
21892-9 Paperbound $2.25

THE BAD CHILD'S BOOK OF BEASTS, MORE BEASTS FOR WORSE CHILDREN, A MORAL ALPHABET, Hilaire Belloc. Three complete humor classics in one volume. Be kind to the frog, and do not call him names . . . and 28 other whimsical animals. Familiar favorites and some not so well known. Illustrated by Basil Blackwell. 156pp.
(USO) 20749-8 Paperbound $1.50

LAST AND FIRST MEN AND STAR MAKER, TWO SCIENCE FICTION NOVELS, Olaf Stapledon. Greatest future histories in science fiction. In the first, human intelligence is the "hero," through strange paths of evolution, interplanetary invasions, incredible technologies, near extinctions and reemergences. Star Maker describes the quest of a band of star rovers for intelligence itself, through time and space: weird inhuman civilizations, crustacean minds, symbiotic worlds, etc. Complete, unabridged. v + 438pp. 21962-3 Paperbound $2.50

THREE PROPHETIC NOVELS, H. G. WELLS. Stages of a consistently planned future for mankind. *When the Sleeper Wakes,* and *A Story of the Days to Come,* anticipate *Brave New World* and *1984,* in the 21st Century; *The Time Machine,* only complete version in print, shows farther future and the end of mankind. All . .ow Wells's greatest gifts as storyteller and novelist. Edited by E. F. Bleiler. x + 335pp. (USO) 20605-X Paperbound $2.50

THE DEVIL'S DICTIONARY, Ambrose Bierce. America's own Oscar Wilde— Ambrose Bierce—offers his barbed iconoclastic wisdom in over 1,000 definitions hailed by H. L. Mencken as "some of the most gorgeous witticisms in the English language." 145pp. 20487-1 Paperbound $1.25

MAX AND MORITZ, Wilhelm Busch. Great children's classic, father of comic strip, of two bad boys, Max and Moritz. Also Ker and Plunk (Plisch und Plumm), Cat and Mouse, Deceitful Henry, Ice-Peter, The Boy and the Pipe, and five other pieces. Original German, with English translation. Edited by H. Arthur Klein; translations by various hands and H. Arthur Klein. vi + 216pp.
20181-3 Paperbound $2.00

PIGS IS PIGS AND OTHER FAVORITES, Ellis Parker Butler. The title story is one of the best humor short stories, as Mike Flannery obfuscates biology and English. Also included, That Pup of Murchison's, The Great American Pie Company, and Perkins of Portland. 14 illustrations. v + 109pp. 21532-6 Paperbound $1.25

THE PETERKIN PAPERS, Lucretia P. Hale. It takes genius to be as stupidly mad as the Peterkins, as they decide to become wise, celebrate the "Fourth," keep a cow, and otherwise strain the resources of the Lady from Philadelphia. Basic book of American humor. 153 illustrations. 219pp. 20794-3 Paperbound $1.50

PERRAULT'S FAIRY TALES, translated by A. E. Johnson and S. R. Littlewood, with 34 full-page illustrations by Gustave Doré. All the original Perrault stories— Cinderella, Sleeping Beauty, Bluebeard, Little Red Riding Hood, Puss in Boots, Tom Thumb, etc.—with their witty verse morals and the magnificent illustrations of Doré. One of the five or six great books of European fairy tales. viii + 117pp. 8⅛ x 11. 22311-6 Paperbound $2.00

OLD HUNGARIAN FAIRY TALES, Baroness Orczy. Favorites translated and adapted by author of the *Scarlet Pimpernel.* Eight fairy tales include "The Suitors of Princess Fire-Fly," "The Twin Hunchbacks," "Mr. Cuttlefish's Love Story," and "The Enchanted Cat." This little volume of magic and adventure will captivate children as it has for generations. 90 drawings by Montagu Barstow. 96pp.
(USO) 22293-4 Paperbound $1.95

POEMS OF ANNE BRADSTREET, edited with an introduction by Robert Hutchinson. A new selection of poems by America's first poet and perhaps the first significant woman poet in the English language. 48 poems display her development in works of considerable variety—love poems, domestic poems, religious meditations, formal elegies, "quaternions," etc. Notes, bibliography. viii + 222pp.

22160-1 Paperbound $2.00

THREE GOTHIC NOVELS: THE CASTLE OF OTRANTO BY HORACE WALPOLE; VATHEK BY WILLIAM BECKFORD; THE VAMPYRE BY JOHN POLIDORI, WITH FRAGMENT OF A NOVEL BY LORD BYRON, edited by E. F. Bleiler. The first Gothic novel, by Walpole; the finest Oriental tale in English, by Beckford; powerful Romantic supernatural story in versions by Polidori and Byron. All extremely important in history of literature; all still exciting, packed with supernatural thrills, ghosts, haunted castles, magic, etc. xl + 291pp.

21232-7 Paperbound $2.50

THE BEST TALES OF HOFFMANN, E. T. A. Hoffmann. 10 of Hoffmann's most important stories, in modern re-editings of standard translations: Nutcracker and the King of Mice, Signor Formica, Automata, The Sandman, Rath Krespel, The Golden Flowerpot, Master Martin the Cooper, The Mines of Falun, The King's Betrothed, A New Year's Eve Adventure. 7 illustrations by Hoffmann. Edited by E. F. Bleiler. xxxix + 419pp.

21793-0 Paperbound $3.00

GHOST AND HORROR STORIES OF AMBROSE BIERCE, Ambrose Bierce. 23 strikingly modern stories of the horrors latent in the human mind: The Eyes of the Panther, The Damned Thing, An Occurrence at Owl Creek Bridge, An Inhabitant of Carcosa, etc., plus the dream-essay, Visions of the Night. Edited by E. F. Bleiler. xxii + 199pp.

20767-6 Paperbound $1.50

BEST GHOST STORIES OF J. S. LEFANU, J. Sheridan LeFanu. Finest stories by Victorian master often considered greatest supernatural writer of all. Carmilla, Green Tea, The Haunted Baronet, The Familiar, and 12 others. Most never before available in the U. S. A. Edited by E. F. Bleiler. 8 illustrations from Victorian publications. xvii + 467pp.

20415-4 Paperbound $3.00

MATHEMATICAL FOUNDATIONS OF INFORMATION THEORY, A. I. Khinchin. Comprehensive introduction to work of Shannon, McMillan, Feinstein and Khinchin, placing these investigations on a rigorous mathematical basis. Covers entropy concept in probability theory, uniqueness theorem, Shannon's inequality, ergodic sources, the E property, martingale concept, noise, Feinstein's fundamental lemma, Shanon's first and second theorems. Translated by R. A. Silverman and M. D. Friedman. iii + 120pp.

60434-9 Paperbound $1.75

SEVEN SCIENCE FICTION NOVELS, H. G. Wells. The standard collection of the great novels. Complete, unabridged. *First Men in the Moon, Island of Dr. Moreau, War of the Worlds, Food of the Gods, Invisible Man, Time Machine, In the Days of the Comet.* Not only science fiction fans, but every educated person owes it to himself to read these novels. 1015pp

20264-X Clothbound $5.00

THE PRINCIPLES OF PSYCHOLOGY, William James. The famous long course, complete and unabridged. Stream of thought, time perception, memory, experimental methods—these are only some of the concerns of a work that was years ahead of its time and still valid, interesting, useful. 94 figures. Total of xviii + 1391pp.
20381-6, 20382-4 Two volumes, Paperbound $8.00

THE STRANGE STORY OF THE QUANTUM, Banesh Hoffmann. Non-mathematical but thorough explanation of work of Planck, Einstein, Bohr, Pauli, de Broglie, Schrödinger, Heisenberg, Dirac, Feynman, etc. No technical background needed. "Of books attempting such an account, this is the best," Henry Margenau, Yale. 40-page "Postscript 1959." xii + 285pp. 20518-5 Paperbound $2.00

THE RISE OF THE NEW PHYSICS, A. d'Abro. Most thorough explanation in print of central core of mathematical physics, both classical and modern; from Newton to Dirac and Heisenberg. Both history and exposition; philosophy of science, causality, explanations of higher mathematics, analytical mechanics, electromagnetism, thermodynamics, phase rule, special and general relativity, matrices. No higher mathematics needed to follow exposition, though treatment is elementary to intermediate in level. Recommended to serious student who wishes verbal understanding. 97 illustrations. xvii + 982pp. 20003-5, 20004-3 Two volumes, Paperbound $6.00

GREAT IDEAS OF OPERATIONS RESEARCH, Jagjit Singh. Easily followed non-technical explanation of mathematical tools, aims, results: statistics, linear programming, game theory, queueing theory, Monte Carlo simulation, etc. Uses only elementary mathematics. Many case studies, several analyzed in detail. Clarity, breadth make this excellent for specialist in another field who wishes background. 41 figures. x + 228pp. 21886-4 Paperbound $2.50

GREAT IDEAS OF MODERN MATHEMATICS: THEIR NATURE AND USE, Jagjit Singh. Internationally famous expositor, winner of Unesco's Kalinga Award for science popularization explains verbally such topics as differential equations, matrices, groups, sets, transformations, mathematical logic and other important modern mathematics, as well as use in physics, astrophysics, and similar fields. Superb exposition for layman, scientist in other areas. viii + 312pp.
20587-8 Paperbound $2.50

GREAT IDEAS IN INFORMATION THEORY, LANGUAGE AND CYBERNETICS, Jagjit Singh. The analog and digital computers, how they work, how they are like and unlike the human brain, the men who developed them, their future applications, computer terminology. An essential book for today, even for readers with little math. Some mathematical demonstrations included for more advanced readers. 118 figures. Tables. ix + 338pp. 21694-2 Paperbound $2.50

CHANCE, LUCK AND STATISTICS, Horace C. Levinson. Non-mathematical presentation of fundamentals of probability theory and science of statistics and their applications. Games of chance, betting odds, misuse of statistics, normal and skew distributions, birth rates, stock speculation, insurance. Enlarged edition. Formerly "The Science of Chance." xiii + 357pp. 21007-3 Paperbound $2.50

PLANETS, STARS AND GALAXIES: DESCRIPTIVE ASTRONOMY FOR BEGINNERS, A. E. Fanning. Comprehensive introductory survey of astronomy: the sun, solar system, stars, galaxies, universe, cosmology; up-to-date, including quasars, radio stars, etc. Preface by Prof. Donald Menzel. 24pp. of photographs. 189pp. 5¼ x 8¼.
21680-2 Paperbound $1.75

TEACH YOURSELF CALCULUS, P. Abbott. With a good background in algebra and trig, you can teach yourself calculus with this book. Simple, straightforward introduction to functions of all kinds, integration, differentiation, series, etc. "Students who are beginning to study calculus method will derive great help from this book." Faraday House Journal. 308pp.
20683-1 Clothbound $2.50

TEACH YOURSELF TRIGONOMETRY, P. Abbott. Geometrical foundations, indices and logarithms, ratios, angles, circular measure, etc. are presented in this sound, easy-to-use text. Excellent for the beginner or as a brush up, this text carries the student through the solution of triangles. 204pp.
20682-3 Clothbound $2.50

BASIC MACHINES AND HOW THEY WORK, U. S. Bureau of Naval Personnel. Originally used in U.S. Naval training schools, this book clearly explains the operation of a progression of machines, from the simplest—lever, wheel and axle, inclined plane, wedge, screw—to the most complex—typewriter, internal combustion engine, computer mechanism. Utilizing an approach that requires only an elementary understanding of mathematics, these explanations build logically upon each other and are assisted by over 200 drawings and diagrams. Perfect as a technical school manual or as a self-teaching aid to the layman. 204 figures. Preface. Index. vii + 161pp. 6½ x 9¼.
21709-4 Paperbound $2.50

THE FRIENDLY STARS, Martha Evans Martin. Classic has taught naked-eye observation of stars, planets to hundreds of thousands, still not surpassed for charm, lucidity, adequacy. Completely updated by Professor Donald H. Menzel, Harvard Observatory. 25 illustrations. 16 x 30 chart. x + 147pp.
21099-5 Paperbound $1.50

MUSIC OF THE SPHERES: THE MATERIAL UNIVERSE FROM ATOM TO QUASAR, SIMPLY EXPLAINED, Guy Murchie. Extremely broad, brilliantly written popular account begins with the solar system and reaches to dividing line between matter and nonmatter; latest understandings presented with exceptional clarity. Volume One: Planets, stars, galaxies, cosmology, geology, celestial mechanics, latest astronomical discoveries; Volume Two: Matter, atoms, waves, radiation, relativity, chemical action, heat, nuclear energy, quantum theory, music, light, color, probability, antimatter, antigravity, and similar topics. 319 figures. 1967 (second) edition. Total of xx + 644pp.
21809-0, 21810-4 Two volumes, Paperbound $5.50

OLD-TIME SCHOOLS AND SCHOOL BOOKS, Clifton Johnson. Illustrations and rhymes from early primers, abundant quotations from early textbooks, many anecdotes of school life enliven this study of elementary schools from Puritans to middle 19th century. Introduction by Carl Withers. 234 illustrations. xxxiii + 381pp.
21031-6 Paperbound $3.50

CATALOGUE OF DOVER BOOKS

THE PHILOSOPHY OF THE UPANISHADS, Paul Deussen. Clear, detailed statement of upanishadic system of thought, generally considered among best available. History of these works, full exposition of system emergent from them, parallel concepts in the West. Translated by A. S. Geden. xiv + 429pp.
21616-0 Paperbound $3.50

LANGUAGE, TRUTH AND LOGIC, Alfred J. Ayer. Famous, remarkably clear introduction to the Vienna and Cambridge schools of Logical Positivism; function of philosophy, elimination of metaphysical thought, nature of analysis, similar topics. "Wish I had written it myself," Bertrand Russell. 2nd, 1946 edition. 160pp.
20010-8 Paperbound $1.50

THE GUIDE FOR THE PERPLEXED, Moses Maimonides. Great classic of medieval Judaism, major attempt to reconcile revealed religion (Pentateuch, commentaries) and Aristotelian philosophy. Enormously important in all Western thought. Unabridged Friedländer translation. 50-page introduction. lix + 414pp.
(USO) 20351-4 Paperbound $3.50

OCCULT AND SUPERNATURAL PHENOMENA, D. H. Rawcliffe. Full, serious study of the most persistent delusions of mankind: crystal gazing, mediumistic trance, stigmata, lycanthropy, fire walking, dowsing, telepathy, ghosts, ESP, etc., and their relation to common forms of abnormal psychology. Formerly *Illusions and Delusions of the Supernatural and the Occult.* iii + 551pp.   20503-7 Paperbound $3.50

THE EGYPTIAN BOOK OF THE DEAD: THE PAPYRUS OF ANI, E. A. Wallis Budge. Full hieroglyphic text, interlinear transliteration of sounds, word for word translation, then smooth, connected translation; Theban recension. Basic work in Ancient Egyptian civilization; now even more significant than ever for historical importance, dilation of consciousness, etc. clvi + 377pp. 6½ x 9¼.
21866-X Paperbound $3.95

PSYCHOLOGY OF MUSIC, Carl E. Seashore. Basic, thorough survey of everything known about psychology of music up to 1940's; essential reading for psychologists, musicologists. Physical acoustics; auditory apparatus; relationship of physical sound to perceived sound; role of the mind in sorting, altering, suppressing, creating sound sensations; musical learning, testing for ability, absolute pitch, other topics. Records of Caruso, Menuhin analyzed. 88 figures. xix + 408pp.
21851-1 Paperbound $3.50

THE I CHING (THE BOOK OF CHANGES), translated by James Legge. Complete translated text plus appendices by Confucius, of perhaps the most penetrating divination book ever compiled. Indispensable to all study of early Oriental civilizations. 3 plates. xxiii + 448pp.   21062-6 Paperbound $3.00

THE UPANISHADS, translated by Max Müller. Twelve classical upanishads: Chandogya, Kena, Aitareya, Kaushitaki, Isa, Katha, Mundaka, Taittiriyaka, Brhadaranyaka, Svetasvatara, Prasna, Maitriyana. 160-page introduction, analysis by Prof. Müller. Total of 670pp. 20992-X, 20993-8 Two volumes, Paperbound $6.50

ADVENTURES OF AN AFRICAN SLAVER, Theodore Canot. Edited by Brantz Mayer. A detailed portrayal of slavery and the slave trade, 1820-1840. Canot, an established trader along the African coast, describes the slave economy of the African kingdoms, the treatment of captured negroes, the extensive journeys in the interior to gather slaves, slave revolts and their suppression, harems, bribes, and much more. Full and unabridged republication of 1854 edition. Introduction by Malcom Cowley. 16 illustrations. xvii + 448pp. 22456-2 Paperbound $3.50

MY BONDAGE AND MY FREEDOM, Frederick Douglass. Born and brought up in slavery, Douglass witnessed its horrors and experienced its cruelties, but went on to become one of the most outspoken forces in the American anti-slavery movement. Considered the best of his autobiographies, this book graphically describes the inhuman treatment of slaves, its effects on slave owners and slave families, and how Douglass's determination led him to a new life. Unaltered reprint of 1st (1855) edition. xxxii + 464pp. 22457-0 Paperbound $2.50

THE INDIANS' BOOK, recorded and edited by Natalie Curtis. Lore, music, narratives, dozens of drawings by Indians themselves from an authoritative and important survey of native culture among Plains, Southwestern, Lake and Pueblo Indians. Standard work in popular ethnomusicology. 149 songs in full notation. 23 drawings, 23 photos. xxxi + 584pp. 6⅝ x 9⅜. 21939-9 Paperbound $4.50

DICTIONARY OF AMERICAN PORTRAITS, edited by Hayward and Blanche Cirker. 4024 portraits of 4000 most important Americans, colonial days to 1905 (with a few important categories, like Presidents, to present). Pioneers, explorers, colonial figures, U. S. officials, politicians, writers, military and naval men, scientists, inventors, manufacturers, jurists, actors, historians, educators, notorious figures, Indian chiefs, etc. All authentic contemporary likenesses. The only work of its kind in existence; supplements all biographical sources for libraries. Indispensable to anyone working with American history. 8,000-item classified index, finding lists, other aids. xiv + 756pp. 9¼ x 12¾. 21823-6 Clothbound $30.00

TRITTON'S GUIDE TO BETTER WINE AND BEER MAKING FOR BEGINNERS, S. M. Tritton. All you need to know to make family-sized quantities of over 100 types of grape, fruit, herb and vegetable wines; as well as beers, mead, cider, etc. Complete recipes, advice as to equipment, procedures such as fermenting, bottling, and storing wines. Recipes given in British, U. S., and metric measures. Accompanying booklet lists sources in U. S. A. where ingredients may be bought, and additional information. 11 illustrations. 157pp. 5⅝ x 8⅛.
(USO) 22090-7 Clothbound $3.50

GARDENING WITH HERBS FOR FLAVOR AND FRAGRANCE, Helen M. Fox. How to grow herbs in your own garden, how to use them in your cooking (over 55 recipes included), legends and myths associated with each species, uses in medicine, perfumes, etc.—these are elements of one of the few books written especially for American herb fanciers. Guides you step-by-step from soil preparation to harvesting and storage for each type of herb. 12 drawings by Louise Mansfield. xiv + 334pp. 22540-2 Paperbound $2.50

MATHEMATICAL PUZZLES FOR BEGINNERS AND ENTHUSIASTS, Geoffrey Mott-Smith. 189 puzzles from easy to difficult—involving arithmetic, logic, algebra, properties of digits, probability, etc.—for enjoyment and mental stimulus. Explanation of mathematical principles behind the puzzles. 135 illustrations. viii + 248pp.

20198-8 Paperbound $1.75

PAPER FOLDING FOR BEGINNERS, William D. Murray and Francis J. Rigney. Easiest book on the market, clearest instructions on making interesting, beautiful origami. Sail boats, cups, roosters, frogs that move legs, bonbon boxes, standing birds, etc. 40 projects; more than 275 diagrams and photographs. 94pp.

20713-7 Paperbound $1.00

TRICKS AND GAMES ON THE POOL TABLE, Fred Herrmann. 79 tricks and games— some solitaires, some for two or more players, some competitive games—to entertain you between formal games. Mystifying shots and throws, unusual caroms, tricks involving such props as cork, coins, a hat, etc. Formerly *Fun on the Pool Table*. 77 figures. 95pp.

21814-7 Paperbound $1.00

HAND SHADOWS TO BE THROWN UPON THE WALL: A SERIES OF NOVEL AND AMUSING FIGURES FORMED BY THE HAND, Henry Bursill. Delightful picturebook from great-grandfather's day shows how to make 18 different hand shadows: a bird that flies, duck that quacks, dog that wags his tail, camel, goose, deer, boy, turtle, etc. Only book of its sort. vi + 33pp. 6½ x 9¼. 21779-5 Paperbound $1.00

WHITTLING AND WOODCARVING, E. J. Tangerman. 18th printing of best book on market. "If you can cut a potato you can carve" toys and puzzles, chains, chessmen, caricatures, masks, frames, woodcut blocks, surface patterns, much more. Information on tools, woods, techniques. Also goes into serious wood sculpture from Middle Ages to present, East and West. 464 photos, figures. x + 293pp.

20965-2 Paperbound $2.00

HISTORY OF PHILOSOPHY, Julián Marias. Possibly the clearest, most easily followed, best planned, most useful one-volume history of philosophy on the market; neither skimpy nor overfull. Full details on system of every major philosopher and dozens of less important thinkers from pre-Socratics up to Existentialism and later. Strong on many European figures usually omitted. Has gone through dozens of editions in Europe. 1966 edition, translated by Stanley Appelbaum and Clarence Strowbridge. xviii + 505pp. 21739-6 Paperbound $3.00

YOGA: A SCIENTIFIC EVALUATION, Kovoor T. Behanan. Scientific but non-technical study of physiological results of yoga exercises; done under auspices of Yale U. Relations to Indian thought, to psychoanalysis, etc. 16 photos. xxiii + 270pp.

20505-3 Paperbound $2.50